# Mathematics for Civil Engineers

By the same author and published by Dunedin Academic Press:

*Mathematical Modelling for Earth Sciences* (2008)
ISBN: 9781903765920

*Introductory Mathematics for Earth Scientists* (2009)
ISBN: 9781906716004

For further details of these and other Dunedin
Earth and Environmental Sciences titles see
www.dunedinacademicpress.co.uk

# Mathematics for Civil Engineers

## AN INTRODUCTION

## Dr Xin-She Yang

Reader, Middlesex University
School of Science and Technology, London

EDINBURGH ◆ LONDON

Published by
Dunedin Academic Press Ltd
*Main Office:* Hudson House, 8 Albany Street, Edinburgh EH1 3QB, Scotland
*London Office:* 352 Cromwell Tower, Barbican, London EC2Y 8NB

www.dunedinacademicpress.co.uk

ISBNs
9781780460840 (Paperback)
9781780465777 (ePub)
9781780465784 (Kindle)

BRITISH LIBRARY CATALOGUING IN PUBLICATION DATA
A catalogue record for this book is available from the British Library

*While all reasonable attempts have been made to ensure the accuracy of information
contained in this publication it is intended for prudent and careful professional and
student use and no liability will be accepted by the author or publishers for any loss,
damage or injury caused by any errors or omissions herein. This disclaimer does not
effect any statutory rights.*

Printed in Great Britain by CPI Antony Rowe

# CONTENTS

# PREFACE

Engineering mathematics is a compulsory course for all undergraduates studying in all disciplines of engineering because it is an essential part of core training in engineering. Students have to develop all necessary mathematical skills and problem-solving practice needed for all other engineering courses such as engineering mechanics, structural analysis, surveying, geotechnics, modelling and simulation, numerical methods and many other advanced courses.

There are many textbooks on engineering mathematics, though most such books are lengthy and generic, not necessarily suitable for students in civil engineering. This book tries to provide a compact and concise approach to introduce mathematics for civil engineers. The main objective of this book is to cover all the major topics in engineering mathematics, and some of the worked examples are chosen for their relevance to civil engineering applications. Thus, the book is more suitable for students to build these key skills in a more confident way.

The main approach and style of this book can be considered as informal, theorem-free and practical. By using an informal and theorem-free approach with more than 150 step-by-step examples, all fundamental mathematics topics required for engineering are covered, and students will gain the same level of fundamental knowledge without going through too many tedious steps of mathematical proofs and analyses. In addition, each chapter is provided with some exercises so as to give readers an opportunity to practise the learned skills, and the answers to these exercises are provided in the Appendix at the end of the book.

*Mathematics for Civil Engineers: An Introduction* aims that undergraduates and civil engineers should rapidly develop all the fundamental knowledge of engineering mathematics. The step-by-step worked examples will help readers to gain more insight and build sufficient confidence in engineering mathematics for problem-solving in real-world applications. In addition, this book will also be useful for graduates in other engineering disciplines, computer science and natural sciences, to refresh their basic mathematical background and skills.

I would like to thank all my students and colleagues who have given valuable feedback and comments on the contents of this book. I also thank the anonymous reviewers for reviewing the full manuscript and providing some good suggestions. Last but not least, I thank my family for all the help and support.

Xin-She Yang
London, September 2017

# CHAPTER 1

# Numbers and Functions

Contents

## Key Points

- Review the basic concepts of real numbers and significant digits.
- Review basic equations and simultaneous equations.
- Review the fundamentals of functions such as power, exponential and logarithm.

The basic requirement of this book is a good understanding of all the basic mathematics such as numbers and equations, taught at school. Here we will first review briefly some of the most relevant and fundamental concepts such as numbers, fractions, real numbers, functions and equations. Then, in the first two chapters, we will introduce some useful concepts such as floating-point notations, significant digits, index notations and polynomials.

## 1. Real Numbers and Significant Digits

Whole numbers such as $-5, 0, 7$ and $100$ are integers, while the positive integers such as $1, 2, 3, ...$ are often called natural numbers. Fractions are ratios of two non-zero integers such as $\frac{2}{3}$ and $\frac{22}{7}$.

In many cases, it may be tedious to write all the numbers such as $1, 2, 3, 4, ..., 100$, so we often use a symbol such as $n$ to denote such numbers. Thus, we can use $n = 1, 2, ..., 100$ to aid our discussions. For example, the ratio of any two non-zero integers $m$ and $n$ in general is a fraction, often written as $m/n$ or $\frac{m}{n}$ where the denominator $n$ should not be zero (that is, $n \neq 0$). For example, $\frac{2}{3}, -\frac{7}{3}$ and $\frac{100}{11}$ are fractions. If $m < n$, the fraction is called a proper fraction.

All the integers and fractions make up the rational numbers. Among all the natural numbers $(1, 2, 3, ...)$, there is a special class of numbers, called prime numbers. A prime number is a positive integer, greater than one, that can only be divisible by 1 and itself. In other words, a prime number $p$ can have only two factors: 1 and itself $p$. It is easy to check that $2, 3, 5, 7, 11, 13, 17$ and $19$ are all prime numbers. The only even prime number is 2, and all the rest prime numbers are odd numbers. Furthermore, numbers such as $\sqrt{2}$ and $\pi$ are called irrational numbers because they cannot be expressed in terms of a fraction.

### Example 1.1

$\pi$ is irrational and we cannot express

$$\pi = 3.14159265358979323846264338327950288...$$

as a ratio or fraction, though the commonly used expression

$$22/7 \approx 3.142857142857...,$$

is just an approximation. Here, we have used the notation '$\approx$' to mean 'is approximately equal to'. Others common approximations of $\pi$ are $311/99$ and $355/113$ because

$$\frac{311}{99} = 3.14141414...,$$

and

$$\frac{355}{113} = 3.14159292035398230088...,$$

which is accurate to the 6th decimal place.

In fact, the decimal digits of an irrational number do not repeat periodically. If any part of the decimal repeats, such as $0.3333333...$, it can always be written as a fraction

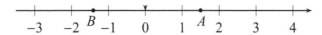

Figure 1.1: Real numbers and their representations (as points) on the number line.

such as 0.3333... = 1/3. Here, 22/7 = 3.142858142857... has a digit sequence 142857 that repeats.

Rational numbers and irrational numbers together make up the real numbers. In mathematics, all the real numbers are often denoted by $\mathbb{R}$ or $\mathfrak{R}$, and a real number corresponds to a unique point or location in the number line (see Fig. 1.1). For example, 3/2 corresponds to point $A$ and $-\sqrt{2}$ corresponds to point $B$.

## 1.1. Notations and Conventions

When approximating $\pi$, we used some notations such as $\approx$ and $\cdots$ without much explanation. We now briefly explain the conventions and notations we often use in engineering mathematics and certainly in this book.

We know that = means equality such as $2 + 3 = 5$, while $\neq$ means 'not equal to' such as $2 + 3 \neq 4$. However, we often use $\approx$ to mean 'is approximately equal to'. For example, we can write $9.99997 \approx 10$ and $\pi \approx 22/7$.

In order to avoid writing all the things out, we often use $n = 1, 2, ..., 10$ to mean exactly $n = 1, 2, 3, 4, 5, 6, 7, 8, 9, 10$. Similarly, when we write $n = 1, 2, 3, \cdots$, we mean all the numbers from 1 up to any large numbers. Basically, we use '$\cdots$' or '...' to mean 'etc'. In addition, we occasionally wish to include the infinity ($\infty$) in our discussion, and there is a positive infinity $+\infty$ and a negative infinity $-\infty$. Any number $n$ is less than $+\infty$, that is, $n < +\infty$. Any number is also greater than $-\infty$ (i.e., $-\infty < n$). Therefore, all numbers are between $-\infty$ and $+\infty$.

For simplicity to denote multiplication, we use $ab$ to mean $a \times b$ or $a * b$ and also $a \cdot b = ab = a * b = a \times b$ where $a$ and $b$ are any two numbers.

Sometimes, to write more compactly, we use $\pm$ to mean either $+$ or $-$. Thus, $2 \pm 3$ means $2 + 3$ or $2 - 3$.

## 1.2. Rounding Numbers and Significant Digits

In engineering, we often have to round up numbers when we do calculations. For example, it is not possible to write out all the digits of $\sqrt{3}$ because it is an irrational number. However, for the first 20 digits of $\sqrt{3}$, we can write

$$\sqrt{3} = 1.7320508075688772935, \tag{1.1}$$

which has 19 decimal places. Here '.' is the decimal point, and the first digit 7 after the decimal point is called first decimal figure or first decimal place. Similarly, '3' in 1.732 is the second decimal figure or second decimal place. Thus, the only digit '9' in the above expression is the 17th decimal place.

In practice, we do not need so many decimal places, and thus we need to carry out the so-called rounding of numbers. For a given number of decimal places, the basic rule for rounding numbers is to look at the number at next decimal place and decide by the following rules:

- If it is less than 5, round down.
- If it is 5 or more, then round up.

## Example 1.2

For example, for $\sqrt{2} = 1.41421356237...$, if we want to round it to 3 decimal places, the current number at the 3rd decimal place is 4. Now we look at the number at the next decimal place, which is 2, so we round it down and we have

$$1.41\underline{4}21356237 \approx 1.414.$$

If we wish to keep 6 decimal places, the current number at the 6th decimal place is 3, and now we look at the next decimal place, which is 5. Thus, we round it up and we have

$$1.41421\underline{3}56237 \approx 1.414214.$$

In science and engineering, we often have to measure a quantity such as temperature and pressure, then the accuracy determines the number of decimal places. Thus, we have to deal with significant digits. If the temperature in a room is 22.15°C, it has four significant digits. When we say the purity of gold is 99.999%, it has five significant digits or five significant figures.

However, care should be taken when dealing with digit 0. For example, 100 has one significant digit, while 100.0 has four significant digits as '.0' signifies something important. Similarly, if we write '100.', it means that it has three significant digits. In addition, 10005 has 5 significant digits, while 0.00005 has only one significant digit. Thus, all non-zero digits are significant, so are the zeros between digits. The zeros to the right of the decimal point are also significant, but zeros to the left of the non-zero digits are not. For example, 1234500 has five significant digits, while 123450.0 has seven significant digits. 0.01234 has four significant digits, while 0.01230 has four significant digits because the final zero in the decimal place is significant.

When adding up numbers with different significant digits, we have to round them to the appropriate number of significant digits. For example,

$$3.1415 + 3.7 = 6.8415, \tag{1.2}$$

is not appropriate because 3.1415 has five significant digits and is accurate to the 4th decimal place, while 3.7 has two significant digits and is accurate to the first decimal place. Thus, the final sum should also be accurate to the first decimal place. We should write

$$3.1415 + 3.7 \approx 6.8. \tag{1.3}$$

## Example 1.3

For multiplication, for example, we have

$$3.1415 \times 0.37 \times 9 = 10.461195, \tag{1.4}$$

but numbers of significant digits of 3.1415, 0.37 and 9 are 5, 2 and 1, respectively. Thus, we have to round the final answer to one significant digit, which means that $10.461195 = 10$. Thus, we have

$$3.1415 \times 0.37 \times 9 \approx 10, \tag{1.5}$$

where we have used $\approx$ to show such rounding and approximation.

It is worth pointing out that the rule of addition and subtraction focuses on the decimal place, while the multiplication and division focus on the number of significant digits. Let us look at another example.

## Example 1.4

If we naively do the calculations, we have

$$(2.71828 - 1.23) \times 3.1415 \div 11 = (1.48828 \times 3.1415)/11$$

$$= 4.67543162/11 = 0.425039238182.$$

This is not the answer with the appropriate number of significant digits.

Since 2.71828 has six significant digits and is accurate to the 5th decimal place, while 1.23 is only accurate to the 2nd decimal place, we should have

$$2.717828 - 1.23 = 1.49.$$

In addition, we have

$$1.49 \times 3.1415 = 4.68 \text{ (three significant digits)}$$

because 3.1415 has five significant digits and 1.49 has three significant digits. Therefore, the answer

$$4.68/11 = 0.4254545...$$

should have only two significant digits because 11 has two significant digits. Finally, we have

$$4.68/11 = 0.43.$$

That is

$$(2.71828 - 1.23) \times 3.1415 \div 11 \approx 0.43.$$

For example, a body of mass $m = 75.0$ kg on the Earth's surface (with the acceleration due to gravity $g = 9.80665$ m/s$^2$), its weight $W$ is

$$W = mg = 75.0 \times 9.80665 = 735.49875 \approx 735,$$

which is about 735 N and should have 3 significant digits.

## 2. Sets

Sometimes, it is much easier to discuss certain concepts using sets and set theory, especially in probability and statistics, which will be introduced in later chapters. Let us first introduce some basic concepts in set theory. A set is any well-defined collection of objects or elements, and the elements are the members or objects in a set. We conventionally use the upper-case letters to denote sets and lower-case letters for elements, and the listed elements are enclosed in the curly brace {}.

## Example 1.5

For example, all the following numbers: $2, 4, 6, 3, -1, 0, 3.7, 8$ form a set and thus can be denoted by

$$\mathscr{A} = \{2, 4, 6, 3, -1, 0, 3.7, 8\},$$

and the number 3.7 is a member of the set, and we can write this membership as

$$3.7 \in \mathscr{A} = \{2, 4, 6, 3, -1, 0, 3.7, 8\}. \tag{1.6}$$

Similarly, we have $2 \in \mathscr{A}$ and $0 \in \mathscr{A}$.

In general, the membership in a set is denoted using $\in$, thus

$$x \in \mathscr{A}, \tag{1.7}$$

means that '$x$ is a member of the set $\mathscr{A}$', while

$$\mathbf{x} \notin \mathscr{A}, \tag{1.8}$$

means that '$x$ is not a member of the set $\mathscr{A}$'.

A special set is the empty set or null set which has no element and is denoted by

$$\emptyset = \{\}, \tag{1.9}$$

which should not be confused with a non-empty set which consists of a single element $\{0\}$. Also, note the special notation $\emptyset$. Therefore, in the previous example, we have $0 \in \mathscr{A}$, but $\emptyset \neq \{0\}$.

We say that $\mathscr{A}$ is a subset of $\mathscr{B}$ if $a \in \mathscr{A}$ implies that $a \in \mathscr{B}$. That is to say that all the members of $\mathscr{A}$ are also members of $\mathscr{B}$. We denote this relationship as

$$\mathscr{A} \subseteq \mathscr{B}. \tag{1.10}$$

If all the members of $\mathscr{A}$ are also members of $\mathscr{B}$, but there exists at least one element $b$ such that $b \in \mathscr{B}$ while $b \notin \mathscr{A}$, we say $\mathscr{A}$ is a *proper* subset of $\mathscr{B}$, and denote this relationship as

$$\mathscr{A} \subset \mathscr{B}. \tag{1.11}$$

When combining sets, we say that $\mathscr{A}$ union $\mathscr{B}$, denoted by

$$\mathscr{A} \cup \mathscr{B}$$

forms a set of all elements that are in $\mathscr{A}$, or $\mathscr{B}$, or both. On the other hand, $\mathscr{A}$ intersect $\mathscr{B}$, written as

$$\mathscr{A} \cap \mathscr{B},$$

is the set of all elements that are in both $\mathscr{A}$ *and* $\mathscr{B}$.

A universal set $\Omega$ is the set that consists of all the elements under consideration. The complement set of $\mathscr{A}$, or *not* $\mathscr{A}$ denoted by $\bar{\mathscr{A}}$, is the set of all the elements that are not in $\mathscr{A}$. The set $\mathscr{A} - \mathscr{B}$ (or $\mathscr{A}$ minus $\mathscr{B}$) is the set of elements that are in $\mathscr{A}$ and not in $\mathscr{B}$, this is equivalent to removing or substracting from $\mathscr{A}$ all the elements that are in $\mathscr{B}$. This leads to

$$\mathscr{A} - \mathscr{B} = \mathscr{A} \cap \bar{\mathscr{B}}, \tag{1.12}$$

and

$$\bar{\mathscr{A}} = \Omega - \mathscr{A}. \tag{1.13}$$

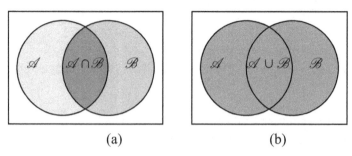

(a)                              (b)

Figure 1.2: Venn diagrams: a) $\mathscr{A} \cap \mathscr{B}$, b) $\mathscr{A} \cup \mathscr{B}$.

## Example 1.6

For two sets

$$\mathscr{A} = \{2, 3, 5, 7\}, \quad \mathscr{B} = \{2, 4, 6, 8, 10\},$$

and a universal set

$$\Omega = \{1, 2, 3, 4, 5, 6, 7, 8, 9, 10\},$$

it is straightforward to check that

$$\mathscr{A} \subset \Omega, \quad \mathscr{B} \subset \Omega, \quad \mathscr{A} - \mathscr{B} = \{3, 5, 7\},$$

$$\mathscr{A} \cup \mathscr{B} = \{2, 3, 4, 5, 6, 7, 8, 10\}, \quad \mathscr{A} \cap \mathscr{B} = \{2\},$$

$$\overline{\mathscr{A}} = \Omega - \mathscr{A} = \{1, 4, 6, 8, 9, 10\},$$

and finally $\overline{\mathscr{A}} \cap \mathscr{A} = \emptyset$.

A better way to represent such abstract mathematical operations of sets is to use the Venn diagrams as the topological representation. Fig. 1.2 represents the intersect $\mathscr{A} \cap \mathscr{B}$ and union $\mathscr{A} \cup \mathscr{B}$, while Fig. 1.3 represents $\mathscr{B} - \mathscr{A}$, $\overline{\mathscr{A}} = \Omega - \mathscr{A}$ and $\mathscr{B} \subset \mathscr{A}$.

Some common sets in mathematics are used so often that they deserve special names or notations. These include:

- $\mathbb{N} = \{1, 2, 3, ...\}$ or $\mathbb{N} = \{0, 1, 2, 3, ...\}$ denotes the set of all natural numbers.
- $\mathbb{Z} = \{..., -3, -2, -1, 0, 1, 2, 3, ...\}$ is the set of all integers.
- $\mathbb{P} = \{2, 3, 5, 7, 11, 13, 17, 19, 23, ...\}$ is the set of all primes.
- $\mathbb{Q} = \{\frac{m}{n} : m, n \in \mathbb{Z}, n \neq 0\} = \{-\frac{1}{2}, 3, -\frac{4}{5}, \frac{22}{7}, ...\}$ is the set of all rational numbers.
- $\mathbb{R}$ is the set of all real numbers consisting of all the rational numbers and all the

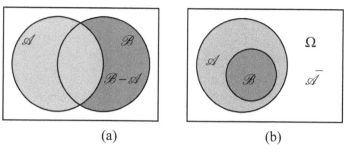

Figure 1.3: Venn diagrams: a) $\mathscr{B} - \mathscr{A}$, b) $\bar{\mathscr{A}} = \Omega - \mathscr{A}$ and $\mathscr{B} \subset \mathscr{A}$.

irrational numbers such as $\sqrt{2}$, $\sqrt{3}$ ad $\pi$.

Sometimes, we may have to deal with division by zero and/or infinity $\infty$ in mathematics. By including the positive infinity $(+\infty)$ and negative infinity $(-\infty)$, the real number system becomes the so-called affinely extended real number system, denoted by $\bar{\mathbb{R}}$. That is $\bar{\mathbb{R}} = \mathbb{R} \cup \{-\infty, +\infty\}$.

All these sets have an infinite number of elements. From the definitions, it is straightforward to check that

$$\mathbb{P} \subset \mathbb{N} \subset \mathbb{Z} \subset \mathbb{Q} \subset \mathbb{R} \subset \bar{\mathbb{R}}. \tag{1.14}$$

## 3. Equations

In engineering mathematics, expressions such as $2x + 3$ often contain an unknown quantity such as $x$, and this unknown quantity is called a variable, which can be any value in the real numbers. To determine the actual value of the variable, a condition is required, and such a condition is often written as an equation.

In general, an equation is a mathematical statement that is written in terms of symbols and an equal sign '='. All the terms on the left of '=' are collectively called the left-hand side (LHS), while those on the right of '=' are called the right-hand side (RHS). For example, $2x + 3 = 5$ is a simple equation with its left-hand side being $2x + 3$ and right-hand side being 5.

An equation has the following properties and thus can be manipulated mathematically as follows:

- Any quantity can be added to, subtracted from or multiplied by both sides of the equation;
- A non-zero quantity can divide both sides;
- Any function such as power (such as $x^2$) and surds (such as $\sqrt{2}$) can be applied to both sides equally.

After such manipulations, an equation may be converted into a completely different equation, though we can obtain the original equation if we can carefully reverse the above procedure, though special care is needed in some cases.

## 3.1. Simple Equation

The idea of these manipulations is to transform the equation to a much simpler form whose solution can be found easily. Let us look at a simple example.

### Example 1.7

For equation

$$3x + 5 = 17 - x,$$

we can first add $+x$ on both sides and we have

$$\underbrace{3x}\ \underbrace{+5}\ \underbrace{+x} = 17\ \underbrace{-x+x,}$$

which becomes

$$4x + 5 = 17.$$

Subtracting 5 from both sides, we have

$$4x + 5 - 5 = 17 - 5, \quad \text{or} \quad 4x = 12.$$

Dividing both sides by 4, we have $x = 3$.

Similarly, from equation

$$x\sqrt{2} = \sqrt{14} + \sqrt{2}, \tag{1.15}$$

we can divide both sides by $\sqrt{2}$ (or multiply by $1/\sqrt{2}$) and we have

$$\frac{x\cancel{\sqrt{2}}}{\cancel{\sqrt{2}}} = \frac{\sqrt{14}}{\sqrt{2}} + \frac{\cancel{\sqrt{2}}}{\cancel{\sqrt{2}}}, \tag{1.16}$$

which gives

$$x = \frac{\sqrt{14}}{\sqrt{2}} + 1 = \frac{\sqrt{2 \times 7}}{\sqrt{2}} + 1 = \frac{\cancel{\sqrt{2}} \times \sqrt{7}}{\cancel{\sqrt{2}}} + 1 = \sqrt{7} + 1.$$

Equations can have many applications and many physical laws and rules are stated using equations. In many applications, when we try to determine some unknown quantities, we have to re-arrange the equation slightly so as to solve the unknown variable or quantity from known quantities. Let us see a simple example.

## Example 1.8

Ohm's law provides a relationship between the current ($I$) through a conductor with a resistance ($R$) to the voltage ($V$) applied across the conductor. That is

$$I = \frac{V}{R}. \tag{1.17}$$

In this equation, the right units are also important. Here, the unit of $I$ is amperes or amps (A), the unit of $V$ is volts (V) and the unit of $R$ is ohms ($\Omega$).

Thus, for $V = 5$ volts and $R = 2.5$ ohms, its current $I$ is

$$I = \frac{V}{R} = \frac{5}{2.5} = 2 \text{ amps} = 2 \text{ A}.$$

If the current required is 10 A, then the voltage for the same conductor/resistor has to increase to

$$V = IR = 10 \times 2.5 = 25 \text{ volts}.$$

In addition, the power dissipated by the resistor can be calculated by

$$P = IV = I \cdot IR = I^2 R,$$

or

$$P = IV = \frac{V}{R} \cdot V = \frac{V^2}{R} = I^2 R.$$

So in the case of $I = 10$ A, the power dissipation is

$$P = IV = 10 \times 25 = 250 \text{ watts}.$$

## 3.2. Simultaneous Equations

The fundamental ideas of mathematical operations can also be applied to a system of equations. In order to solve two simultaneous equations for the unknowns $x$ and $y$

$$\begin{cases} 2x & +y & = 5, \\ x & -3y & = -1, \end{cases} \tag{1.18}$$

we first multiply both sides of the first equation by 3, we have

$$3 \times (2x + y) = 3 \times (5), \tag{1.19}$$

and we get

$$6x + 3y = 15. \tag{1.20}$$

Now we add this equation (thinking of the same quantity on both sides) to the

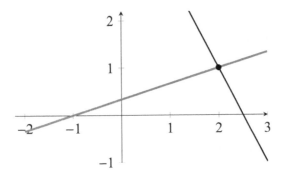

Figure 1.4: The solution corresponds to the cross point by two lines.

second equation of the system. We have

$$
\begin{array}{rll}
6x & +3y & = 15 \\
x & -3y & = -1 \\
\hline
7x & +0 & = 14.
\end{array}
\tag{1.21}
$$

This means that

$$
7x = 14, \quad \text{or} \quad x = \frac{14}{7} = 2.
\tag{1.22}
$$

Now substitute this solution $x = 2$ into one of the equations (say, the first); we have

$$
2 \times (2) + y = 5,
\tag{1.23}
$$

or

$$
4 + y = 5.
\tag{1.24}
$$

This simply leads to $y = 5 - 4 = 1$. Thus, we finally get $x = 2$ and $y = 1$. As each equation corresponds to a line in the two-dimensional (2D) Cartesian plane, the cross point $(2,1)$ of the two lines is the solution we just obtained (see Fig. 1.4).

In general, for $n$ unknowns or variables, we need a system of $n$ equations. In this case we usually have to deal with a system of linear equations, often in terms of matrix forms, which will be introduced later in this book.

## 3.3. Inequality

The linear equations we have discussed so far are essentially equalities. In an equality, the left-hand side is always equal to the right-hand side. In the case when the left-hand side is greater than or less than the right-hand side, we have to deal with an inequality. For example, $x > 2$ is a simple inequality because it states that the value of $x$ must be greater than 2. Thus, $x = 10, 10^9, 3.14$ and 2.2 all satisfy the inequality. In addition, the values 2.01 and 2.00001 are also valid, but $x = 2$ does not satisfy the inequality $x > 2$. If we wish to include 2 in the inequality, we can write

$$x \geq 2, \tag{1.25}$$

which means that $x$ is 'greater than' or 'equal to' 2. Similarly, $x < 0$ is an inequality that states all the values of $x$ must be negative.

Inequalities can be manipulated in almost the same ways as those for equations. We can add or subtract a term on both sides and we can also multiply both sides by the same positive number. From $x \geq 2$, we can have

$$x + 7 \geq 2 + 7, \quad \text{or} \quad x + 7 \geq 9. \tag{1.26}$$

Similarly, we also have

$$3(x + 7) \geq 3 \times 9, \tag{1.27}$$

which leads to

$$3(x + 7) \geq 27. \tag{1.28}$$

However, when we multiply a negative number, the inequality sign must be reversed. That is, ' $>$' (or ' $<$') becomes ' $<$' (or ' $>$'). For example, from $x \geq 2$, we have

$$(-3) \times x \leq (-3) \times 2, \tag{1.29}$$

which means that

$$-3x \leq -6. \tag{1.30}$$

We can also take the square of both sides, thus we have

$$(x)^2 \geq (2)^2, \tag{1.31}$$

which gives

$$x^2 \geq 4. \tag{1.32}$$

Care should be taken in this case. For example, from (1.30), we have to reverse the inequality side because the square operation in this case is equivalent to multiplying by a negative number (itself). Thus, we have

$$(-3x)^2 \geq (-6)^2, \tag{1.33}$$

or

$$9x^2 \geq 36. \tag{1.34}$$

Though we can obtain $x^2 \geq 4$ from $x \geq 2$, they are not exactly equivalent because we can also get $x^2 \geq 4$ from $x \leq -2$. From

$$x^2 \geq 4, \tag{1.35}$$

we can take the square root of both sides, we have two cases:

$$x \geq 2, \quad \text{or} \quad x \leq -2. \tag{1.36}$$

## Example 1.9

Let us simplify the following inequality

$$(x - 3)^2 + 5 \geq 9.$$

Adding $-5$ on both sides, we have

$$(x - 3)^2 \geq 4,$$

whose square roots become

$$(x - 3) \geq 2, \quad \text{or} \quad (x - 3) \leq -2.$$

So they become

$$x \geq 5, \quad \text{or} \quad x \leq 1.$$

## 4. Functions

Loosely speaking, a function is a quantity (say $y$) which varies with another independent quantity $x$ in a deterministic way. The area $y$ of a circle with a radius $r$ is simply $\pi r^2$. To use the most generic notation for an independent variable $x$, we now use $x$ to represent the radius, so we have

$$y = \pi x^2, \tag{1.37}$$

which is an example of simple quadratic functions.

For any given value of $x$, there is a unique corresponding value of $y$. By varying $x$ smoothly, we can vary $y$ in such a manner that the point $(x, y)$ will trace out a curve on the $x$-$y$ plane (see Fig. 1.5). Thus, $x$ is called the independent variable, and $y$ is called the dependent variable or function. Sometimes, in order to emphasize the relationship as a function, we use $f(x)$ to express a generic function, showing that it is a function of $x$. This can also be written as $y = f(x)$.

For any real number $x$, there is a corresponding unique value $y = f(x) = \pi x^2$. Although a negative radius is meaningless physically, it can be valid mathematically.

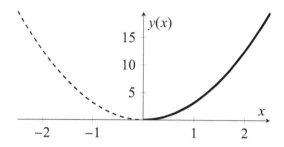

Figure 1.5: The graph of function $y = f(x) = \pi x^2$.

This relationship is a one-to-one mapping from $x$ to $y$ for $x \geq 0$. That is to say, for a given value of $x \geq 0$ such as $x = 1$, there is a unique value of $y$ that corresponds to each $x$. When $x = 1$, we have $y = \pi x^2 = \pi$. However, mathematically speaking, $x = -1$ can also correspond to the same $y = \pi$, though $x = -1$ as a radius is meaningless (see the dashed curve in Fig. 1.5). This means the mapping from $x$ to $y$ is not a one-to-one mapping if negative values of $x$ are allowed.

Strictly speaking, functions such as $x^2$ and $x^3$ are power functions. We will discuss them in more detail when index notations and polynomials are introduced later in the next chapter.

## 4.1. Domain and Range

The domain of a function is the set of numbers $x$ for which the function $f(x)$ is defined validly. If a function is defined over a range $a \leq x \leq b$, we say its domain is $[a, b]$ which is called a closed interval. If $a$ and $b$ are not included, we have $a < x < b$ which is denoted as $(a, b)$, and we call this interval an open interval. If $b$ is included, while $a$ is not, we have $a < x \leq b$, and we often write this half-open and half-closed interval as $(a, b]$. Thus, the domain of function $f(x) = x^2$ is the whole set of all real numbers $\mathbb{R}$, so we have

$$f(x) = x^2, \qquad -\infty < x < +\infty. \tag{1.38}$$

Here the notation $\infty$ means infinity. In this case, the domain of $f = x^2$ is all the real numbers, which can be simply written as $\mathbb{R}$. That is $x \in \mathbb{R}$.

All the values that a function can take for a given domain form the range of the function. Thus, the range of $y = f(x) = x^2$ is $0 \leq y < +\infty$ or $[0, +\infty)$. Here, the closed bracket '[' means that the value (here 0) is included as part of the range, while the open round bracket ')' means that the value (here $+\infty$) is not part of the range.

As we have seen earlier when we discussed the area of a circle, there is some difference between the domain of physical quantities and mathematical quantities. As most physical quantities are non-negative, it is important in engineering to ensure the quantities are meaningful. Some values may have mathematical meaning, but may not have any real physical meaning such as negative radius.

For example, the dewpoint or dewpoint temperature ($T_d$) is an important concept in weather forecasting, and it is a special temperature to which the moist air must be cooled so as to reach saturation (forming dew) under the conditions of constant pressure and constant water content. It is closely related to the moisture or water content in the air. Relative humidity $R$ is the ratio of the partial pressure of water vapor in the mixture of air and water to the equilibrium water vapor pressure, which is a percentage between 0 to 100%.

## Example 1.10

When the relative humidity $R$ is above 50%, there is an approximate linear function to estimate the dewpoint temperature for given dry air temperature $T$ and $R$. We have

$$T_d \approx T - \frac{(100 - R)}{5}.$$

As the fraction (the second term) on the right-hand side of the above equation is always positive for $R < 100$, this means that the dewpoint temperature is always less than the air temperature. When $R = 100\%$ (that is the air is saturated with water vapor), we have $T_d = T$ for $R = 100$.

For example, for $T = 30°C$ and $R = 60\%$, we have

$$T_d \approx 30 - \frac{(100 - 60)}{5} = 30 - \frac{40}{5} = 30 - 8 = 22,$$

which is 22°C. When the dewpoint is above 20°C, it will be uncomfortable for most people. In practice, the most comfortable dewpoints for most people range from 10°C to 16°C.

It is worth pointing out that the above formula becomes significantly inaccurate when $R < 50\%$. For example, for $T = 15°C$ and $R = 10\%$, we have

$$T_d \approx 15 - \frac{(100 - 10)}{5} = 15 - 18 = -3,$$

which is ridiculously inaccurate as it is below the freezing point.

Sometimes, it is more convenient to write a function in a concise mathematical language. For example, for $f(x) = \frac{1}{2}x^2 + 2x + 3$, we can write

$$f : x \mapsto \frac{1}{2}x^2 + 2x + 3, \qquad x \in \mathbb{R}, \qquad (1.39)$$

which is called a mapping. This means that $f(x)$ is a function of $x$ and the actual role of this function is to turn any real number $x$ (input) in the domain into another number (output) in the corresponding range: $\frac{1}{2}x^2 + 2x + 3$.

## Example 1.11

When we interpret the meaning of a function, we should think of it as a relationship between its argument $x$ (the input) and another quantity (the output). Thus, the input can be another expression when appropriate.

For example, for a function

$$y = f(x) = \frac{1}{2}x^2 + 2x + 3,$$

when $x = 2$, we have

$$f(2) = \frac{1}{2} \times 2^2 + 2 \times 2 + 3 = 2 + 4 + 3 = 9.$$

Similarly, for $x = \sqrt{2}$, we have

$$y = f(\sqrt{2}) = \frac{1}{2}(\sqrt{2})^2 + 2\sqrt{2} + 3 = \frac{1}{2} \times 2 + 2\sqrt{2} + 3 = 2\sqrt{2} + 4.$$

However, if we replace $x$ by $a - 2$, we have

$$f(x) = f(a-2) = \frac{1}{2}(a-2)^2 + 2(a-2) + 3 = \frac{1}{2}(a^2 - 4a + 4) + 2a - 4 + 3$$

$$= \frac{a^2}{2} \underbrace{-2a}_{} + 2 + \underbrace{2a}_{} - 1 = \frac{a^2}{2} + 1.$$

In addition, if $x$ is replaced by $-a$ (that is $x = -a$), we have

$$f(-a) = \frac{1}{2}(-a)^2 + 2(-a) + 3 = \frac{a^2}{2} - 2a + 3.$$

Therefore, a function should be considered as a relationship in the most general sense, which will become useful when discussing Fourier and Laplace transforms as well as convolution integrals, which will be introduced in later chapters.

## 4.2. Linear Function and Modulus Function

The simplest general function is probably the linear function

$$y = kx + c, \tag{1.40}$$

where $k$ and $c$ are real constants. For example, when $k = 1$ and $c = 0$, we have $y = x$, which corresponds to a simple straight line through the origin $(0,0)$ with a gradient

$k = 1.$

## Example 1.12

The relationship between the Fahrenheit (°F) and degrees Celsius (°C) for temperature can be written as a linear function

$$F = \frac{9}{5}C + 32 = 1.8C + 32,$$

which gives the intercept of 32 and the gradient of 1.8. The above equation can be rewritten as

$$C = \frac{F - 32}{1.8} = \frac{5(F - 32)}{9},$$

to convert from Fahrenheit to degrees Celsius.

The normal human body temperature is about $C = 36.8°C$, what is the temperature in Fahrenheit? From the above equation, we have

$$F = 1.8C + 32 = 1.8 \times 36.8 + 32 = 98.24,$$

which is about 98°F. Obviously, the temperature of a human body will vary slightly, depending on where the measurement is taken.

Now let us ask if there is a temperature at which both the Fahrenheit and Celsius readings will be the same?

From $F = 1.8C + 32$, if the readings are the same (i.e., $F = C$), we have

$$F = 1.8F + 32,$$

which gives

$$(1 - 1.8)F = 32, \quad \text{or} \quad -0.8F = 32,$$

which leads to

$$F = \frac{32}{-0.8} = -40.$$

That is, the reading at -40°C is the same as -40°F.

A special function is the modulus function $|x|$ which is defined by

$$|x| = \begin{cases} x & \text{if } x \geq 0, \\ -x & \text{if } x < 0. \end{cases} \tag{1.41}$$

That is to say, $|x|$ is always non-negative. For example, $|5| = 5$ and $|-5| = 5$.

It is worth pointing out that the modulus function is not linear in general, even though it shares many similarity with linear functions.

The modulus function has the following properties

$$|a \times b| = |a| \times |b|, \qquad a, b \in \mathbb{R}, \tag{1.42}$$

which means that $|a^2| = |a|^2 = a^2 \geq 0$ for any $a \in \mathbb{R}$. Similarly, for any real numbers $a$ and $b \neq 0$, we have

$$\left|\frac{a}{b}\right| = \frac{|a|}{|b|}. \tag{1.43}$$

## Example 1.13

For example, when $a = -3$, $b = 4$ and $c = 5$, we have

$$|a| = |-3| = 3, \quad |b| = |4| = 4.$$

so we have

$$|a \times b| = |-3 \times 4| = |-12| = 12, \quad |a| \times |b| = 3 \times 4 = 12,$$

which indeed means that

$$|a \times b| = |a| \times |b|.$$

Similarly, we have

$$\left|\frac{a}{b}\right| = \left|\frac{-3}{4}\right| = \frac{3}{4},$$

which is the same as $|a|/|b| = 3/4$. It is also straightforward to show that

$$|abc| = |a|\,|b|\,|c| = 3 \times 4 \times 5 = 60.$$

But for addition in general, we have $|a + b| \neq |a| + |b|$. Here, the sign $\neq$ means that the left-hand side is not equal to the right-hand side.

In fact, there is an inequality in this case: $|a + b| \leq |a| + |b|$. The sign $\leq$ means less than or equal to. Obviously, the equality holds in the above case when either $a = 0$ or $b = 0$. We will discuss this in detail in the vector analysis in later chapters.

From the previous example, we know that

$$|a + b| = |-3 + 4| = |1| = 1, \quad |a| + |b| = |-3| + |4| = 3 + 4 = 7,$$

which gives that $|a| + |b| > |a + b|$. Similarly, we have

$$|a + b + c| = |-3 + 4 + 5| = |6| = 6, \quad |c| = |5| = 5,$$

so we have

$$|a| + |b| + |c| = 3 + 4 + 5 = 12 > |a + b + c| = 6.$$

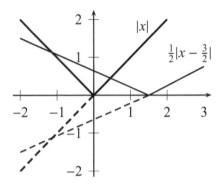

Figure 1.6: Graphs of modulus functions.

Two examples of modulus functions $f(x) = |x|$ and $f(x) = \frac{1}{2}|x - \frac{3}{2}|$ are shown as solid lines in Fig. 1.6. The dashed lines correspond to $x$ and $\frac{1}{2}(x - \frac{3}{2})$ (without the modulus operator), respectively.

### 4.3. Power Functions

When we calculate the area of a circle, we have used $x^2$ that is a power function. In general, a power function can be defined as

$$f(x) = x^p, \tag{1.44}$$

where $p$ is a constant, called the exponent of the power function. We usually read $x^p$ as $x$ to the $p$th power or simply $x$ to the $p$. In the case when $p$ is a positive integer, it becomes the familiar powers $x$, $x^2$, $x^3$ as we learned in school. In this case, $x$ can be any real numbers, thus the domain is $x \in \mathbb{R}$ if $p$ is a positive integer.

Many useful relationships between physical quantities can be expressed by using power functions. For example, the surface area and volume of a sphere with a radius $r$ are $4\pi r^2$ and $4\pi r^3/3$, respectively.

### Example 1.14

The well-known Stefan-Boltzmann law for black-body radiation can be written as

$$E = \sigma T^4,$$

where $E$ is the energy radiated per unit surface area per unit time (for all wavelengths), $T$ is the absolute temperature in Kelvin (K) and $\sigma = 5.67 \times 10^{-8}$ W/(m$^2$

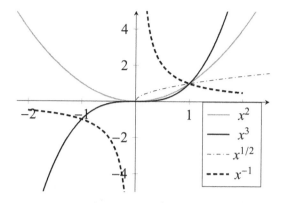

Figure 1.7: Graphs of power functions $x^p$ with $p = 2, 3, 1/2$ and $-1$.

K$^4$) is the Stefan-Boltzmann constant.

In general, $p$ can be any real number. For example, when $p = 1/2$, we have the square root function

$$f(x) = x^{1/2} = \sqrt{x}. \tag{1.45}$$

In this case, its domain is $x \geq 0$ because a negative number does not have a root in the real domain. The graphs of some power functions with different $p$ values are given in Fig. 1.7.

## Example 1.15

As an example, we now try to calculate the time taken for an object to fall freely from a height of $h$ to the ground. If we assume the object is released with a zero velocity at $h$, we know that the distance $h$ it travels can be determined by

$$h = \frac{1}{2}gt^2, \tag{1.46}$$

where $t$ is the time taken, and $g = 9.8$ m/s$^2$ is the acceleration due to gravity. Now we have

$$t = \sqrt{2h/g}. \tag{1.47}$$

The velocity just before it hits the ground is $v = gt$ or

$$v = g \sqrt{\frac{2h}{g}} = \sqrt{2gh}. \tag{1.48}$$

For example, the time taken for an object to fall from $h = 20$ m to the ground is about $t = \sqrt{2 \times 20/9.8} \approx 2$ seconds.

The velocity as it approaches the ground is

$$v \approx \sqrt{2 \times 9.8 \times 20} \approx 19.8 \text{ m/s}.$$

Obviously, as $h$ increases, $t$ will also increase, so does the velocity. Interestingly, the leaking velocity of water from a leaking water tank can be approximated by $\sqrt{2gh}$ where $h$ is the depth of the water column above the orifice, thus the discharge $Q$ can be calculated by $Q = a\sqrt{2gh}$ where $a$ is the area of the leaking orifice.

In many applications, the inverse-square laws with exponent $p = -2$ are specially relevant where a physical quantity varies with the distance $r$ in the form of $r^{-2}$ or $1/r^2$. Good examples of inverse-square laws are Newton's law of gravitation, Coulomb's law of electrostatics, and the variation of light intensity with distance.

## 4.4. Exponentials and Logarithms

An exponential function can be written in the following generic form:

$$f(x) = b^x, \tag{1.49}$$

where $b > 0$ is a known or given real constant, called the base, while the independent variable $x$ is called the exponent. In the context of real numbers, we have to limit our discussion here to $b > 0$.

An interesting extension for $f(x) = b^x$ is that

$$f(-x) = b^{-x} = \frac{1}{b^x} = (\frac{1}{b})^x. \tag{1.50}$$

In this case, we have to be careful. When $b < 0$, for some values of $x$, $b^x$ will be meaningless. For example, if $b = -3$ and $x = 1/2$, then $(-3)^{1/2} = \sqrt{-3}$ would be meaningless. Similarly, say, when $b = 0$ and $x = -5$, then $b^x = 0^{-5} = 1/0^5$ would also be meaningless.

A widely used exponential function, called natural exponential function, is when the base $b = e = 2.7182818284...$ is a special constant $e$ (often called Euler's number). Thus, we have

$$y = e^x = \exp(x). \tag{1.51}$$

Here $e = 2.718281828459045235360287...$ is an irrational number.

## Example 1.16

For example, the discharge of a capacitor with the initial charge $Q_0$ will be governed by

$$Q(t) = Q_0 e^{-x/\tau},$$

where $\tau$ is the time constant. In addition, the barometric formula for pressure variation $P$ with height $z$ is approximately

$$P = P_0 e^{-z/\bar{h}},$$

where constant $\bar{h}$ is the scale height (about 8.4 km) and $P_0$ is the pressure on the Earth's surface $(z = 0)$.

In addition, the well-known Boltzmann distribution is an exponential function

$$P = A e^{-E/k_B T},$$

where $P$ is the probability of the state with energy $E$. $T$ is the absolute temperature of the system and $k_B = 1.38 \times 10^{-23}$ J/K is the Boltzmann constant. Here, $A$ is a normalization constant.

---

In addition, the ratio of two exponential functions $a^x$ and $b^x$ is

$$\frac{a^x}{b^x} = (\frac{a}{b})^x. \tag{1.52}$$

Now suppose we know the value of $b > 0$ and the value of the function $y = b^x$, the question is if we can determine $x$ given $b$ and $y$. This is essentially to calculate the inverse of the exponential function so as to find $x$ that satisfies

$$b^x = y. \tag{1.53}$$

The notation for such inverse is

$$x = \log_b y, \tag{1.54}$$

where $b$ is again the base. That is to say, $x$ is the logarithm of $y$ to the base $b$. As discussed earlier, $y = b^x > 0$ is always positive for $b > 0$, the logarithm is only possible for $y > 0$ and $b > 0$. Since $b^0 = 1$, we have $0 = \log_b 1$. From $b^1 = b$, we have $1 = \log_b b$. In addition, we can also combine the above two equations, and we have

$$b^{\log_b y} = y. \tag{1.55}$$

## Example 1.17

Claude Shannon's information theory laid the foundation for modern digital communications. The key ideas are Shannon's <u>bit</u> and entropy. In order to transmit a series of 0s and 1s, it is useful to know the information contents they contain. For example, with a simple 'yes' (1) or 'no' (0), each digit has a 50-50 chance to appear. That is, 1 may appear $p = 1/2$ of times, so is 0. Shannon's information is defined as

$$I = -\log_2 p.$$

Here, the base is 2.

If you ask your friend whether he or she will come to a party, the information content of a simple answer 'yes' (or 'no') is exactly $I = -\log_2 \frac{1}{2} = 1$ (bit).

It is worth pointing out that the input argument of a logarithm function cannot be zero or negative. That is, the domain of

$$y = \log_b(x), \quad b > 0, \tag{1.56}$$

is $x > 0$. If $x > b$, then $y > 1$. In addition, $x = 1$ gives $y = \log_b(1) = 0$. In the case $0 < x < 1$, then $y < 0$. The range of $y$ is thus $-\infty < y < +\infty$.

## Example 1.18

For a function $y = \log_2(x + 1)$, its domain can be derived from the condition $x + 1 > 0$. That is $x > -1$. Similarly, for the following function

$$y = \log_b(x^2 - 4), \quad b > 0,$$

it requires that

$$x^2 - 4 > 0,$$

which leads to

$$x > 2, \quad \text{or} \quad x < -2.$$

Thus, the domain of this function is $(-\infty, -2) \cup (2, +\infty)$.

Since, for any real numbers $p, q$ and $b > 0$, we can define two new variables $u = b^p$ and $v = b^q$, thus we have

$$y = b^p b^q = uv = b^{p+q}. \tag{1.57}$$

From the definition of the logarithm, we now have

$$p + q = \log_b y = \log_b(uv). \tag{1.58}$$

Since $p = \log_b u$ and $q = \log_b v$, we finally obtain

$$\log_b(uv) = p + q = \log_b u + \log_b v. \tag{1.59}$$

This is the multiplication rule. In a special case when $u = v$, we have

$$\log_b(u \cdot u) = \log_b(u^2) = \log_b u + \log_b u = 2 \log_b u. \tag{1.60}$$

If we replace $v$ by $u^2$, we can get

$$\log_b(u \cdot u^2) = \log_b u + 2 \log_b u = 3 \log_b u. \tag{1.61}$$

In fact, if we follow a similar procedure, we finally obtain

$$\log_b u^n = n \log_b u, \quad n = 1, 2, 3, ..., \tag{1.62}$$

which can be extended to any value of $n$, including negative values and real numbers. In general, we have

$$\log_b[u(x)^{f(x)}] = f(x) \log_b[u(x)], \tag{1.63}$$

as long as the domains are valid (i.e., $u^f > 0$ and $u > 0$). For example, we have

$$\log_b 2^x = x \log_b 2, \tag{1.64}$$

and

$$\log_b a^{-x^2} = -x^2 \log_b a, \quad a > 0. \tag{1.65}$$

Even though any base $b > 0$ is valid, some bases are convenient for calculating logarithms and thus are more widely used than others. For historical reasons and because of the decimal systems, base $b = 10$ is widely used and we write $\log_{10} u$.

Another special base is the base $e$ for natural or Napierian logarithms where

$$e = 2.7182818284..., \tag{1.66}$$

and in this case, we simply write the logarithm as ln, using a special notation to distinguish from the common logarithms (log).

$$\ln u \equiv \log_e u. \tag{1.67}$$

The reasons why $e = 2.71828...$ is a natural choice for the base are many, and readers can refer to some textbooks about the history of mathematics. One of the reasons is that its corresponding exponential function

$$y = e^x. \tag{1.68}$$

Sometimes, it is written explicitly as $\exp(x)$ or $\exp[x]$.

Obviously, functions such as $\exp(x)$, $\exp(-x)$ and $\exp(-x^2)$ appear in a wide range of real-world applications. The graphs of some exponential functions and $\ln(x)$ are

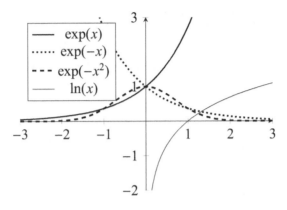

Figure 1.8: Exponential functions and logarithm.

shown in Fig. 1.8.

The well-known formula for entropy $S$ discovered by Ludwig Boltzmann can be written as

$$S = k_B \ln W,$$

where $k_B = 1.38064852 \times 10^{-23}$ J/K is the Boltzmann constant and $W$ is the probability in terms of state permutations.

In fact, natural logarithms have many applications in science and engineering, from earthquakes and number theory to music and fractals.

## Example 1.19

The widely used Decibel (dB) is a logarithm-based unit for expressing the ratio of a physical quantity relative to a reference level. For example, the acoustic sound pressure level is defined as

$$L = 20 \log_{10} \left( \frac{A}{A_0} \right) \text{dB},$$

where $A_0$ is the reference sound pressure and $A_0 = 20 \times 10^{-6}$ (i.e., 20 $\mu$Pa or 20 micropascals) is used in air. Thus, for a pressure level of 90 dB, we have

$$20 \log_{10}(\frac{A}{A_0}) = 90 \text{ dB},$$

which gives

$$\log_{10}(\frac{A}{A_0}) = 4.5, \quad \text{or} \quad \frac{A}{A_0} = 10^{4.5},$$

which gives the sound pressure level as

$$A = 10^{4.5} A_0 = 20 \times 10^{-6} \times 10^{4.5} = 20 \times 10^{-1.5} \approx 0.632 \text{ Pa.}$$

It is worth pointing out that the above definition of dB is for field quantities ($A$). If the power is concerned, the definition will be different by a factor of 2 because the power is proportional to the square of the field quantity (i.e., $A^2$). In this case, we have

$$r = 10 \log_{10} \left( \frac{P}{P_0} \right) \text{dB,} \qquad (1.69)$$

where $P_0$ is the power at the reference level. Even though dB is a logarithmic unit, it is in essence a dimensionless unit because it is the logarithm of a ratio.

## Example 1.20

For different applications, different reference levels can be used. For example, in describing the power of WiFi signals, $P_0 = 1$ miliwatt is often used. In order to show the different reference, dB units have different variations. If $P_0 = 1$ miliwatt (mW), the unit becomes dBm. Thus, we have

$$r = 10 \log_{10} \frac{P}{1 \text{ mW}},$$

or

$$P = 10^{r/10} \text{ mW.}$$

Thus, for a radio transmission power 25 dB, we have

$$P = 10^{25/10} = 10^{2.5} = 316.2 \text{ mW.}$$

while, for a WiFi signal strength of -20 dBm, its power is

$$P = 10^{-20/10} = 10^{-2} \text{ mW} = 10 \text{ } \mu\text{W.}$$

The calculations of logarithms are not straightforward, so they were commonly listed in various mathematical tables before computers became widely used.

Sometimes, the calculation of logarithm becomes simpler if we can change its base. The so-called change-of-base formula can be written as

$$\log_b(x) = \frac{\log_k(x)}{\log_k(b)}, \qquad (1.70)$$

where $k > 0$ is the new base. This formula changes the calculation of $\log_b(x)$ from the base $b$ to base $k$. In fact, we can use any base that is convenient to our calculation.

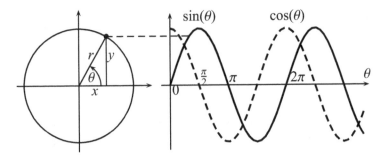

Figure 1.9: A sine function $\sin(\theta)$ (solid) and a cosine function $\cos(\theta)$ (dashed).

Thus, we have

$$\log_b(x) = \frac{\log_2(x)}{\log_2(b)} = \frac{\log_{10}(x)}{\log_{10}(b)} = \frac{\ln(x)}{\ln b}. \tag{1.71}$$

## Example 1.21

For example, to calculate $\log_2(10)$, we can change it to base $e$, and we have

$$\log_2(10) = \frac{\ln(10)}{\ln(2)} = \frac{2.30259}{0.69315} \approx 3.32193,$$

where we have used a calculator to estimate $\ln(2)$ and $\ln(10)$.

## 4.5. Trigonometrical Functions

Trigonometrical functions are widely used in engineering. Referring to the variables in Fig. 1.9, the basic sine and cosine functions are defined as

$$\sin\theta = \frac{y}{r}, \qquad \cos\theta = \frac{x}{r}. \tag{1.72}$$

Since $|x| \le r$ and $|y| \le r$ for any triangle, we have

$$-1 \le \sin\theta \le 1, \qquad 1 \le \cos\theta \le 1. \tag{1.73}$$

From the graphs of $\sin\theta$ and $\cos\theta$ in Fig. 1.9, we know that

$$\sin 0 = \sin\pi = 0, \quad \cos 0 = \cos 2\pi = 1, \quad \sin\frac{\pi}{2} = 1, \tag{1.74}$$

and

$$\cos\pi = -1, \quad \sin\frac{3\pi}{2} = -1. \tag{1.75}$$

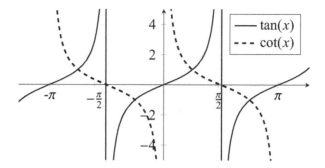

Figure 1.10: Graphs of tan($x$) and cot($x$).

In fact, the sine curve becomes a cosine curve if you shift to the left by $\pi/2$. That is

$$\sin(\theta + \frac{\pi}{2}) = \cos \theta. \tag{1.76}$$

Similarly, we have $\cos(\theta + \frac{\pi}{2}) = -\sin \theta$. If we add $2\pi$ or $360°$ to the angle $\theta$, we will reach the same point; we then have

$$\sin(2\pi + \theta) = \sin(\theta), \qquad \cos(2\pi + \theta) = \cos \theta. \tag{1.77}$$

This is equivalent to saying that both $\sin \theta$ and $\cos \theta$ functions have a period of $2\pi$ or $360°$. In addition, if we replace $\theta$ by $-\theta$, then from the definition $\sin \theta = y/r$, the point at $(x, y)$ becomes $(x, -y)$ with the same $r$. This means that

$$\sin(-\theta) = \frac{-y}{r} = -\sin(\theta), \tag{1.78}$$

which suggests that $\sin \theta$ is any odd function. A similar argument suggests that $\cos(-\theta) = \cos(\theta)$, which is an even function.

Other trigonometrical functions can be defined using the basic sin and cos functions. For example, the $\tan \theta$ can be defined as

$$\tan \theta = \frac{y}{x} = \frac{\sin \theta}{\cos \theta}. \tag{1.79}$$

Similarly, we have

$$\cot \theta = \frac{1}{\tan \theta} = \frac{\cos \theta}{\sin \theta}, \qquad \operatorname{cosec} \theta = \frac{1}{\sin \theta}, \qquad \sec \theta = \frac{1}{\cos \theta}. \tag{1.80}$$

The graphs of tan($x$) and cot($x$) are given in Fig. 1.10.

## Example 1.22

In geotechnical engineering, the Mohr-Coulomb equation

$$\tau = \sigma' \tan\phi + c_h,$$

links the shear strength $\tau$ of soil to the effective stress $\sigma' = \sigma - p$ where $\sigma$ is the total normal stress and $p$ is the pore water pressure. Here, $\phi$ is the angle of internal friction, and $c_h$ is the so-called cohesion. The equivalent friction coefficient is thus $\mu = \tan\phi$.


In trigonometry, we often write either $\sin(\theta)$ or $\sin\theta$ as long as there is no ambiguity. We also write the power as $\sin^n \theta \equiv (\sin\theta)^n$ for clarity and simplicity.

Some functions can have a certain symmetry when $x$ is replaced by $-x$. For example, we know that $f(x) = x^2$ will remain the same when $x$ is replaced by $-x$. That is

$$f(-x) = f(x), \tag{1.81}$$

and function is symmetric about the $y$-axis. In this case, we call such functions even functions. It is straightforward to check that $\cos(x)$, $x^2$, $|x|$, $x^4$ and $\exp(-x^2)$ are all even functions.

With other functions such as $g(x) = x^3$, the function value will become the original value multiplied by $-1$ if replace $x$ by $-x$. For example, we have $g(-x) = (-x)^3 = -x^3 = -g(x)$. Such functions are called odd functions, and they are symmetric when rotated around the origin $(0, 0)$ by $180°$. It is easy to check that $x$, $\sin(x)$, and $x^5$ are all odd functions. They all satisfy

$$g(-x) = -g(x). \tag{1.82}$$

Examples of both even and odd functions are shown in Fig. 1.11.

In most cases, functions such as $\sqrt{x}$ and $\exp(x)$ do not satisfy $f(-x) = \pm f(x)$, which means that they are neither even or odd.


## Example 1.23

Capillary action is an interesting phenomenon which is responsible for the rise of a liquid column in a thin tube. For a tube of radius $r$ filled with a liquid such as water or mercury, the height $h$ of the liquid column is governed by

$$h = \frac{2\gamma \cos\theta}{\rho g r},$$

where $\gamma$ is the surface tension of the liquid-air interface, $\rho$ is the liquid density, and

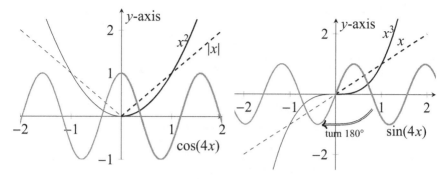

Figure 1.11: Even functions (left) and odd functions (right).

$g$ is the acceleration due to gravity. In addition, $\theta$ is the contact angle defined as the angle from the solid surface to the tangent plane of liquid surface at the contact. When $\theta$ is less than 90° (i.e., hydrophilic) such as for water-glass contact, the surface meniscus is concave. When $\theta$ is greater than 90° (i.e., hydrophobic) such as mercury-glass contact, the surface is convex. Even for the same liquid, the contact angle will vary with the solid surface. For example, the contact angle of water on clean glass is about 0°, while the contact angle of water on Teflon is about 108°.

For a glass tube filled with water in the open air, we have $\rho = 1000$ kg/m$^3$, $g = 9.8$ m/s$^2$, $\gamma = 0.073$ N/m, and $\theta = 0°$, we have

$$h = \frac{2 \times 0.073 \times \cos 0}{1000 \times 9.8r} \approx \frac{1.5 \times 10^{-5}}{r} \text{ m.}$$

Thus, for a tube of radius $r = 1$ cm, the rise of the water column inside the tube is about $h \approx 1.5$ mm. If $r$ is reduced to $r = 1$ mm, we have $h \approx 15$ mm. If $r = 0.01$ mm (or 10 $\mu$m), then $h \approx 1500$ mm or 1.5 metres.

Capillary action, together with osmosis, provides a water and nutrition transport mechanism in trees via their tube-like xylem vessels.

In many cases, we have to calculate the angle from a known function value. For example, we know that $\sin 30° = \sin(\pi/6) = 0.5$. If we are asked to find the angle $\theta$ so that $\sin \theta = 1/2$, we are dealing with the inverse of a sine function. In general, if we know the value of $y$ in $y = \sin(x)$, we can write it as

$$x = \sin^{-1}(y) = \arcsin(y). \tag{1.83}$$

Thus, we have

$$x = \sin^{-1}(0.5) = 30° = \frac{\pi}{6}. \tag{1.84}$$

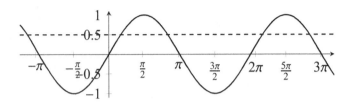

Figure 1.12: Sin(x) and the inverse of $x = \sin^{-1}(0.5)$.

However, we have to be careful here. From Fig. 1.9, we know that there are many possible $x$ values for a given $y$ (for example, by drawing a horizontal line at a fixed value $y = 0.5$) and this is because the period of $\sin(x)$ is $2\pi$. Thus, the above inverse function can have an infinite number of solutions when shifted by $2\pi n$ (where $n = \pm 1, \pm 2, ...$). Therefore, we have

$$x = \frac{\pi}{6} + 2\pi n, \quad n = \pm 1, \pm 2, ..., \tag{1.85}$$

for the same $\sin x = 0.5$. This means that the inverse sine function is not a one-to-one mapping (see Fig. 1.12).

If we want to get a one-to-one mapping for $x = \sin^{-1}(y)$, we have to focus on the range $\pi/2 \leq x \leq \pi/2$ for $-1 \leq y \leq +1$. In this case, $[-\pi/2, \pi/2]$ is called the range of the principal values. Similarly, for $y = \cos(x)$, we have

$$x = \cos^{-1}(y), \tag{1.86}$$

whose range of principal values is

$$0 \leq x \leq \pi. \tag{1.87}$$

Other inverse functions of trigonometrical functions can be defined in a similar manner. For example, the inverse tangent can be defined by

$$x = \tan^{-1}(y), \quad -\frac{\pi}{2} < x < +\frac{\pi}{2}, \quad -\infty < y < +\infty. \tag{1.88}$$

## 4.6. Composite Functions

Sometimes, it is necessary to do mathematical calculations using composite functions. Loosely speaking, a composite function is a function of a function. For example, if we have two functions: $f(x) = x^2 + 1$ and $g(x) = \sin(x)$, we can calculate

$$f(g(x)) = [g(x)]^2 + 1 = (\sin(x))^2 + 1 = \sin^2 x + 1, \tag{1.89}$$

which essentially takes $g(x)$ as the input for $f(x)$. This is a composite function.

In general, we can essentially define a composite function

$$(f \circ g)(x) = f\big(g(x)\big), \tag{1.90}$$

which applies $g$ first as a function of $x$, then applies $f$ as a function of $g(x)$. This notation highlights that order of the application of the operators ($f$ or $g$), and the operators on the right apply first. It is worth pointing out that $(f \circ g)$ is also written as $fg$ in some textbooks.

Similarly, we can define

$$(g \circ f)(x) = g\big(f(x)\big). \tag{1.91}$$

In general, for functions $f(x)$, $g(x)$ and $h(x)$, we have defined

$$(h \circ g \circ f)(x) = h\bigg(g\big(f(x)\big)\bigg). \tag{1.92}$$

Obviously, there are other possible permutations, depending on the order of composition. Higher-order composite functions can also be defined in a similar manner. In addition, a function can be composite itself. For example,

$$(f \circ f) = f(f(x)), \quad (f \circ f \circ f) = f(f(f(x))). \tag{1.93}$$

It is straightforward to check that

$$f \circ (g \circ h) = (f \circ g) \circ h = f \circ g \circ h. \tag{1.94}$$

## Example 1.24

For functions $f(x) = x^3$, $g(x) = x + 2$ and $h(x) = \exp(x)$, we have

$$(f \circ g) = f(x + 2) = (x + 2)^3 = x^3 + 6x^2 + 12x + 8,$$

$$(g \circ f) = g(x^3) = x^3 + 2, \quad (f \circ f) = (x^3)^3 = x^9,$$

and

$$(g \circ g) = (x + 2) + 2 = x + 4, \quad (g \circ g \circ g) = ((x + 2) + 2) + 2 = x + 6.$$

In addition, we have

$$(h \circ g \circ f)(x) = (h \circ g)(x^3) = h\big(x^3 + 2\big) = e^{x^3 + 2},$$

$$(f \circ g \circ h)(x) = (f \circ g)(e^x) = f\big(e^x + 2\big) = (e^x + 2)^3, \quad (h \circ h)(x) = e^{e^x}.$$

The composition of functions provides a way to decompose complex functions and such decomposition can become convenient when calculating derivatives of complex functions using the chain rules to be introduced in later chapters. For example, for a function

$$\phi(x) = \exp(\sqrt{2x + 3}), \tag{1.95}$$

we can define $f(x) = \exp(x)$, $g(x) = \sqrt{x}$ and $h(x) = 2x + 3$, and we have

$$\phi(x) = \exp(\sqrt{2x + 3}) = (f{\circ}g{\circ}h)(x). \tag{1.96}$$

To do a proper composition, we have to be careful about the domains and ranges of functions. For example, $f(x) = \ln(x)$ and $g(x) = x^2 - 9$, the domain of $f(x)$ is $x > 0$, while the domain for $g(x)$ is all real numbers. However, the domain of $(f{\circ}g)(x) = \ln(x^2 - 9)$ is $x^2 - 9 > 0$, that is $x > 3$ and $x < -3$.

In addition, it is worth pointing out that it is not always possible to carry out a composition of two functions. For example, $f(x) = \ln(x)$ and $g(x) = -|x|$. Thus, $(f{\circ}g) = f\big(g(x)\big) = \ln(-|x|)$ is not valid because for any $x$, the value of $g(x)$ is always negative or zero, which means that $f(g(x))$ is not defined for any $x$ in the real domain. In this case, the composition is meaningless. However, $(g{\circ}f)(x) = -|\ln x|$ is defined for all $x$ except for $x = 0$.

## Exercises

**1.1.** Solve the following equations:
- $2x + 3 = 7$.
- $3x + 2 = 11$.
- $7x + \pi = 0$.
- $x + 2y = 5$ and $y - x = 1$.
- $2x + 3x = 7$ and $2x + y = 5$.

**1.2.** Round up the following using significant figures:
- Round up 19820 to 1 significant figure.
- What is 0.0012345 to 2 significant figures?
- What is $19.77 \times 8.0$?
- What is $2.017 + 3.14 + 1.12345$?

**1.3.** Find the domains and ranges of the following functions:
- $f(x) = (x - 1)^2$.
- $f(x) = |x - 2| + (|x + 2| + 1)^2$.
- $f(x) = \ln(x^2 - 1)$.
- $f(x) = 2^{-x} + 3$.
- $f(x) = \ln(2^{-x^2})$.

**1.4.** Use $\sin^2 x + \cos^2 x = 1$ and $\sin(2x) = 2 \sin x \cos x$ to show that $\sin(2x) = \frac{2 \tan x}{1 + \tan^2 x}$.

# CHAPTER 2

# Equations and Polynomials

Contents

---

## Key Points

- Review the basic concepts concerning index notations.
- Introduce the binomial expansions.
- Review the basics about quadratic equations and introduce polynomials and roots.

---

## 1. Index Notation

Before we proceed to study polynomials, let us first introduce the notations of indices. For the discussion of the product or multiplication of $2 \times 2 \times 2 \times 2 \times 2 \times 2 \times 2 \times 2$ (that is, 2 multiplied by itself 8 times), it is more compact to write it as

$$2^8 \equiv 2 \times 2 \times 2 \times 2 \times 2 \times 2 \times 2 \times 2. \tag{2.1}$$

Here, we use '$\equiv$' to mean the definition, as the left-hand side and the right-hand side are always equal in this case. The compact index notation is not only just to save time and space, but it also looks better. Imagine writing $2^{100}$ using the multiplication symbol '$\times$' 99 times.

For higher-order products such as $a \times a \times a \times a \times a \times a \times a$ or $aaaaaaa$, it might be more economical to write it as a simpler form $a^7$. In general, we use the index form to express the product

$$a^n = \overbrace{a \times a \times ... \times a}^{n}, \tag{2.2}$$

where $a$ is called the base, and $n$ is called the index or exponent. Conventionally, we write $a \times a = aa = a^2$, $a \times a \times a = aaa = a^3$ and so on and so forth. For example, if

$n = 100$, it is obviously advantageous to use the index form. A very good feature of index notation is the multiplication rule

$$a^n \times a^m = \overbrace{a \times a \times \ldots \times a}^{n} \times \overbrace{a \times a \times \ldots \times a}^{m} = \overbrace{a \times a \times \ldots \times a}^{n+m} = a^{n+m}.$$

Thus, the product of factors with the same base can easily be carried out by adding their indices. If we interpret $a^{-n} = 1/a^n$, we then have the division rule

$$a^n \div a^m = \frac{a^n}{a^m} = a^{n-m}. \tag{2.3}$$

In the special case when $n = m$, it requires that $a^0 = 1$ because any non-zero number divided by itself should be equal to 1.

## Example 2.1

In order to calculate $2^5 \times 2^7 \times 3^3 \times 3^4$, we have

$$2^5 \times 2^7 \times 3^3 \times 3^4 = 2^{5+7} \times 3^{3+4} = 2^{12} \times 3^7 = 4096 \times 2187 = 8957952.$$

Similarly, we can calculate

$$2^{-5} \times 2^9 \times 4^5 = 2^{-5+9} \times (2^2)^5 = 2^{-5+9+10} = 16384,$$

and

$$2^{-11} \times 2^3 \times 2^7 = 2^{-11+3+7} = 2^{-1} = \frac{1}{2} = 0.5.$$

If we replace $a$ by $a \times b$ and use $a \times b = b \times a$, we get the factor rule

$$(a \times b)^n = a^n \times b^n = a^n b^n. \tag{2.4}$$

Similarly, using $a^m$ to replace $a$ in the expression (2.2), it is straightforward to verify the following power-on-power rule

$$(a^m)^n = a^{n \times m} = a^{nm}. \tag{2.5}$$

Here, the exponents $m$ and $n$ are integers and they can be positive or negative or zero. Fractional exponents can be treated by certain conventions. For example, the fraction exponent $1/n$ means to take the $n$th root of $a$. That is,

$$a^{\frac{1}{n}} = a^{1/n} = \sqrt[n]{a}. \tag{2.6}$$

Thus, we have

$$a^{\frac{m}{n}} = a^{m/n} = \sqrt[n]{a^m} = (\sqrt[n]{a})^m, \tag{2.7}$$

which gives the following examples:

$$2^{\frac{4}{3}} = \sqrt[3]{2^4} = \sqrt[3]{16}, \quad 4^{-2/3} = \frac{1}{4^{2/3}} = \frac{1}{\sqrt[3]{4^2}} = \frac{1}{\sqrt[3]{16}}. \tag{2.8}$$

However, the exponents can also be generalized to be any real values such as 0.5 or -1/2, so $a^{0.5} = \sqrt{a}$ and $a^{-1/2} = 1/\sqrt{a}$. Therefore, in the rest of this book, we will treat the exponents in the most general sense; that is, the exponents can be real numbers.

Now we can revisit the power function

$$f(x) = x^p. \tag{2.9}$$

There are four cases:
- $p > 1$: The power $x^p$ increases faster as $x$ becomes large.
- $p = 1$: It becomes a linear function or a straight line $x$.
- $0 < p < 1$: The power increases more slowly compared to the increase of $x$.
- $p < 0$: It decreases to zero as $x$ becomes large.

Care should be taken when $p$ is a fraction, especially a negative fraction. For example, if $p = -2/3$, we have

$$f(x) = x^{-2/3} = \frac{1}{x^{2/3}} = \frac{1}{\sqrt[3]{x^2}}, \tag{2.10}$$

whose domain can be any real number, except for $x = 0$. This means that this function is not defined at $x = 0$. Similarly, when $p = 1/4$, we have

$$f(x) = x^{1/4} = \sqrt[4]{x}, \tag{2.11}$$

whose domain is $x \geq 0$ because we cannot find an even root for a negative number.

## Example 2.2

The surface-area-to-volume ratio $R_{sv}$ is a very important factor in engineering and chemistry. For a sphere of radius $a$, its surface area is $S = 4\pi a^2$ and its volume is $V = 4\pi a^3/3$, so the surface-to-volume ratio is

$$R_{sv} = \frac{S}{V} = \frac{4\pi a^2}{\frac{4\pi a^3}{3}} = \frac{3}{a}.$$

Similarly, for a cube of side length $a$, its volume is $V = a^3$ and surface area is $S = 6a^2$, so the ratio is simply

$$R_{sv} = 6a^2/a^3 = 6/a.$$

As an exercise, we leave you to check that most object shapes will have a surface-to-volume ratio that is inverse proportional to its size.

We may wonder what happens when $p = 0$. For consistency in mathematics, it is necessary for $x^0 = 1$ for any value of $x \neq 0$. In addition, in a very special case when $x = 0$, we have $0^0$. In this case, $0^0$ is undefined (that is, its value is not defined with a single fixed value).

One way to look at $0^0$ is that for any non-zero $a$, $a^0 = 1$ as mentioned earlier. Another way to look at it is that for $0^2 = 0$, $0^1 = 0$ or $0^a = 0$ for any $a > 0$. Thus, it seems that either $0^0 = 0$ or $0^0 = 1$ is possible. In fact, some mathematicians suggest that $0^0 = 1$ can allow certain formulas to be expressed more compactly. Though a mathematical curiosity, $0^0$ does not make any difference in practice because it rarely occurs in any real-world applications. Many phenomena in nature obey the so-called power laws. Both gravitation and electromagnetism obey the inverse-square law. For example, Newton's inverse-square law of gravitation has an exponent $-2$. In reality, many phenomena can have various exponents.

## Example 2.3

For example, the buckling load $P$ of a slender column such as a cylindrical rod with a length $L$ is given by

$$P = \frac{\pi^2 EI}{L^2},$$

where $E$ is the modulus of elasticity (or Young's modulus) and $I$ is the minimum area moment of inertia. This axial buckling load is also called Euler's critical load. In general, this relationship should be extended to

$$P = \frac{\pi^2 EI}{(K_e L)^2},$$

where $K_e$ is an effective length factor depending on the boundary support conditions. For a simple support column, $K_e = 1$ can be used. If the area of cross section is $A$, the critical stress is

$$\sigma = \frac{P}{A} = \frac{\pi^2 EI}{A(K_e L)^2} = \frac{\pi^2 E}{(K_e L)^2 / (I/A)}.$$

If we define $r = \sqrt{I/A}$ as the radius of gyration and $s = K_e L / r$ as the slenderness ratio, we have

$$\sigma = \frac{\pi^2 E}{s^2}.$$

It is an inverse square law in terms of $s$ or $L$.

## 2. Binomial Expansions

From the discussions in the previous chapters, we know $(a + b)^0 = 1$ and $(a + b)^1 = (a + b) = a + b$. We also know that

$$(a + b)^2 = (a + b)(a + b) = a(a + b) + b(a + b) = a^2 + 2ab + b^2. \qquad (2.12)$$

If we follow a similar procedure to expand, we can deal with higher-order expansions. For example,

$$(a + b)^3 = (a + b)(a + b)^2 = (a + b)(a^2 + 2ab + b^2),$$

$$= a(a^2 + 2ab + b^2) + b(a^2 + 2ab + b^2) = a^3 + 3a^2b + 3ab^2 + b^3. \qquad (2.13)$$

Similarly, we have

$$(a + b)^4 = a^4 + 4a^3b + 6a^2b^2 + 4ab^3 + b^4, \qquad (2.14)$$

$$(a + b)^5 = a^5 + 5a^4b + 10a^3b^2 + 10a^2b^3 + 5ab^4 + b^5. \qquad (2.15)$$

Now the question is: do we have a general expression for $(a + b)^n$ for any $n \geq 1$? In this case, we need the binomial theorem. Before we introduce the binomial theorem, we first introduce some other relevant concepts such as factorials.

For any positive integer $n$, the factorial $n$ (or $n$ factorial), denoted by $n!$, is the product of all the $n$ natural numbers from 1 to $n$. That is

$$n! = n \times (n - 1) \times (n - 2) \times ... \times 1. \qquad (2.16)$$

For example, we have $1! = 1, 2! = 2 \times 1 = 2, 3! = 3 \times 2 \times 1 = 6$, and $5! = 5 \times 4 \times 3 \times 2 \times 1 = 120$. Similarly, $10! = 10 \times 9 \times ... \times 2 \times 1 = 3628800$. From the above definition, we can easily obtain a recursive formula

$$(n + 1)! = n! \times (n + 1), \qquad (n = 1, 2, 3, ...). \qquad (2.17)$$

In fact, the above equation is still valid for $n = 0$. Now you may wonder what is the value of $0!$? For consistency and simplicity in writing mathematical expressions, it requires that we define $0! = 1$. In fact, it is also possible to define the factorial of negative integers, but it involves more difficult concepts such as $\Gamma$ function and $\Gamma$ integrals. Anyway, factorials of negative numbers such as $(-5)!$ are rarely used in sciences and engineering, so we do not need to study them further.

The combinatorial coefficient or binomial coefficient is often written in the form

$$\binom{n}{r} \equiv {}^nC_r \equiv \frac{n!}{r!(n - r)!}, \qquad (2.18)$$

where $n, r \geq 0$ are integers. Here the symbol '$\equiv$' means 'it is defined as' or 'exactly the same as'. In some literature, the notation ${}^nC_r$ is also widely used. To write the expression explicitly, we have

$$\binom{n}{r} = \frac{n(n-1)(n-2)...(n-r+1)}{1 \times 2 \times ... \times (r-1) \times r}. \tag{2.19}$$

## Example 2.4

For example, we can calculate $\binom{7}{5}$ by using either

$$\binom{7}{5} = \frac{7 \times 6 \times 5 \times 4 \times 3}{1 \times 2 \times 3 \times 4 \times 5} = \frac{2520}{120} = 21, \quad \text{or} \quad \binom{7}{5} = \frac{7!}{5!(7-5)!} = \frac{5040}{120 \times 2} = 21.$$

This number also represents the number of choices or combinations of choosing (for example) 5 among 7 candidates to form a team.
    Similarly, we have

$$\binom{9}{3} = 84, \quad \binom{20}{3} = 1140, \quad \binom{100}{2} = 4950, \quad \binom{100}{5} = 75287520.$$

The binomial expansion in general can be written as

$$(a+b)^n = \binom{n}{0}a^n + \binom{n}{1}a^{n-1}b + \binom{n}{2}a^{n-2}b^2 + ... + \binom{n}{n-1}ab^{n-1} + \binom{n}{n}b^n. \tag{2.20}$$

In the special case $a = 1$ and $b = x$, we have

$$(1+x)^n = 1 + nx + ... + nx^{n-1} + x^n. \tag{2.21}$$

In the case of very small $x$ (i.e., $|x| \ll 1$ where '$\ll$' means 'much less than'), we have an approximate expression

$$(1+x) \approx 1 + nx, \tag{2.22}$$

if we ignore all the higher order terms (because $|x^2| \ll |x|$ and $|x^3| \ll |x^2|$ if $|x| \ll 1$). This approximation can be useful for error estimation.

## Example 2.5

We know the volume of a sphere is $V = 4\pi r^3/3$. If the radius $r$ is $R = 0.50$ meters with an error of $\delta = 10$ mm, we have $r = R \pm \delta r$. The error in length measurement is $E = \delta r/R = 0.010/0.5 = 2\%$. What is the error in the volume calculated?
    From

$$V = \frac{4\pi}{3}r^3 = \frac{4\pi}{3}R^3(1+E)^3 = V_0(1+E)^3 \approx V_0(1+3E),$$

where

$$V_0 = \frac{4\pi}{3}R^3 = \frac{4 \times 3.14159}{3} \times 0.5^3 \approx 0.524 \text{ m}^3.$$

Thus, the error in volume can be estimated by

$$\frac{(V - V_0)}{V_0} = 3E = 3 \times \frac{2}{100} = 0.06,$$

which is about 6%.

---

## 3. Floating Point Numbers

In science and engineering, we often have to deal with very large or very small numbers, and thus floating point notations become a very useful tool. For example, the mass of the Earth is about $5.972 \times 10^{24}$ kg, while the mass of a proton is about $m_p \approx 1.67262 \times 10^{-27}$ kg.

In general, a floating-point number has three parts: significand, base and exponent. Here, the mass of the Earth

$$\text{significand} \times \text{base}^{\text{exponent}} = 5.972 \times 10^{24}, \tag{2.23}$$

where 5.972 is the significand or significant digits, 10 is the base and 24 is the exponent. Such notations are also called the scientific notation, standard form or standard index form. Floating-point numbers provide a convenient way to deal with any real numbers (large or small), and thus are widely used in scientific computing and engineering applications.

It is worth pointing out that the base 10 is usually written as $E$ in the scientific computing and numeric outputs in many programming languages. For example, $5.972 \times 10^{24}$ is often written as $5.972E + 24$ or $5.972E24$.

In addition, depending on the precision or accuracy of the calculations, there is a trade-off using such notations. For example, we know the basic charge of an electron is

$$q = (1.602176487 \pm 0.000000040) \times 10^{-19} \tag{2.24}$$

which has 10 significant digits. However, if we are concerned with the basic estimates, we can approximate the above simply as

$$q = 1.602 \times 10^{-19}, \quad \text{or even} \quad 1.6 \times 10^{-19}, \tag{2.25}$$

depending on the precision we may want.

## Example 2.6

The black-body radiation of an object at an absolute temperature $T$ emits the energy $E$ per unit area per unit time, which obeys the Stefan-Boltzmann law

$$E_b = \sigma T^4,$$

where $\sigma = 5.67 \times 10^{-8}$ W K$^{-4}$m$^{-2}$ is the Stefan-Boltzmann constant. For example, a typical human body has a skin surface area of about $A = 1.5$ m$^2$. The normal body temperature $T_h = 36.8°C$ or $T_h = 273 + 36.8 = 309.8$ K (Kelvin) in a room with a constant room temperature $T_0 = 20°C$ or $T_0 = 273 + 20 = 293$ K. The skin tempera-ture $T_s$ in contact with clothes will be about $T_s \approx 32.0°C$ or $T_s = 273 + 32.0 = 305$ K. Thus, total energy per unit time radiated by an adult human body is

$$E = A(\sigma T_s^4 - \sigma T_0^4) = A\sigma(T_s^4 - T_0^4)$$

$$= 1.5 \times 5.67 \times 10^{-8} \times (305^4 - 293^4) \approx 109 \text{ J/s} = 109 \text{ Watts.}$$

This is equivalent to the power of more than four 25-watt light bulbs.

## 4. Quadratic Equations

When an equation involves one unknown variable $x$, it may not always be linear. Some terms may involve $x^2$, in which case, we have to deal with quadratic functions or equations. Quadratic functions are widely used in many applications.

In general, a quadratic function $y = f(x)$ can be written as

$$f(x) = ax^2 + bx + c, \tag{2.26}$$

where the coefficients $a, b$ and $c$ are real numbers. Here we can assume $a \neq 0$. If $a = 0$, it reduces to the case of linear functions that we have discussed earlier. Two examples of quadratic functions are shown in Fig. 2.1 where both $a > 0$ and $a < 0$ are shown, respectively. Depending on the combination of the values of $a$, $b$, and $c$, the curve may cross the $x$-axis twice, once (just touch), and not at all. The points at which the curve crosses the $x$-axis are the roots of $f(x) = 0$.

As we can assume $a \neq 0$, we have

$$x^2 + \frac{b}{a}x + \frac{c}{a} = 0. \tag{2.27}$$

Of course, we can use factorization to find the solution in many cases. For example, we can factorize $x^2 - 5x + 6 = 0$ as

$$x^2 - 5x + 6 = (x - 2)(x - 3) = 0, \tag{2.28}$$

and the solution is thus either $(x - 2) = 0$, or $(x - 3) = 0$. That is $x = 2$ or $x = 3$. How-

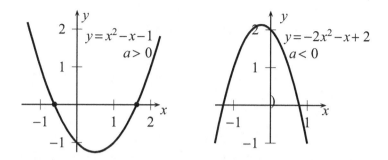

Figure 2.1: Graphs of quadratic functions.

ever, sometimes it is difficult (or even impossible) to factorize even seemingly simple expressions such as $x^2 - x - 1$. In this case, it might be better to address the quadratic equation in a generic manner using the so-called complete-square method.

In general, completing the square of the expression can be used to solve this equation, and we have

$$x^2 + \frac{b}{a}x + \frac{c}{a} = \left(x + \frac{b}{2a}\right)^2 - \frac{b^2}{4a^2} + \frac{c}{a} = \left(x + \frac{b}{2a}\right)^2 - \frac{(b^2 - 4ac)}{4a^2} = 0, \qquad (2.29)$$

which leads to

$$\left(x + \frac{b}{2a}\right)^2 = \frac{(b^2 - 4ac)}{4a^2}. \qquad (2.30)$$

Taking the square root of the above equation, we have either

$$x + \frac{b}{2a} = +\sqrt{\frac{b^2 - 4ac}{4a^2}}, \quad \text{or} \quad x + \frac{b}{2a} = -\sqrt{\frac{b^2 - 4ac}{4a^2}}. \qquad (2.31)$$

Using $\sqrt{4a^2} = +2a$ or $-2a$, we have

$$x = -\frac{b}{2a} \pm \frac{\sqrt{b^2 - 4ac}}{2a} = \frac{-b \pm \sqrt{b^2 - 4ac}}{2a}. \qquad (2.32)$$

In some books, the expression $b^2 - 4ac$ is often denoted by $\Delta$. That is $\Delta = b^2 - 4ac$.
- If $\Delta = b^2 - 4ac < 0$, there is no solution which corresponds to the case where the curve does not cross the $x$-axis at all.
- In the special case of $b^2 - 4ac = 0$, we have a single solution $x = -b/(2a)$, which corresponds to the case where the curve just touches the $x$-axis.
- If $b^2 - 4ac > 0$, we have two distinct real roots.

Now let us look at an example about the golden ratio.

## Example 2.7

The golden ratio is related to the following equation:

$$x^2 - x - 1 = 0.$$

Thus, we have $a = 1$, $b = -1$ and $c = -1$. Since $\Delta = b^2 - 4ac = (-1)^2 - 4 \times 1 \times (-1) = 5 > 0$, it has two different real roots. We have

$$x_{1,2} = \frac{-(-1) \pm \sqrt{\Delta}}{2} = \frac{1 \pm \sqrt{5}}{2}.$$

So the two roots are $(1 - \sqrt{5})/2$ and $(1 + \sqrt{5})/2$ which are marked as solid circles on the graph in Fig. 2.1.

The positive root is the famous golden ratio

$$\phi = \frac{1 + \sqrt{5}}{2} = 1.6180339887...,$$

and this ratio appears in many natural phenomena, from flower petals and DNA structures to star formation and spiral galaxies. Many design applications in engineering and architecture also use the golden ratio.

## 5. Polynomials and Roots

The functions we have discussed so far are a special case of polynomials. In particular, a quadratic function $ax^2 + bx + c$ is a special case of polynomials. In general, a polynomial $p(x)$ is an explicit expression, often written as a sum or linear combination, which consists of multiples of non-negative integer powers of $x$. Therefore, we often write a polynomial in the descending order (in terms of the power $x^n$)

$$p(x) = a_n x^n + a_{n-1} x^{n-1} + ... + a_1 x + a_0, \qquad (n \geq 0), \qquad (2.33)$$

where $a_i (i = 0, 1, ..., n)$ are known constants. The coefficient $a_n$ is often called the leading coefficient, and the highest power $n$ of $x$ is called the degree of the polynomial. Each partial expression such as $a_n x^n$ and $a_{n-1} x^{n-1}$ is called a term. The term $a_0$ is called the constant term. Obviously, the simplest polynomial is a constant (degree 0).

For example, $2x^3 + x^2 - 4x + 5$ is a polynomial of degree 3 with a leading coefficient 2. It has four terms. The expression

$$-x^5 - 2x^4 + 3x^2 + 1$$

is a polynomial of the fifth degree with a leading coefficient $-1$. It has only four terms as the terms $x^3$ and $x$ do not appear in the expression (their coefficients are zero).

When adding (or subtracting) two polynomials, we have to add (or subtract) the corresponding terms (with the same power $n$ of $x$), and the degree of resulting polynomial may have any degree from 0 to the highest degree of two polynomials. However, when multiplying two polynomials, the degree of the resulting polynomial is the sum of the degrees of the two polynomials. For example,

$$(a_n x^n + a_{n-1} x^{n-1} + ... + a_0)(b_m x^m + b_{m-1} x^{m-1} + ... + b_0)$$

$$= a_n b_m x^{n+m} + (a_n b_{m-1} + a_{n-1} b_m) x^{n+m-1} + ... + a_0 b_0. \tag{2.34}$$

## Example 2.8

If we add $x^3 + 2x^2 + 1$ with $2x^3 - 3x^2 + 4x - 5$, we have

$$(x^3 + 2x^2 + 1) + (2x^3 - 3x^2 + 4x - 5) = (1x^3 + 2x^2 + 0x + 1) + (2x^3 - 3x^3 + 4x - 5)$$

$$= (1 + 2)x^3 + (2 - 3)x^2 + (0 + 4)x + (1 - 5) = 3x^3 + (-1)x^2 + 4x - 4 = 3x^3 - x^2 + 4x - 4,$$

which is a cubic polynomial. However, if we subtract the second polynomial from the first, we have

$$(x^3 + 2x^2 + 1) - (2x^3 - 3x^2 + 4x - 5)$$

$$= (1 - 2)x^3 + [2 - (-3)]x^2 + (0 - 4)x + [1 - (-5)] = -x^3 + 5x^2 - 4x + 6,$$

which is cubic. Similarly, the product of $x^4 - 2x - 1$ and $-x^5 - x^2 + 2$ is a polynomial of degree 9 because

$$(x^4 - 2x - 1)(-x^5 - x^2 + 2) = -x^9 + x^6 + x^5 + 2x^4 + 2x^3 + x^2 - 4x - 2,$$

has 9 as the exponent in the highest-order term (i.e., $x^9$).

For a given polynomial such as

$$f(x) = x^3 - 4x = 0, \tag{2.35}$$

the value $x_*$ that satisfies the above equation is called a root. We use $*$ to highlight that this value is a special value of $x$. That is $x_*^3 - 4x_* = 0$. Obviously, for this simple equation, we can guess that $x_* = 2$ is a solution or root because $2^3 - 4 \times 2 = 0$. It is also easy to check that $x_* = 0$ is also a root. Now the question is: have we found all the roots for this simple problem? If not, how many are left? If yes, how can we be sure?

If we look at this problem more carefully, we realize that $x_* = -2$ is also a root because $(-2)^3 - 4(-2) = -8 + 8 = 0$. Now there are three roots:

$$x_* = 0, \quad x_* = +2, \quad x_* = -2. \tag{2.36}$$

If we try even harder, it seems that we cannot find another root.

Another way to look this problem is to factorize the polynomial. Since $f(x) = x^3 - 4x = x(x^2 - 4) = x(x + 2)(x - 2)$, we have $x(x + 2)(x - 2) = 0$, which gives three roots: $x = 0$, $x = -2$ and $x = 2$.

Now a natural question is: how to find the roots of a polynomial in general? We know that there exists an explicit formula for finding the roots of a quadratic equation. Analytical forms are also possible for a cubic, and a quartic function (polynomial of degree 4), though quite complicated. In some special cases, it may be possible to find a factor via factorization or even by educated guess; then some of the solutions can be found. For example, for $f(x) = x^8 - 1$, we can have

$$f(x) = x^8 - 1 = (x - 1)(x + 1)(x^2 + 1)(x^4 + 1) = 0, \qquad (2.37)$$

but to figure out such factors is not easy. For example, it is not straightforward to find all the factor such that

$$x^8 - 16 = (x^2 - 2)(x^2 + 2)(x^2 - 2x + 2)(x^2 + 2x + 2).$$

In general, factorization of high-order polynomials is not always possible. In most cases, numerical methods such as Newton's root-finding algorithm are needed.

## Exercises

**2.1.** Simplify the following expressions:
- $2^2 \times 2^7 / 16^2$
- $3^{-3} \times 3^4 \times 100^0$
- $\sqrt{2}(\sqrt{7})^3 / \sqrt{14}$
- $2^9 2^{-7} - 3^4 3^{-3} + \pi^0$
- $\log_2(4) + \log_2(9) - \log_2(18)$.

**2.2.** Show that $\begin{pmatrix} 6 \\ 4 \end{pmatrix} = \begin{pmatrix} 5 \\ 3 \end{pmatrix} + \begin{pmatrix} 5 \\ 4 \end{pmatrix}$.

**2.3.** Show that $x^6 - 1 = (x - 1)(x + 1)(x^2 - x + 1)(x^2 + x + 1)$, then solve $x^6 - 1 = 0$.

**2.4.** Show that $x^4 - 81 = (x + 3)(x - 3)(x^2 + 9)$ and find the roots of $x^4 - 81 = 0$.

**2.5.** Find the roots of $x^x = x$.

# CHAPTER 3

# Vectors and Matrices

Contents

---

## Key Points

- Introduce the basic concepts of vectors, together with vector algebra such as the dot product, cross product and triple vector product.
- Introduce the basic concepts of matrices and matrix algebra.
- Introduce the eigenvalues and eigenvectors of a square matrix.
- Introduce briefly the main concepts of tensors.

---

Many quantities such as forces, displacements and velocities are vectors, and they are widely used in engineering. We will first introduce the basic concepts of vectors and vector algebra and then introduce matrices and tensors.

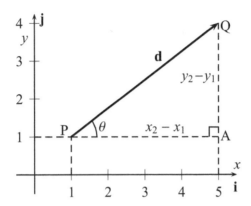

Figure 3.1: The displacement from $P(x_1, y_1)$ to point $Q(x_2, y_2)$.

## 1. Vectors

A vector is a quantity with both direction and magnitude. Typical examples in engineering are forces, velocities and displacements.

Suppose a robot moves from a point P at $(x_1, y_1)$ to another point Q at $(x_2, y_2)$, then its displacement is represented by a vector $\mathbf{d} = \overrightarrow{PQ}$ (see Fig. 3.1).

For the displacement vector, we need the magnitude (or the length or distance) between P and Q, and also the direction or angle $\theta$ to determine the vector uniquely. Since the coordinates of two points P and Q are given, the distance between P and Q can be calculated using the Cartesian distance. The length or magnitude of $\mathbf{d}$ can conveniently be written as $d \equiv \|\mathbf{d}\| \equiv |\mathbf{d}|$, and we have

$$PQ = |\overrightarrow{PQ}| = \|\mathbf{d}\| = |\mathbf{d}| = \sqrt{(x_2 - x_1)^2 + (y_2 - y_1)^2}. \tag{3.1}$$

The magnitude of a vector is also called the modulus of the vector. Here we follow the conventions of using a single letter in italic form to denote the magnitude while using the same letter in bold type to denote the vector itself.

The direction of the vector is represented by the angle $\theta$ from the $x$-axis. We have

$$\tan \theta = \frac{y_2 - y_1}{x_2 - x_1}. \tag{3.2}$$

Conventionally, we often write a vector using bold font $\mathbf{d}$, rather than $d$. In many books, vectors are also written in the overhead arrow form such as $\overrightarrow{PQ}$ or simply $\overrightarrow{d}$. The notation $\overrightarrow{PQ}$ signifies that the vector is pointing from P to Q. Here we will use the bold-type notations as they are more popularly used in mathematics. The components of the vector **d** are $x_2 - x_1$ along the x-axis and $y_2 - y_1$ along the y-axis. This provides a way to write the vector as

$$\mathbf{d} = \overrightarrow{PQ} = \begin{pmatrix} x_2 - x_1 \\ y_2 - y_1 \end{pmatrix}. \tag{3.3}$$

Here we write the vector as a column, called a column vector. In general, a vector **u** in an $n$–dimensional space can be written as

$$\mathbf{u} = \begin{pmatrix} u_1 \\ u_2 \\ u_3 \\ \vdots \\ u_n \end{pmatrix}, \tag{3.4}$$

where $u_i (i = 1, 2, ..., n)$ are its components. This vector can also be written as a row vector by transposing a column into a row:

$$\mathbf{u} = \begin{pmatrix} u_1, & u_2, & ..., & u_n \end{pmatrix}^T, \tag{3.5}$$

where $T$ denotes the transpose operation. The length or modulus of **u** is

$$\|\mathbf{u}\| = \sqrt{u_1^2 + u_2^2 + ... + u_n^2}. \tag{3.6}$$

## Example 3.1

Reading from the graph shown in Fig. 3.1, we know that P is at $(1, 1)$ and Q is at $(5, 4)$. The displacement can be represented in the mathematical form as

$$\mathbf{d} = \begin{pmatrix} x_2 - x_1 \\ y_2 - y_1 \end{pmatrix} = \begin{pmatrix} 5 - 1 \\ 4 - 1 \end{pmatrix} = \begin{pmatrix} 4 \\ 3 \end{pmatrix}.$$

Therefore, the distance PQ or the magnitude $d$ of the displacement **d** is

$$d = |\mathbf{d}| = \sqrt{(5 - 1)^2 + (4 - 1)^2} = \sqrt{4^2 + 3^2} = 5.$$

The angle $\theta$ is given by

$$\tan \theta = \frac{4 - 1}{5 - 1} = \frac{3}{4} = 0.75,$$

or

$$\theta = \tan^{-1} 0.75 \approx 36.87°.$$

For real-world problems, the unit of length must be given, either in metres or any other suitable units. Vectors and trigonometrical functions are widely used in all disciplines of engineering, especially surveying and engineering mechanics.

It is worth pointing out that the vector $\overrightarrow{QP}$ is pointing the opposite direction of $\overrightarrow{PQ}$, and we thus have

$$\overrightarrow{QP} = \begin{pmatrix} x_1 - x_2 \\ y_1 - y_2 \end{pmatrix} = \begin{pmatrix} (-1)(x_2 - x_1) \\ (-1)(y_2 - y_1) \end{pmatrix} = -\begin{pmatrix} x_2 - x_1 \\ y_2 - y_1 \end{pmatrix} = -\overrightarrow{PQ} = -\mathbf{d}.$$

In general for any real number $\beta \neq 0$ and a vector $\mathbf{v} = \begin{pmatrix} a \\ b \end{pmatrix}$, we have

$$\beta \mathbf{v} = \beta \begin{pmatrix} a \\ b \end{pmatrix} = \begin{pmatrix} \beta a \\ \beta b \end{pmatrix}. \tag{3.7}$$

A vector whose magnitude is 1 is called a unit vector. So all the following vectors are unit vectors

$$\mathbf{i} = \begin{pmatrix} 1 \\ 0 \end{pmatrix}, \quad \mathbf{j} = \begin{pmatrix} 0 \\ 1 \end{pmatrix}, \quad \mathbf{w} = \begin{pmatrix} \cos\theta \\ \sin\theta \end{pmatrix}. \tag{3.8}$$

For any $\theta$, we know $|\mathbf{w}| = \sqrt{\cos^2\theta + \sin^2\theta} = 1$ due to the identity $\sin^2\theta + \cos^2\theta = 1$. The vectors $\mathbf{i}$ and $\mathbf{j}$ are the unit vectors along $x$-axis and $y$-axis directions, respectively.

Since a vector has a magnitude and a direction, any two vectors with the same magnitude and direction should be equal since there is no other constraint. This means that we can shift and move both ends of a vector by any same amount in any direction, and we still have the same vector. In other words, if two vectors are equal, they must have the same magnitude and direction. Mathematically, their corresponding components must be equal.

The addition of two vectors is a vector whose components are simply the addition of their corresponding components. If we define the subtraction of any two vectors $\mathbf{u}$ and $\mathbf{v}$ as

$$\mathbf{u} - \mathbf{v} = \mathbf{u} + (-\mathbf{v}), \tag{3.9}$$

where $-\mathbf{v}$ is obtained by reversing the direction of $\mathbf{v}$. In general, we have

$$\mathbf{v}_1 \pm \mathbf{v}_2 = \begin{pmatrix} a_1 \\ b_1 \end{pmatrix} \pm \begin{pmatrix} a_2 \\ b_2 \end{pmatrix} = \begin{pmatrix} a_1 \pm a_2 \\ b_1 \pm b_2 \end{pmatrix}. \tag{3.10}$$

The addition of any two vectors $\mathbf{u}$ and $\mathbf{v}$ is commutative, that is

$$\mathbf{v}_1 + \mathbf{v}_2 = \mathbf{v}_2 + \mathbf{v}_1. \tag{3.11}$$

This is because each of its components is commutative: $a_1 + a_2 = a_2 + a_1$ and $b_1 + b_2 = b_2 + b_1$. Similarly, as the addition of scalars is associative (i.e., $a_1 + (a_2 + a_3) = (a_1 + a_2) + a_3$), then the addition of vectors is associative as well. That is

$$\mathbf{v}_1 + (\mathbf{v}_2 + \mathbf{v}_3) = (\mathbf{v}_1 + \mathbf{v}_2) + \mathbf{v}_3. \tag{3.12}$$

So far we have only focused on the vectors in a two-dimensional plane; we can easily extend our discussion to any higher-dimensional vectors. In general, for any two vectors

$$\mathbf{u} = \begin{pmatrix} u_1 \\ u_2 \\ \vdots \\ u_n \end{pmatrix}, \quad \mathbf{v} = \begin{pmatrix} v_1 \\ v_2 \\ \vdots \\ v_n \end{pmatrix}, \tag{3.13}$$

we have

$$\mathbf{u} \pm \mathbf{v} = \begin{pmatrix} u_1 \pm v_1 \\ u_2 \pm v_2 \\ \vdots \\ u_n \pm v_n \end{pmatrix}. \tag{3.14}$$

The addition and substraction of two vectors are element-wise operations. However, the multiplication of two vectors can be a bit tricky.

## 2. Vector Products

The product of two vectors can be either a scalar or a vector, depending on the way we carry out the multiplications.

### 2.1. Dot Product

The scalar product of two vectors $\mathbf{F}$ and $\mathbf{d}$ is defined as

$$\mathbf{F} \cdot \mathbf{d} = |\mathbf{F}| \, |\mathbf{d}| \, \cos\theta = Fd\cos\theta, \tag{3.15}$$

where $d = |\mathbf{d}|$ and $F = |\mathbf{F}|$ are the norms or magnitudes of $\mathbf{d}$ and $\mathbf{F}$, respectively. This rather odd definition has some physical meaning. For example, we know that the work $W$ done by a force $f$ to move an object a distance $s$, is simply $W = fs$ on the condition that the force is applied along the direction of movement. If a force $\mathbf{F}$ is applied at an angle $\theta$ related to the displacement $\mathbf{d}$ (see Fig. 3.2), we first have to decompose or project the force $\mathbf{F}$ onto the displacement direction so that the component actually acts on the object along the direction of $\mathbf{d}$ is $\mathbf{F}_\| = F\cos\theta$. So the actual work done becomes

$$W = \mathbf{F}_\| d = Fd\cos\theta, \tag{3.16}$$

which means that the amount of work $W$ is the scalar product

$$W = \mathbf{F} \cdot \mathbf{d}. \tag{3.17}$$

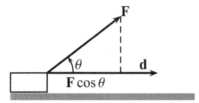

Figure 3.2: Work done $W$ by a force $\mathbf{F}$ to move an object in the direction of displacement $\mathbf{d}$ is $W = \mathbf{F} \cdot \mathbf{d} = Fd\cos\theta$.

Here the $\cdot$ symbol denotes such a scalar product. From such notations, the scalar product of two vectors is also called the dot product or inner product.

If we intend to compute in terms of their components

$$\mathbf{F} = \begin{pmatrix} f_1 \\ f_2 \\ f_3 \end{pmatrix}, \qquad \mathbf{d} = \begin{pmatrix} d_1 \\ d_2 \\ d_3 \end{pmatrix}, \tag{3.18}$$

the dot product can be calculated by

$$\mathbf{F} \cdot \mathbf{d} = f_1 d_1 + f_2 d_2 + f_3 d_3. \tag{3.19}$$

Since $\cos 90° = 0$, when the scalar product is zero, it suggests that the two vectors are perpendicular to each other; sometimes we also say they are orthogonal. So for the unit vectors $\mathbf{i}, \mathbf{j}$, and $\mathbf{k}$, we have

$$\mathbf{i} \cdot \mathbf{i} = \mathbf{j} \cdot \mathbf{j} = \mathbf{k} \cdot \mathbf{k} = 1, \quad \text{or} \quad \mathbf{i} \cdot \mathbf{j} = \mathbf{j} \cdot \mathbf{k} = \mathbf{k} \cdot \mathbf{i} = 0. \tag{3.20}$$

These basic properties can easily be verified by using the formula (3.19).

If we know the dot product, we can use it to determine the angle $\theta$, and we have

$$\cos \theta = \frac{\mathbf{F} \cdot \mathbf{d}}{Fd}. \tag{3.21}$$

## Example 3.2

To test if the following vectors are orthogonal or not

$$\mathbf{u} = \begin{pmatrix} 1 \\ 2 \\ 3 \end{pmatrix}, \quad \mathbf{v} = \begin{pmatrix} 2 \\ 2 \\ -2 \end{pmatrix},$$

we can calculate their dot product

$$\mathbf{u} \cdot \mathbf{v} = 1 \cdot 2 + 2 \cdot 2 + 3 \cdot (-2) = 2 + 4 - 6 = 0.$$

As their dot product is zero, these two vectors are perpendicular to each other.

The dot product has some interesting properties. From its definition, it is easy to see that $\mathbf{F} \cdot \mathbf{d} = \mathbf{d} \cdot \mathbf{F}$. Another interesting property is the distributive law:

$$\mathbf{F} \cdot (\mathbf{d} + \mathbf{s}) = \mathbf{F} \cdot \mathbf{d} + \mathbf{F} \cdot \mathbf{s}. \tag{3.22}$$

## Example 3.3

If a horse is towing a boat with a similar configuration as shown in Fig. 3.2, the force is 400 N with an angle of 30° degrees in the direction of travel. The boat travels for 1 km=1000 m at the speed of 3.6 km/hour. What is the work done? What is the output power of the horse?

The work done $W$ is

$$W = \mathbf{F} \cdot \mathbf{d} = 400 \times 1000 \cos 30° = 4 \times 10^5 \times \frac{\sqrt{3}}{2} \approx 3.464 \times 10^5 \text{ Joules.}$$

The speed of the boat (and the horse) is 3.6 km/hour or $v = 3.6 \times 1000/3600 = 1$ m/s. Thus, the output power of the horse is

$$P = \mathbf{F} \cdot \mathbf{v} = 400 \times 1 \cos 30° = 400 \times \frac{\sqrt{3}}{2} \approx 346.4 \text{ watts.}$$

In a special case when calculating the dot product of a vector with itself, we have

$$\mathbf{u} \cdot \mathbf{u} = \|\mathbf{u}\|^2 = u_1^2 + u_2^2 + \dots + u_n^2. \tag{3.23}$$

## 2.2. Cross Product

The vector product, also called the cross product or outer product, of two vectors $\mathbf{u}$ and $\mathbf{v}$ forms another vector $\mathbf{Q}$. The definition can be written as

$$\mathbf{u} \times \mathbf{v} = |\mathbf{u}| \, |\mathbf{v}| \, \sin\theta \, \mathbf{n} = uv \sin\theta \, \mathbf{n}, \tag{3.24}$$

where $\mathbf{n}$ is the unit vector, and $\mathbf{Q}$ points to the direction of $\mathbf{n}$ which is perpendicular to both vectors $\mathbf{u}$ and $\mathbf{v}$, forming a right-handed system (see Fig. 3.3). In addition, $\theta$ is the angle between $\mathbf{u}$ and $\mathbf{v}$, and $u = |\mathbf{u}|$ and $v = |\mathbf{v}|$ are the magnitudes of $\mathbf{u}$ and $\mathbf{v}$, respectively.

In many books, the notation $\mathbf{u} \wedge \mathbf{v}$ is also used, that is

$$\mathbf{u} \wedge \mathbf{v} \equiv \mathbf{u} \times \mathbf{v}. \tag{3.25}$$

The right-handed system suggests that, if we change the order of the product, there is

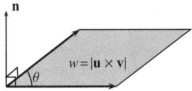

Figure 3.3: The direction of $\mathbf{u} \times \mathbf{v}$ points to the direction along $\mathbf{n}$ while the magnitude $\mathbf{Q} = |\mathbf{u} \times \mathbf{v}| = |u||v| \sin\theta$ is the area of the shaded region.

a sign change. That is, $\mathbf{v} \times \mathbf{u} = -\mathbf{u} \times \mathbf{v}$.

Though the vector product is a vector, its magnitude has a geometrical meaning. That is, the magnitude is the area of the shaded parallelogram shown in Fig. 3.3.

Using $\mathbf{u} = \left( u_1 \quad u_2 \quad u_3 \right)^T$, and $\mathbf{v} = \left( v_1 \quad v_2 \quad v_3 \right)^T$ where the superscript $T$ means the transpose which turns a column vector into a row vector or vice versa, we can write the vector product in terms of their components

$$\mathbf{u} \times \mathbf{v} = \begin{vmatrix} \mathbf{i} & \mathbf{j} & \mathbf{k} \\ u_1 & u_2 & u_3 \\ v_1 & v_2 & v_3 \end{vmatrix} = \begin{pmatrix} u_2 v_3 - u_3 v_2 \\ u_3 v_1 - u_1 v_3 \\ u_1 v_2 - u_2 v_1 \end{pmatrix}$$

$$= (u_2 v_3 - u_3 v_2)\mathbf{i} + (u_3 v_1 - u_1 v_3)\mathbf{j} + (u_1 v_2 - u_2 v_1)\mathbf{k}. \qquad (3.26)$$

Here, we have used $|\cdot|$ as a notation, though it is the determinant of a matrix, and we will introduce this concept later in this chapter.

## Example 3.4

A robotic arm rotates around an axis with an angular velocity $\omega = (1, 0, 1)^T$, what is the linear velocity at the position $\mathbf{r} = (2, 3, 4)^T$?

The velocity $\mathbf{v}$ can be calculated by

$$\mathbf{v} = \omega \times \mathbf{r} = \begin{pmatrix} 1 \\ 0 \\ 1 \end{pmatrix} \times \begin{pmatrix} 2 \\ 3 \\ 4 \end{pmatrix} = \begin{vmatrix} \mathbf{i} & \mathbf{j} & \mathbf{k} \\ 1 & 0 & 1 \\ 2 & 3 & 4 \end{vmatrix}$$

$$= (0 \times 4 - 1 \times 3)\mathbf{i} + (1 \times 2 - 1 \times 4)\mathbf{j} + (1 \times 3 - 0 \times 2)\mathbf{k} = -3\mathbf{i} - 2\mathbf{j} + 3\mathbf{k} = \begin{pmatrix} -3 \\ -2 \\ 3 \end{pmatrix}.$$

So the velocity points towards the $(-3, -2, 3)^T$ direction.

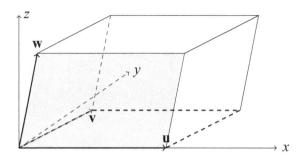

Figure 3.4: Triple vector product and volume $V = \mathbf{w} \cdot (\mathbf{u} \times \mathbf{v})$.

## 2.3. Triple Product of Vectors

For three different vectors $\mathbf{u}$, $\mathbf{v}$ and $\mathbf{w}$, it is possible to form a triple product with some physical meaning.

For example, the product $\mathbf{w} \cdot (\mathbf{u} \times \mathbf{v})$ is the volume $V$ of a parallelepiped with sides $\mathbf{u}$, $\mathbf{v}$ and $\mathbf{w}$. That is

$$V = \mathbf{w} \cdot (\mathbf{u} \times \mathbf{v}). \tag{3.27}$$

Since $|\mathbf{u} \times \mathbf{v}|$ is the area of the (base) parallelogram formed by $\mathbf{u}$ and $\mathbf{v}$, and the component of $\mathbf{w}$ in the direction of $\mathbf{u} \times \mathbf{v}$ is the height, the triple product is indeed the volume (see Fig. 3.4).

Careful readers may already wonder why we choose $\mathbf{u}$ and $\mathbf{v}$ to form the base parallelogram? In fact, there are two other ways to calculate the triple product with the same result. We have

$$\mathbf{u} \cdot (\mathbf{v} \times \mathbf{w}) = \mathbf{w} \cdot (\mathbf{u} \times \mathbf{v}) = \mathbf{v} \cdot (\mathbf{w} \times \mathbf{u}), \tag{3.28}$$

which is often referred to as the cyclic permutation.

### Example 3.5

Let us calculate the volume of the parallelepiped formed by the following three vectors

$$\mathbf{u} = \begin{pmatrix} 4 \\ 0 \\ 0 \end{pmatrix}, \quad \mathbf{v} = \begin{pmatrix} 2 \\ 1 \\ 0 \end{pmatrix}, \quad \mathbf{w} = \begin{pmatrix} 1/2 \\ 0 \\ 5/2 \end{pmatrix}.$$

First, we form the cross product

$$\mathbf{u} \times \mathbf{v} = \begin{pmatrix} u_2 v_3 - u_3 v_2 \\ u_3 v_1 - u_1 v_3 \\ u_1 v_2 - u_2 v_1 \end{pmatrix} = \begin{pmatrix} 0 \times 0 - 0 \times 1 \\ 0 \times 2 - 4 \times 0 \\ 4 \times 1 - 0 \times 2 \end{pmatrix} = \begin{pmatrix} 0 \\ 0 \\ 4 \end{pmatrix}.$$

So $|\mathbf{u} \times \mathbf{v}| = 4$, the area of the base formed by $\mathbf{u}$ and $\mathbf{v}$ is 4, as shown in Fig. 3.4. The volume is the triple product

$$\mathbf{w} \cdot (\mathbf{u} \times \mathbf{v}) = \begin{pmatrix} 1/2 \\ 0 \\ 5/2 \end{pmatrix} \cdot \begin{pmatrix} 0 \\ 0 \\ 4 \end{pmatrix} = \frac{1}{2} \times 0 + 0 \times 0 + \frac{5}{2} \times 4 = 10.$$

Here we have only covered the very basics of the vector algebra and there are more complex manipulations of vectors including other triple products, differentiation and integrals. We will introduce such more advanced concepts in later chapters.

## 3. Matrix Algebra

Matrices can be considered as an extension of vectors. In engineering simulations and scientific computations, matrices are widely used.

## 3.1. Matrix, Addition and Multiplication

A matrix is a rectangular array of numbers. For example, the following arrays:

$$\mathbf{A} = \begin{pmatrix} 1 & 2 & 3 \\ 4 & 5 & 6 \end{pmatrix}, \quad \mathbf{B} = \begin{pmatrix} 1 & 2 \\ 2.3 & 5.6 \end{pmatrix}, \quad \mathbf{C} = \begin{pmatrix} 1 & 0 & 2.2 & 3.1 \\ 7 & -1.5 & -0.01 & 8.9 \end{pmatrix}, \quad (3.29)$$

are all matrices. In most mathematics textbooks, a matrix is denoted by an uppercase letter in bold font. The size of a matrix is determined by its numbers of rows and columns. Since matrix $\mathbf{A}$ has 2 rows and 3 columns, the size of $\mathbf{A}$ is thus $2 \times 3$ (or 2 by 3). Similarly, the size of $\mathbf{B}$ is 2 by 2, and the size of $\mathbf{C}$ is $2 \times 4$.

Each number in a matrix is called an entry or element, and the location of the entry is given by its row and column numbers. We usually use a lower case letter with subscripts to denote such entries. For the above matrix $\mathbf{A}$, we can write

$$\mathbf{A} = \begin{pmatrix} 1 & 2 & 3 \\ 4 & 5 & 6 \end{pmatrix} = \begin{pmatrix} a_{11} & a_{12} & a_{13} \\ a_{21} & a_{22} & a_{23} \end{pmatrix}, \quad (3.30)$$

and we have $a_{11} = 1$, $a_{12} = 2$, $a_{13} = 3$, $a_{21} = 4$, $a_{22} = 5$ and $a_{23} = 6$. Use the same convention of notations, we can say that $b_{21}$ of the 2nd row and 1st column of $\mathbf{B}$ is $b_{21} = 2.3$, and $c_{23} = -0.01$ of $\mathbf{C}$.

In general, for a matrix with $m$ rows and $n$ columns, arranged in a rectangular array, we can write it as

$$
\mathbf{A} = \begin{pmatrix} a_{11} & a_{12} & \cdots & a_{1n} \\ a_{21} & a_{22} & \cdots & a_{2n} \\ \vdots & \vdots & \ddots & \vdots \\ a_{m1} & a_{m2} & \cdots & a_{mn} \end{pmatrix},
\tag{3.31}
$$

where $a_{ij}(i = 1, 2, ..., m; j = 1, 2, ..., n)$ are numbers. $m$ is the number of rows and $n$ is the number of columns. In this case, the matrix $\mathbf{A}$ is said to have a size of $m \times n$. In mathematics, we usually write the matrix in bold font such as $\mathbf{A}$ and write its element in lower case such as $a_{11}$. Sometimes, in order to show its elements, we may can write

$$
\mathbf{A} = [a_{ij}], \quad (i = 1, 2, ..., m; \ j = 1, 2, ..., n),
\tag{3.32}
$$

where we use $[a_{ij}]$ to explicitly show the elements of $\mathbf{A}$.

If the number of rows ($m$) is equal to the number of columns $n$ (i.e., $m = n$), we say the matrix is a square matrix.

The transpose of a matrix can be formally defined as

$$
\mathbf{A}^T = [a_{ij}]^T = [a_{ji}], \text{ for all } (i, j).
\tag{3.33}
$$

Sometimes, it makes a statement simpler if we write $\forall(i, j)$ to mean for all values of $(i = 1, 2, ..., m; \ j = 1, 2, ..., n)$.

If a matrix is a square matrix satisfying

$$
\mathbf{A}^T = \mathbf{A}, \text{ or } a_{ij} = a_{ji}, \quad \forall(i, j),
\tag{3.34}
$$

it is called a symmetric matrix. The expression $\mathbf{A}^T = \mathbf{A}$ emphasize the overall form, while $a_{ij} = a_{ji}$ focuses on the properties of elements, though both forms are essentially identical and thus used interchangeably.

## Example 3.6

For matrices

$$
\mathbf{A} = \begin{pmatrix} 2 & -2 & 0 \\ 7 & 8 & -9 \end{pmatrix}, \quad \mathbf{B} = \begin{pmatrix} 2 & -1 \\ -1 & 3 \end{pmatrix},
$$

their sizes are 2 by 3, and 2 by 2, respectively. Their transposes are

$$
\mathbf{A}^T = \begin{pmatrix} 2 & 7 \\ -2 & 8 \\ 0 & -9 \end{pmatrix}, \quad \mathbf{B}^T = \begin{pmatrix} 2 & -1 \\ -1 & 3 \end{pmatrix}.
$$

Since $\mathbf{B}^T = \mathbf{B}$, matrix $\mathbf{B}$ is symmetric.

In principle, each element $a_{ij}$ of a matrix $\mathbf{A}$ can be a real or a complex number. If all elements are real numbers, the matrix is called a real matrix. If at least one element contains a complex number (to be introduced later in this book), the matrix is called complex. For the moment, we will focus mainly on real matrices.

Matrix addition and subtraction are possible only if both matrices are of the same size. If $\mathbf{A} = [a_{ij}]$ and $\mathbf{G} = [g_{ij}]$ are the same size of $m \times n$, their sum $\mathbf{S} = [s_{ij}]$ can be obtained by adding their corresponding entries. That is

$$s_{ij} = a_{ij} + g_{ij}, \quad \forall(i, j), \tag{3.35}$$

where $\mathbf{S}$ has the same size of $m \times n$. Their differences also form a matrix $\mathbf{M} = [m_{ij}]$

$$m_{ij} = a_{ij} - g_{ij}, \tag{3.36}$$

which also has the same size. If a matrix can be multiplied by a scalar $\beta \neq 0$, we have

$$\beta\mathbf{A} = [\beta a_{ij}], \quad \forall(i, j). \tag{3.37}$$

## Example 3.7

For $\mathbf{A} = \begin{pmatrix} 1 & 2 & 3 \\ 4 & 5 & 6 \end{pmatrix}$ and $\mathbf{G} = \begin{pmatrix} 1 & 1 & 7 \\ 2 & -2 & 1 \end{pmatrix}$, we have

$$\mathbf{A} + \mathbf{G} = \begin{pmatrix} 1 & 2 & 3 \\ 4 & 5 & 6 \end{pmatrix} + \begin{pmatrix} 1 & 1 & 7 \\ 2 & -2 & 1 \end{pmatrix}$$

$$= \begin{pmatrix} 1+1 & 2+1 & 3+7 \\ 4+2 & 5+(-2) & 6+1 \end{pmatrix} = \begin{pmatrix} 2 & 3 & 10 \\ 6 & 3 & 7 \end{pmatrix},$$

and

$$\mathbf{A} - \mathbf{G} = \begin{pmatrix} 1 & 2 & 3 \\ 4 & 5 & 6 \end{pmatrix} - \begin{pmatrix} 1 & 1 & 7 \\ 2 & -2 & 1 \end{pmatrix} = \begin{pmatrix} 1-1 & 2-1 & 3-7 \\ 4-2 & 5-(-2) & 6-1 \end{pmatrix} = \begin{pmatrix} 0 & 1 & -4 \\ 2 & 7 & 5 \end{pmatrix}.$$

In addition, we have

$$7\mathbf{A} = 7\begin{pmatrix} 1 & 2 & 3 \\ 4 & 5 & 6 \end{pmatrix} = \begin{pmatrix} 7\times1 & 7\times2 & 7\times3 \\ 7\times4 & 7\times5 & 7\times6 \end{pmatrix} = \begin{pmatrix} 7 & 14 & 21 \\ 28 & 35 & 42 \end{pmatrix}.$$

Matrix multiplication requires that the number of columns of the first matrix is equal to the number of rows of the second matrix. If $\mathbf{A} = [a_{ij}]$ is an $m \times n$ matrix, and

$\mathbf{B} = [b_{jk}]$ is an $n \times p$ matrix, then $\mathbf{C} = \mathbf{AB}$ is an $m \times p$ matrix. We have

$$\mathbf{C} = [c_{ik}] = \begin{pmatrix} \boxed{c_{11}} & c_{12} & \cdots & c_{1p} \\ c_{21} & c_{22} & \cdots & c_{2p} \\ \vdots & \vdots & \ddots & \vdots \\ c_{m1} & c_{m2} & \cdots & c_{mp} \end{pmatrix} = \mathbf{AB} = [a_{ij}][b_{jk}]$$

$$= \begin{pmatrix} \boxed{a_{11}} & \boxed{a_{12}} & \boxed{\cdots} & \boxed{a_{1n}} \\ a_{21} & a_{22} & \cdots & a_{2n} \\ \vdots & \vdots & \ddots & \vdots \\ a_{m1} & a_{m2} & \cdots & a_{mn} \end{pmatrix} \begin{pmatrix} \boxed{b_{11}} & b_{12} & \cdots & b_{1p} \\ \boxed{b_{21}} & b_{22} & \cdots & b_{2p} \\ \boxed{\vdots} & \vdots & \ddots & \vdots \\ \boxed{b_{n1}} & b_{n2} & \cdots & b_{np} \end{pmatrix}, \tag{3.38}$$

where

$$c_{ik} = a_{i1}b_{1k} + a_{i2}b_{2k} + a_{i3}b_{3k} + \ldots + a_{in}b_{nk} = \sum_{j=1}^{n} a_{ij}b_{jk}. \tag{3.39}$$

Here, the entry $c_{ik}$ is obtained by the sum of multiplying the $i$th row of $\mathbf{A}$ by the corresponding entry in the $k$th column of $\mathbf{B}$. For example, we have

$$c_{11} = \sum_{j=1}^{n} a_{1j}b_{j1} = a_{11}b_{11} + a_{12}b_{21} + \ldots + a_{1n}b_{n1}, \tag{3.40}$$

and these entries are highlighted in boxes as Eq.(3.38). Similarly, we have

$$c_{23} = \sum_{j=1}^{n} a_{2j}b_{j3} = a_{21}b_{13} + a_{22}b_{23} + \ldots + a_{2n}b_{n3}, \tag{3.41}$$

and so on. Let us look at an example.

## Example 3.8

For two matrices

$$\mathbf{A} = \begin{pmatrix} a_{11} & a_{12} \\ a_{21} & a_{22} \end{pmatrix} = \begin{pmatrix} 1 & 2 \\ 3 & 4 \end{pmatrix}, \quad \mathbf{B} = \begin{pmatrix} b_{11} & b_{12} \\ b_{21} & b_{22} \end{pmatrix} = \begin{pmatrix} 5 & 6 \\ 7 & 8 \end{pmatrix},$$

where $a_{11} = 1$, $b_{11} = 5$, ..., $b_{22} = 8$. For the multiplication of these two matrices, we have

$$\mathbf{C} = \begin{pmatrix} c_{11} & c_{12} \\ c_{21} & c_{22} \end{pmatrix} = \mathbf{AB} = \begin{pmatrix} a_{11} & a_{12} \\ a_{21} & a_{22} \end{pmatrix} \begin{pmatrix} b_{11} & b_{12} \\ b_{21} & b_{22} \end{pmatrix}.$$

So we have

$$c_{11} = \sum_{j=1}^{2} a_{1j}b_{j1} = a_{11}b_{11} + a_{12}b_{21} = 1 \times 5 + 2 \times 7 = 19.$$

$$c_{12} = \sum_{j=1}^{2} a_{1j}b_{j2} = a_{11}b_{12} + a_{12}b_{22} = 1 \times 6 + 2 \times 8 = 22.$$

$$c_{21} = \sum_{j=1}^{2} a_{2j}b_{j1} = a_{21}b_{11} + a_{22}b_{21} = 3 \times 5 + 4 \times 7 = 43.$$

$$c_{22} = \sum_{j=1}^{2} a_{2j}b_{j2} = a_{21}b_{12} + a_{22}b_{22} = 3 \times 6 + 4 \times 8 = 50.$$

Therefore, we have

$$\mathbf{C} = \mathbf{AB} = \begin{pmatrix} 1 & 2 \\ 3 & 4 \end{pmatrix} \begin{pmatrix} 5 & 6 \\ 7 & 8 \end{pmatrix} = \begin{pmatrix} 19 & 22 \\ 43 & 50 \end{pmatrix}.$$

When a scalar $\alpha$ multiplies a matrix $\mathbf{A}$, the result is the matrix with each of $\mathbf{A}$'s elements multiplying by $\alpha$. For example,

$$\alpha \mathbf{A} = \alpha \begin{pmatrix} a & b \\ c & d \\ e & f \end{pmatrix} = \begin{pmatrix} \alpha a & \alpha b \\ \alpha c & \alpha d \\ \alpha e & \alpha f \end{pmatrix}. \tag{3.42}$$

Generally speaking, the addition of two matrices is commutative

$$\mathbf{P} + \mathbf{Q} = \mathbf{Q} + \mathbf{P}. \tag{3.43}$$

The addition of three matrices is associative, that is

$$(\mathbf{P} + \mathbf{Q}) + \mathbf{A} = \mathbf{P} + (\mathbf{Q} + \mathbf{A}). \tag{3.44}$$

However, matrices multiplication is not commutative. That is $\mathbf{AB} \neq \mathbf{BA}$.

## Example 3.9

For example, for two matrices

$$\mathbf{A} = \begin{pmatrix} 1 & 2 \\ 3 & 4 \end{pmatrix}, \quad \mathbf{B} = \begin{pmatrix} 5 & 6 \\ 7 & 8 \end{pmatrix},$$

we have

$$\mathbf{D} = \begin{pmatrix} d_{11} & d_{12} \\ d_{21} & d_{22} \end{pmatrix} = \mathbf{BA} = \begin{pmatrix} b_{11} & b_{12} \\ b_{21} & b_{22} \end{pmatrix} \begin{pmatrix} a_{11} & a_{12} \\ a_{21} & a_{22} \end{pmatrix}.$$

Following the same procedure as in the previous example, we have

$$d_{11} = \sum_{j=1}^{2} b_{1j} a_{j1} = b_{11} a_{11} + b_{12} \times a_{21} = 5 \times 1 + 6 \times 3 = 23.$$

Similarly, we have

$$d_{12} = 34, \quad d_{21} = 31, \quad d_{22} = 46.$$

Thus, we have

$$\mathbf{D} = \mathbf{BA} = \begin{pmatrix} 23 & 34 \\ 31 & 46 \end{pmatrix}.$$

From the previous example, we know that

$$\mathbf{AB} = \begin{pmatrix} 19 & 22 \\ 43 & 50 \end{pmatrix},$$

we can conclude that

$$\mathbf{AB} \neq \mathbf{BA}.$$

In general, if the product of two matrices $\mathbf{A}$ and $\mathbf{B}$ exists, their transposes have the following properties:

$$(\mathbf{AB})^T = \mathbf{B}^T \mathbf{A}^T. \tag{3.45}$$

It leaves an exercise to use the above examples to show that this relationship is true.

There are two special matrices: the zero matrix and the identity matrix. A zero matrix is a matrix whose every element is zero. For example, we have $1 \times 4$ zero matrix as $\mathbf{O} = \begin{pmatrix} 0 & 0 & 0 & 0 \end{pmatrix}$.

If the diagonal elements are 1's and all the other elements are zeros, it is called a unit matrix or an identity matrix. For example, the $3 \times 3$ unit matrix can be written as

$$\mathbf{I} = \begin{pmatrix} 1 & 0 & 0 \\ 0 & 1 & 0 \\ 0 & 0 & 1 \end{pmatrix}. \tag{3.46}$$

For an $n \times n$ identity matrix, we can write it as

$$\mathbf{I} = \begin{pmatrix} 1 & 0 & 0 & \cdots & 0 \\ 0 & 1 & 0 & \cdots & 0 \\ 0 & 0 & 1 & \cdots & 0 \\ \vdots & \vdots & \vdots & \ddots & 0 \\ 0 & 0 & 0 & \cdots & 1 \end{pmatrix}. \tag{3.47}$$

If a matrix $\mathbf{A}$ is the same size as the unit matrix, then it is commutative with $\mathbf{I}$. That is

$$\mathbf{IA} = \mathbf{AI} = \mathbf{A}. \tag{3.48}$$

## Example 3.10

For $\mathbf{A} = \begin{pmatrix} 11 & -2 \\ -3 & 5 \end{pmatrix}$, we can check that

$$\mathbf{AI} = \begin{pmatrix} 11 & -2 \\ -3 & 5 \end{pmatrix} \begin{pmatrix} 1 & 0 \\ 0 & 1 \end{pmatrix} = \begin{pmatrix} 11 & -2 \\ -3 & 5 \end{pmatrix}.$$

Similarly, we have

$$\mathbf{IA} = \begin{pmatrix} 1 & 0 \\ 0 & 1 \end{pmatrix} \begin{pmatrix} 11 & -2 \\ -3 & 5 \end{pmatrix} = \begin{pmatrix} 11 & -2 \\ -3 & 5 \end{pmatrix},$$

which indeed means that

$$\mathbf{AI} = \mathbf{IA}.$$

## 3.2. Transformation and Inverse

When a point $P(x, y)$ (say, a point on a robotic arm) is rotated by an angle $\theta$, it becomes its corresponding point $P'(x', y')$ (see Fig. 3.5). The relationship between the old co-ordinates $(x, y)$ and the new coordinates $(x', y')$ can be derived by using trigonometry.

The new coordinates at the new location $P'(x', y')$ can be written in terms of the coordinates at the original location $P(x, y)$ and the angle $\theta$. From basic geometry, we know that $\angle A'P'Q = \theta = \angle A'OW$. Since $x = OA = OA'$, $y = AP = A'P'$, and $x' = OS = OW - SW = OW - QA'$, we have

$$x' = OW - QA' = x \cos \theta - y \sin \theta. \tag{3.49}$$

Similarly, we have

$$y' = SP' = SQ + QP' = WA' + QP' = x \sin \theta + y \cos \theta. \tag{3.50}$$

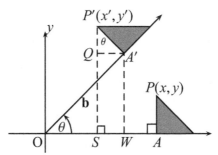

Figure 3.5: The rotational transformation.

The above two equations can be written in a compact form

$$\begin{pmatrix} x' \\ y' \end{pmatrix} = \begin{pmatrix} \cos\theta & -\sin\theta \\ \sin\theta & \cos\theta \end{pmatrix} \begin{pmatrix} x \\ y \end{pmatrix},$$

(3.51)

where the matrix for the rotation or transformation is

$$\mathbf{R}_\theta = \begin{pmatrix} \cos\theta & -\sin\theta \\ \sin\theta & \cos\theta \end{pmatrix}.$$

(3.52)

Therefore, for any point $(x, y)$, its new coordinates after rotating by an angle $\theta$ can be obtained by

$$\begin{pmatrix} x' \\ y' \end{pmatrix} = \mathbf{R}_\theta \begin{pmatrix} x \\ y \end{pmatrix}.$$

(3.53)

## Example 3.11

For a vector **u** such as a part of a robotic arm, linking point A=(1,2) to B=(2,1), both points A and B rotate 45° anticlockwise in the Cartesian coordinate system. What are the new coordinates of A and B? What is the new vector **u**′?

The rotation matrix is

$$\mathbf{R}_{45} = \begin{pmatrix} \cos 45° & -\sin 45° \\ \sin 45° & \cos 45° \end{pmatrix} = \begin{pmatrix} \sqrt{2}/2 & -\sqrt{2}/2 \\ \sqrt{2}/2 & \sqrt{2}/2 \end{pmatrix} = \frac{\sqrt{2}}{2}\begin{pmatrix} 1 & -1 \\ 1 & 1 \end{pmatrix}.$$

The new point $A'$ after rotating from A is

$$\frac{\sqrt{2}}{2}\begin{pmatrix} 1 & -1 \\ 1 & 1 \end{pmatrix}\begin{pmatrix} 1 \\ 2 \end{pmatrix} = \frac{\sqrt{2}}{2}\begin{pmatrix} 1\times 1 + (-1)\times 2 \\ 1\times 1 + 1\times 2 \end{pmatrix} = \frac{\sqrt{2}}{2}\begin{pmatrix} -1 \\ 3 \end{pmatrix}.$$

Point B will become point $B'$ at

$$\frac{\sqrt{2}}{2}\begin{pmatrix} 1 & -1 \\ 1 & 1 \end{pmatrix}\begin{pmatrix} 2 \\ 1 \end{pmatrix} = \frac{\sqrt{2}}{2}\begin{pmatrix} 1\times 2 + (-1)\times 1 \\ 1\times 2 + 1\times 1 \end{pmatrix} = \frac{\sqrt{2}}{2}\begin{pmatrix} 1 \\ 3 \end{pmatrix}.$$

Before the rotation, vector $\mathbf{u}$ is

$$\mathbf{u} = \overrightarrow{AB} = \begin{pmatrix} 2 \\ 1 \end{pmatrix} - \begin{pmatrix} 1 \\ 2 \end{pmatrix} = \begin{pmatrix} 1 \\ -1 \end{pmatrix},$$

which points 45° downwards to the right. Now the new vector $\mathbf{u}'$ after the rotation becomes

$$\mathbf{u}' = \overrightarrow{A'B'} = \frac{\sqrt{2}}{2}\begin{pmatrix} 1 \\ 3 \end{pmatrix} - \frac{\sqrt{2}}{2}\begin{pmatrix} -1 \\ 3 \end{pmatrix} = \frac{\sqrt{2}}{2}\begin{pmatrix} 2 \\ 0 \end{pmatrix},$$

which points to the right horizontally.

If point $P$ is rotated by $\theta + \psi$, we have

$$\begin{pmatrix} x'' \\ y'' \end{pmatrix} = \begin{pmatrix} \cos(\theta + \psi) & -\sin(\theta + \psi) \\ \sin(\theta + \psi) & \cos(\theta + \psi) \end{pmatrix}\begin{pmatrix} x \\ y \end{pmatrix} = \mathbf{R}_{\theta+\psi}\begin{pmatrix} x \\ y \end{pmatrix}, \tag{3.54}$$

which can also be achieved by two steps: first by rotating $\theta$ to get $P'(x', y')$ and then rotating by $\psi$ from $P'(x', y')$ to $P''(x'', y'')$. This is to say

$$\begin{pmatrix} x'' \\ y'' \end{pmatrix} = \mathbf{R}_\psi\begin{pmatrix} x' \\ y' \end{pmatrix} = \mathbf{R}_\psi\mathbf{R}_\theta\begin{pmatrix} x \\ y \end{pmatrix}. \tag{3.55}$$

Combining with (3.54), we have

$$\mathbf{R}_{\theta+\psi} = \mathbf{R}_\psi\mathbf{R}_\theta, \tag{3.56}$$

or

$$\begin{pmatrix} \cos(\theta + \psi) & -\sin(\theta + \psi) \\ \sin(\theta + \psi) & \cos(\theta + \psi) \end{pmatrix} = \begin{pmatrix} \cos\psi & -\sin\psi \\ \sin\psi & \cos\psi \end{pmatrix}\begin{pmatrix} \cos\theta & -\sin\theta \\ \sin\theta & \cos\theta \end{pmatrix}$$

$$= \begin{pmatrix} \cos\psi\cos\theta - \sin\psi\sin\theta & -[\cos\psi\sin\theta + \sin\psi\cos\theta] \\ \sin\psi\cos\theta + \cos\psi\sin\theta & \cos\psi\cos\theta - \sin\psi\sin\theta \end{pmatrix}, \tag{3.57}$$

which is another way of deriving the sine and cosine of the addition of two angles. Thus, this equation implies that

$$\sin(\theta + \psi) = \sin\theta\cos\psi + \cos\theta\sin\psi, \tag{3.58}$$

and

$$\cos(\theta + \psi) = \cos\theta\cos\psi - \sin\theta\sin\psi. \tag{3.59}$$

In a special case of $\theta = \psi$, they become the so-called double angle formulae:

$$\sin(2\theta) = 2\sin\theta\cos\theta, \quad \cos(2\theta) = \cos^2\theta - \sin^2\theta = 2\cos^2\theta - 1, \tag{3.60}$$

where we have used $\sin^2\theta + \cos^2\theta = 1$.

In a special case when first rotating by $\theta$, followed by rotating back by $-\theta$, a point $P(x, y)$ should reach its original point. That is

$$\begin{pmatrix} x \\ y \end{pmatrix} = \begin{pmatrix} 1 & 0 \\ 0 & 1 \end{pmatrix}\begin{pmatrix} x \\ y \end{pmatrix} = \mathbf{R}_{-\theta}\mathbf{R}_{\theta}\begin{pmatrix} x \\ y \end{pmatrix}, \tag{3.61}$$

which means that

$$\mathbf{R}_{-\theta}\mathbf{R}_{\theta} = \begin{pmatrix} 1 & 0 \\ 0 & 1 \end{pmatrix} = \mathbf{I}. \tag{3.62}$$

In other words, $\mathbf{R}_{-\theta}$ is the inverse of $\mathbf{R}_{\theta}$. That is to say

$$\mathbf{R}_{-\theta} = \begin{pmatrix} \cos(-\theta) & -\sin(-\theta) \\ \sin(-\theta) & \cos(-\theta) \end{pmatrix} = \begin{pmatrix} \cos\theta & \sin\theta \\ -\sin\theta & \cos\theta \end{pmatrix}, \tag{3.63}$$

is the inverse of

$$\mathbf{R}_{\theta} = \begin{pmatrix} \cos\theta & -\sin\theta \\ \sin\theta & \cos\theta \end{pmatrix}. \tag{3.64}$$

In general, the inverse $\mathbf{A}^{-1}$ of a square matrix $\mathbf{A}$, if it exists, is defined by

$$\mathbf{A}^{-1}\mathbf{A} = \mathbf{A}\mathbf{A}^{-1} = \mathbf{I}, \tag{3.65}$$

where $\mathbf{I}$ is a unit matrix which is the same size as $\mathbf{A}$.

It is worth pointing out that a matrix is not invertible if its determinant $\det(\mathbf{A})$ is zero, and $\mathbf{A}^{-1}$ does not exist. We will formally define the determinant on the next page. Let us look at a simple example first.

### Example 3.12

For example, a $2 \times 2$ matrix $\mathbf{A}$ and its inverse $\mathbf{A}^{-1}$

$$\mathbf{A} = \begin{pmatrix} a & b \\ c & d \end{pmatrix}, \quad \mathbf{A}^{-1} = \begin{pmatrix} \alpha & \beta \\ \gamma & \kappa \end{pmatrix},$$

can be related by

$$\mathbf{A}\mathbf{A}^{-1} = \begin{pmatrix} a & b \\ c & d \end{pmatrix}\begin{pmatrix} \alpha & \beta \\ \gamma & \kappa \end{pmatrix} = \begin{pmatrix} a\alpha + b\gamma & a\beta + b\kappa \\ c\alpha + d\gamma & c\beta + d\kappa \end{pmatrix} = \begin{pmatrix} 1 & 0 \\ 0 & 1 \end{pmatrix} = \mathbf{I}.$$

This means that

$$a\alpha + b\gamma = 1, \quad a\beta + b\kappa = 0, \quad c\alpha + d\gamma = 0, \quad c\gamma + d\kappa = 1.$$

These four equations will solve the four unknowns $\alpha, \beta, \gamma$ and $\kappa$. After some simple rearrangement and calculations, we have

$$\alpha = \frac{d}{\Delta}, \quad \beta = \frac{-b}{\Delta}, \quad \gamma = \frac{-c}{\Delta}, \quad \kappa = \frac{a}{\Delta},$$

where $\Delta = ad - bc$ is the determinant of $\mathbf{A}$. Therefore, the inverse becomes

$$\mathbf{A}^{-1} = \frac{1}{ad - bc} \begin{pmatrix} d & -b \\ -c & a \end{pmatrix}.$$

It is straightforward to verify that $\mathbf{A}^{-1}\mathbf{A} = \begin{pmatrix} 1 & 0 \\ 0 & 1 \end{pmatrix}$.

In the case of

$$\mathbf{R}_\theta = \begin{pmatrix} \cos\theta & -\sin\theta \\ \sin\theta & \cos\theta \end{pmatrix},$$

we have

$$\mathbf{R}^{-1} = \frac{1}{\cos\theta\cos\theta - (-\sin\theta)\sin\theta} \begin{pmatrix} \cos\theta & \sin\theta \\ -\sin\theta & \cos\theta \end{pmatrix} = \begin{pmatrix} \cos\theta & \sin\theta \\ -\sin\theta & \cos\theta \end{pmatrix},$$

where we have used $\cos^2\theta + \sin^2\theta = 1$. This is the same as (3.63).

We have seen that some special combination of the elements such as the determinant $\Delta = ad - bc$ is very important. We now try to define it more generally.

The determinant of an $n \times n$ square matrix $\mathbf{A} = [a_{ij}]$ is a number which can be obtained by a cofactor expansion either by row or by column

$$\det(\mathbf{A}) \equiv |A| = \sum_{j=1}^{n} (-1)^{i+j} a_{ij} |\mathbf{M}_{ij}|, \tag{3.66}$$

where $|\mathbf{M}_{ij}|$ is the cofactor or the determinant of a minor matrix $\mathbf{M}$ of $\mathbf{A}$, obtained by deleting row $i$ and column $j$. This is a recursive relationship. For example, $\mathbf{M}_{12}$ of a $3 \times 3$ matrix is obtained by deleting the first row and the second column

$$\begin{vmatrix} \overline{a_{11}} & -\overline{a_{12}}- & \overline{a_{13}} \\ a_{21} & a_{22} & a_{23} \\ a_{31} & a_{32} & a_{33} \end{vmatrix} \implies |\mathbf{M}|_{12} = \begin{vmatrix} a_{21} & a_{23} \\ a_{31} & a_{33} \end{vmatrix}. \tag{3.67}$$

Obviously, the determinant of a $1 \times 1$ matrix $|a_{11}| = a_{11}$ is the number itself. The determinant of a $2 \times 2$ matrix

$$\det(\mathbf{A}) = \begin{vmatrix} a_{11} & a_{12} \\ a_{21} & a_{22} \end{vmatrix} = a_{11}a_{22} - a_{12}a_{21}. \tag{3.68}$$

The determinant of a $3 \times 3$ matrix is given by $\det(\mathbf{A})$ or

$$\begin{vmatrix} a_{11} & a_{12} & a_{23} \\ a_{21} & a_{22} & a_{23} \\ a_{31} & a_{32} & a_{33} \end{vmatrix} = (-1)^{1+1}a_{11}\begin{vmatrix} a_{22} & a_{23} \\ a_{32} & a_{33} \end{vmatrix} + (-1)^{1+2}a_{12}\begin{vmatrix} a_{21} & a_{23} \\ a_{31} & a_{33} \end{vmatrix}$$

$$+(-1)^{1+3}a_{13}\begin{vmatrix} a_{21} & a_{22} \\ a_{31} & a_{32} \end{vmatrix} = a_{11}(a_{22}a_{33} - a_{32}a_{23})$$

$$-a_{12}(a_{21}a_{33} - a_{31}a_{23}) + a_{13}(a_{21}a_{32} - a_{31}a_{22}). \tag{3.69}$$

Here we used the expansion along the first row $i = 1$. We can also expand it along any other rows or columns, and the results are the same. As the determinant of a matrix is a scalar or a simple number, it is not difficult to understand the following properties

$$\det(\mathbf{AB}) = \det(\mathbf{A})\det(\mathbf{B}), \qquad \det(\mathbf{A}^T) = \det(\mathbf{A}). \tag{3.70}$$

If $\det(\mathbf{A}) = 0$, the square matrix is called singular and its inverse does not exist. For example, $\mathbf{B} = \begin{pmatrix} 1 & 2 \\ 2 & 4 \end{pmatrix}$, its determinant is $\det(\mathbf{B}) = 1 \times 4 - 2 \times 2 = 0$, so its inverse does not exist.

There are many applications of determinants. The inverse of a matrix exists only if $\det(\mathbf{A}) \neq 0$. Here we will use it to calculate the inverse $\mathbf{A}^{-1}$ using

$$\mathbf{A}^{-1} = \frac{\text{adj}(\mathbf{A})}{\det(\mathbf{A})} = \frac{\mathbf{B}^T}{\det(\mathbf{A})}, \qquad \mathbf{B} = \left[(-1)^{i+j}|M_{ij}|\right], \tag{3.71}$$

where the matrix $\mathbf{B}^T$ is called the adjoint of matrix $\mathbf{A}$ with the same size as $\mathbf{A}$, and $i, j = 1, ..., n$. Each part of the element $\mathbf{B}$ is expressed in terms of a cofactor so that $b_{ij} = (-1)^{i+j}|M_{ij}|$. $\mathbf{B}$ itself is called the cofactor matrix, while $\text{adj}(\mathbf{A}) = \mathbf{B}^T$ is sometimes used to denote the adjoint matrix. This seems too complicated, so let us compute the inverse of a $3 \times 3$ matrix as an example.

## Example 3.13

In order to compute the inverse of

$$\mathbf{A} = \begin{pmatrix} 1 & 1 & -2 \\ 1 & 0 & 2 \\ 2 & 1 & 1 \end{pmatrix},$$

we first construct its adjoint matrix $\mathbf{B}^T$ with

$$\mathbf{B} = [b_{ij}] = \left[(-1)^{i+j}|M_{ij}|\right].$$

The first element $b_{11}$ can be obtained by

$$b_{11} = (-1)^{1+1}\begin{vmatrix} 0 & 2 \\ 1 & 1 \end{vmatrix} = (-1)^2 \times (0 \times 1 - 2 \times 1) = -2.$$

The element $b_{12}$ is

$$b_{12} = (-1)^{1+2}\begin{vmatrix} 1 & 2 \\ 2 & 1 \end{vmatrix} = -1 \times (1 \times 1 - 2 \times 2) = 3,$$

while the element $b_{21}$ is

$$b_{21} = (-1)^{2+1}\begin{vmatrix} 1 & -2 \\ 1 & 1 \end{vmatrix} = (-1)^3 \times (1 \times 1 - 1 \times (-2)) = -3.$$

Following a similar procedure, we have $\mathbf{B}$ and its transpose $\mathbf{B}^T$ as

$$\mathbf{B} = \begin{pmatrix} -2 & 3 & 1 \\ -3 & 5 & 1 \\ 2 & -4 & -1 \end{pmatrix}, \quad \text{or} \quad \mathbf{B}^T = \begin{pmatrix} -2 & -3 & 2 \\ 3 & 5 & -4 \\ 1 & 1 & -1 \end{pmatrix}.$$

Then, the determinant of $\mathbf{A}$ is

$$\det(\mathbf{A}) = \begin{vmatrix} 1 & 1 & -2 \\ 1 & 0 & 2 \\ 2 & 1 & 1 \end{vmatrix} = 1 \times \begin{vmatrix} 0 & 2 \\ 1 & 1 \end{vmatrix} - 1 \times \begin{vmatrix} 1 & 2 \\ 2 & 1 \end{vmatrix} + (-2) \times \begin{vmatrix} 1 & 0 \\ 2 & 1 \end{vmatrix}$$

$$= 1 \times (0 \times 1 - 2 \times 1) - 1 \times (1 \times 1 - 2 \times 2) - 2 \times (1 \times 1 - 2 \times 0)$$

$$= 1 \times (-2) - 1 \times (-3) - 2 \times 1 = -1.$$

Finally, the inverse becomes

$$\mathbf{A}^{-1} = \frac{\mathbf{B}^T}{\det(\mathbf{A})} = \frac{1}{-1}\begin{pmatrix} -2 & -3 & 2 \\ 3 & 5 & -4 \\ 1 & 1 & -1 \end{pmatrix} = \begin{pmatrix} 2 & 3 & -2 \\ -3 & -5 & 4 \\ -1 & -1 & 1 \end{pmatrix}.$$

This result will be used in the next example.

## 4. System of Linear Equations

A system of linear equations can be written as a large matrix equation, and the solution of such a linear system will become straightforward if the inverse of a square matrix is used. Let us demonstrate this by an example. For a linear system consisting of three simultaneous equations, we have

$$a_{11}x + a_{12}y + a_{13}z = b_1,$$

$$a_{21}x + a_{22}y + a_{23}z = b_2,$$

$$a_{31}x + a_{32}y + a_{33}z = b_3,$$

which can be written compactly as

$$\begin{pmatrix} a_{11} & a_{12} & a_{13} \\ a_{21} & a_{22} & a_{23} \\ a_{31} & a_{32} & a_{33} \end{pmatrix} \begin{pmatrix} x \\ y \\ z \end{pmatrix} = \begin{pmatrix} b_1 \\ b_2 \\ b_3 \end{pmatrix}, \tag{3.72}$$

or more compactly as

$$\mathbf{Au = b}, \tag{3.73}$$

where $\mathbf{u} = \begin{pmatrix} x & y & z \end{pmatrix}^T$. By multiplying $\mathbf{A}^{-1}$ on both sides, we have

$$\mathbf{A}^{-1}\mathbf{Au} = \mathbf{A}^{-1}\mathbf{b}, \quad \text{or} \quad \mathbf{Iu} = \mathbf{A}^{-1}\mathbf{b}. \tag{3.74}$$

Therefore, its solution can be written as

$$\mathbf{u} = \mathbf{A}^{-1}\mathbf{b}.$$

Now let us look at an example of the solution of a simple linear system.

## Example 3.14

In order to solve the following system

$$x + y - 2z = -14, \quad x + 2z = 15, \quad 2x + y + z = 8,$$

we first write it as $\mathbf{Au = b}$, or

$$\begin{pmatrix} 1 & 1 & -2 \\ 1 & 0 & 2 \\ 2 & 1 & 1 \end{pmatrix} \begin{pmatrix} x \\ y \\ z \end{pmatrix} = \begin{pmatrix} -14 \\ 15 \\ 8 \end{pmatrix}.$$

We know from the earlier example that the inverse of $\mathbf{A}^{-1}$ is

$$\mathbf{A}^{-1} = \begin{pmatrix} 2 & 3 & -2 \\ -3 & -5 & 4 \\ -1 & -1 & 1 \end{pmatrix},$$

we now have $\mathbf{u} = \mathbf{A}^{-1}\mathbf{b}$ or

$$\begin{pmatrix} x \\ y \\ z \end{pmatrix} = \begin{pmatrix} 2 & 3 & -2 \\ -3 & -5 & 4 \\ -1 & -1 & 1 \end{pmatrix} \begin{pmatrix} -14 \\ 15 \\ 8 \end{pmatrix} = \begin{pmatrix} 2 \times (-14) + 3 \times 15 + (-2) \times 8 \\ -3 \times (-14) + (-5) \times 15 + 4 \times 8 \\ -1 \times (-14) + (-1) \times 15 + 1 \times 8 \end{pmatrix} = \begin{pmatrix} 1 \\ -1 \\ 7 \end{pmatrix},$$

which gives a unique set of solutions $x = 1, y = -1$ and $z = 7$.

In general, a linear system of $n$ equations for $n$ unknowns can be written in the compact form as

$$\begin{pmatrix} a_{11} & a_{12} & \cdots & a_{1n} \\ a_{21} & a_{22} & \cdots & a_{2n} \\ \vdots & \vdots & & \\ a_{n1} & a_{n2} & \cdots & a_{nn} \end{pmatrix} \begin{pmatrix} u_1 \\ u_2 \\ \vdots \\ u_n \end{pmatrix} = \begin{pmatrix} b_1 \\ b_2 \\ \vdots \\ b_n \end{pmatrix}, \tag{3.75}$$

or simply

$$\mathbf{A}\mathbf{u} = \mathbf{b}. \tag{3.76}$$

Its solution can be obtained by inverse

$$\mathbf{u} = \mathbf{A}^{-1}\mathbf{b}. \tag{3.77}$$

However, this is not always possible. If the determinant $\det(\mathbf{A}) = 0$, the system becomes degenerate or ill-posed, and there may an infinite number of solutions. For example, the system of $x + 2y = 5$ and $2x + 4y = 10$ can have an infinite number of solutions because

$$\mathbf{A} = \begin{pmatrix} 1 & 2 \\ 2 & 4 \end{pmatrix}, \tag{3.78}$$

has no inverse, and the two equations are not linearly independent. Multiplying the first equation by 2 gives the second equation. These two equations overlap as a single line on the Cartesian plane. On the other hand, the following system of $x + 2y = 5$ and $2x + 4y = 11$ does not have any solution, which corresponds to two parallel lines that do not cross each other.

You may wonder how you can get the inverse $\mathbf{A}^{-1}$ of a larger matrix $\mathbf{A}$ more efficiently. For large systems, direct inverse is not a good option. There are many other more efficient methods to obtain the solutions, including the powerful Gauss-Jordan elimination, matrix decomposition, and iteration methods. Interested readers can refer to more advanced literature for details.

## 5.  Eigenvalues and Eigenvectors

A special case of a linear system $\mathbf{Au} = \mathbf{b}$ is when $\mathbf{b} = \lambda\mathbf{u}$, and this becomes an eigenvalue problem.

### 5.1.  Eigenvalues and Eigenvectors of a Matrix

An eigenvalue $\lambda$ and corresponding eigenvector $\mathbf{u}$ of a square matrix $\mathbf{A}$ satisfy

$$\mathbf{Au} = \lambda\mathbf{u}. \tag{3.79}$$

Obviously, we can multiply both sides by an identity matrix $\mathbf{I}$ of the same size as that of $\mathbf{A}$, and we have

$$\mathbf{IAu} = \mathbf{I}\lambda\mathbf{u}, \quad \text{or} \quad \mathbf{Au} = \lambda\mathbf{Iu}, \tag{3.80}$$

where we have used $\mathbf{IA} = \mathbf{A}$ and $\mathbf{I}\lambda = \lambda\mathbf{I}$. The above equation can also be written as

$$(\mathbf{A} - \lambda\mathbf{I})\mathbf{u} = 0. \tag{3.81}$$

In order to satisfy this equation, we can either have $\mathbf{u} = 0$ (a trivial solution) or

$$\mathbf{A} - \lambda\mathbf{I} = 0.$$

More formally, any nontrivial solution requires that

$$\det|\mathbf{A} - \lambda\mathbf{I}| = 0. \tag{3.82}$$

In civil engineering, especially in structural analysis, eigenvalue problems are related to the natural frequencies of basic vibration modes of structures. In this context, the eigenvector corresponding to an eigenvalue provides information about the mode shapes of vibrations in structures. Therefore, eigenvalue problems are very useful for analyzing responses of bridges and buildings subject to time-varying loads such as winds, earthquake and probabilistic loads.

In addition, eigenvalues and eigenvectors can be used to detect any flaws such as cracks that may develop in structures under dynamic loading. Non-destructive testings often use the change in eigenvalues to figure out subtle changes in critical components such as turbine blades and high-precision structures.

### Example 3.15

Let us first look at a simple example. For a $2 \times 2$ matrix

$$\mathbf{A} = \begin{pmatrix} 1 & 7 \\ 7 & 1 \end{pmatrix},$$

its eigenvalues can be obtained by solving

$$\det|\mathbf{A} - \lambda\mathbf{I}| = \begin{vmatrix} 1-\lambda & 7 \\ 7 & 1-\lambda \end{vmatrix} = (1-\lambda)(1-\lambda) - 7 \times 7 = (1-\lambda)^2 - 49 = 0,$$

which can be written as

$$(1-\lambda)^2 = 7^2, \quad \text{or} \quad 1-\lambda = \pm 7,$$

which gives two eigenvalues

$$\lambda_1 = -6, \quad \lambda_2 = 8.$$

Here, we have seen that a $2 \times 2$ matrix can have two eigenvalues. In general, for a real matrix $\mathbf{A}$ of size $n \times n$, we have

$$\mathbf{A} = \begin{pmatrix} a_{11} & a_{12} & \cdots & a_{1n} \\ a_{21} & a_{22} & \cdots & a_{2n} \\ \vdots & \vdots & \ddots & \vdots \\ a_{n1} & a_{n2} & \cdots & a_{nn} \end{pmatrix}, \tag{3.83}$$

its eigenvalues can be determined by

$$\begin{vmatrix} a_{11}-\lambda & a_{12} & \cdots & a_{1n} \\ a_{21} & a_{22}-\lambda & \cdots & a_{2n} \\ \vdots & \vdots & \ddots & \vdots \\ a_{n1} & a_{n2} & \cdots & a_{nn}-\lambda \end{vmatrix} = 0, \tag{3.84}$$

which is equivalent to a polynomial of order $n$. In general, the characteristic equation has $n$ solutions ($\lambda_i, i = 1, 2, ...n$), though it is usually not easy to find them if $n$ is large.

In many applications, eigenvalues are estimated using numerical methods or by certain approximation techniques. In practice, analytical methods are rarely used to calculate them accurately for large-scale problems related to structures.

## Example 3.16

For a simple $2 \times 2$ matrix

$$\mathbf{A} = \begin{pmatrix} 1 & 5 \\ 2 & 4 \end{pmatrix},$$

its eigenvalues can be determined by

$$\begin{vmatrix} 1-\lambda & 5 \\ 2 & 4-\lambda \end{vmatrix} = 0,$$

or

$$(1-\lambda)(4-\lambda) - 2 \times 5 = 0,$$

which is equivalent to

$$(\lambda + 1)(\lambda - 6) = 0.$$

Thus, the eigenvalues are $\lambda_1 = -1$ and $\lambda_2 = 6$. The trace of $\mathbf{A}$ is $\text{tr}(\mathbf{A}) = a_{11} + a_{22} = 1 + 4 = 5 = \lambda_1 + \lambda_2$. This relationship can be generalized, as we will see later.

For an eigenvalue $\lambda$, there is an eigenvector $\mathbf{u}$ that corresponds to it so that they satisfy

$$\mathbf{Au} = \lambda\mathbf{u}. \tag{3.85}$$

Interestingly, we can multiply a non-zero scalar or number $\beta \neq 0$ and we have

$$\mathbf{Au}\beta = \lambda\mathbf{u}\beta, \tag{3.86}$$

which can be written as

$$\mathbf{A}(\beta\mathbf{u}) = \lambda(\beta\mathbf{u}). \tag{3.87}$$

This means that $\beta\mathbf{u}$ is also an eigenvector that corresponds to the same eigenvalue $\lambda$. That is to say that there could be an infinite number of eigenvectors if we choose different values of $\beta$. In the other words, we can only determine the direction of the eigenvector $\mathbf{u}$, not its length uniquely.

Therefore, it is customary to impose an extra condition that the length of the eigenvector is unity, and in this case, the eigenvector can be determined uniquely. Even with the unity requirement, there are still two vectors $\mathbf{u}$ and $-\mathbf{u}$ (or the same direction but with opposite unit vectors) that correspond to the same eigenvalue, and some care should be taken.

## Example 3.17

In order to obtain the eigenvector for each eigenvalue in the previous example, we assume

$$\mathbf{v} = \begin{pmatrix} v_1 \\ v_2 \end{pmatrix}.$$

For the eigenvalue $\lambda_1 = -1$, we plug this into

$$(\mathbf{A} - \lambda\mathbf{I})\mathbf{v} = 0,$$

and we have

$$\begin{pmatrix} 1-(-1) & 5 \\ 2 & 4-(-1) \end{pmatrix}\begin{pmatrix} v_1 \\ v_2 \end{pmatrix} = 0, \qquad \begin{pmatrix} 2 & 5 \\ 2 & 5 \end{pmatrix}\begin{pmatrix} v_1 \\ v_2 \end{pmatrix} = 0,$$

which is equivalent to

$$2v_1 + 5v_2 = 0, \qquad \text{or} \qquad v_1 = -\frac{5}{2}v_2.$$

This equation has infinite solutions; each corresponds to the vector parallel to the unit eigenvector. As the eigenvector should be normalized so that its modulus is unity, this additional condition requires

$$v_1^2 + v_2^2 = 1,$$

which means

$$(\frac{-5v_2}{2})^2 + v_2^2 = 1, \qquad \text{or} \qquad v_2^2 = \frac{4}{29},$$

which gives two solutions

$$v_2 = \pm\frac{2}{\sqrt{29}}.$$

If we choose the positive root, we have $v_2 = 2/\sqrt{29}$ and $v_1 = -5/\sqrt{29}$. Thus, we have the first set of eigenvalue and eigenvector

$$\lambda_1 = -1, \qquad \mathbf{v}_1 = \begin{pmatrix} -\frac{5}{\sqrt{29}} \\ \frac{2}{\sqrt{29}} \end{pmatrix}. \tag{3.88}$$

Similarly, the second eigenvalue $\lambda_2 = 6$ gives

$$\mathbf{v}_2 = \begin{pmatrix} \frac{\sqrt{2}}{2} \\ \frac{\sqrt{2}}{2} \end{pmatrix}.$$

Eigenvalues have interesting connections with their original matrix. The trace of any square matrix is defined as the sum of its diagonal elements, i.e.,

$$\mathrm{tr}(\mathbf{A}) = \sum_{i=1}^{n} a_{ii} = a_{11} + a_{22} + \ldots + a_{nn}. \tag{3.89}$$

The sum of all the eigenvalues of a square matrix $\mathbf{A}$ is equivalent to the trace of $\mathbf{A}$.

That is

$$\text{tr}(\mathbf{A}) = a_{11} + a_{22} + \ldots + a_{nn} = \sum_{i=1}^{n} \lambda_i = \lambda_1 + \lambda_2 + \ldots + \lambda_n. \tag{3.90}$$

In addition, the eigenvalues are also related to the determinant by

$$\det(\mathbf{A}) = \lambda_1 \lambda_2 \ldots \lambda_n = \prod_{i=1}^{n} \lambda_i. \tag{3.91}$$

## Example 3.18

For the previous example, we know that

$$\mathbf{A} = \begin{pmatrix} 1 & 5 \\ 2 & 4 \end{pmatrix},$$

its trace and determinant of $\mathbf{A}$ are

$$\text{tr}(\mathbf{A}) = 1 + 4 = 5, \quad \det(\mathbf{A}) = 1 \times 4 - 2 \times 5 = -6.$$

The sum of the eigenvalues is

$$\sum_{i=1}^{2} \lambda_i = -1 + 6 = 5 = \text{tr}(\mathbf{A}),$$

while the product of the eigenvalues is

$$\prod_{i=1}^{2} \lambda_i = -1 \times 6 = -6 = \det(\mathbf{A}).$$

Indeed, the above relationships about eigenvalues are true.

When a real square matrix is symmetric, its eigenvalues are all real. If two eigenvalues $\lambda_1$ and $\lambda_2$ are distinct (i.e., $\lambda_1 \neq \lambda_2$ and $\lambda_1, \lambda_2 \in \mathbb{R}$), their corresponding non-zero eigenvectors $\mathbf{u}_1$ and $\mathbf{u}_2$ are orthogonal to each other. From the basic definitions, we have

$$\mathbf{A}\mathbf{u}_1 = \lambda_1 \mathbf{u}_1, \quad \mathbf{A}\mathbf{u}_2 = \lambda_2 \mathbf{u}_2. \tag{3.92}$$

By multiplying the first equation by $\mathbf{u}_2^T$ on the both sides, we have

$$\mathbf{u}_2^T \mathbf{A}\mathbf{u}_1 = \mathbf{u}_2^T \lambda_1 \mathbf{u}_1 = \lambda_1 \mathbf{u}_2^T \mathbf{u}_1. \tag{3.93}$$

Here it is in fact a pre-multiplication because $\mathbf{u}_2^T$ is applied as the first factor on both sides. Since $\mathbf{A}$ is real and symmetric (i.e., $\mathbf{A}^T = \mathbf{A}$) and $\mathbf{u}^T \mathbf{A} = (\mathbf{A}^T \mathbf{u})^T$, we have

$$\mathbf{u}_2^T \mathbf{A}\mathbf{u}_1 = (\mathbf{A}^T \mathbf{u}_2)^T \mathbf{u}_1 = (\mathbf{A}\mathbf{u}_2)^T \mathbf{u}_1 = (\lambda_2 \mathbf{u}_2)^T \mathbf{u}_1 = \lambda_2 \mathbf{u}_2^T \mathbf{u}_1, \tag{3.94}$$

where we have used $\mathbf{A}\mathbf{u}_2 = \lambda_2\mathbf{u}_2$. Therefore, the above two equations give

$$\lambda_1\mathbf{u}_2^T\mathbf{u}_1 = \lambda_2\mathbf{u}_2^T\mathbf{u}_1, \tag{3.95}$$

or

$$(\lambda_1 - \lambda_2)\mathbf{u}_2^T\mathbf{u}_1 = 0, \tag{3.96}$$

which implies $\mathbf{u}_2^T\mathbf{u}_1 = 0$ if $\lambda_1 \neq \lambda_2$. That is, two eigenvectors with two distinct eigenvalues are orthogonal to each other.

On the other hand, a matrix can have a repeated eigenvalue. For example, a $2 \times 2$ identity matrix

$$\mathbf{I} = \begin{pmatrix} 1 & 0 \\ 0 & 1 \end{pmatrix}, \tag{3.97}$$

has a repeated eigenvalue of 1. In this case, any two-dimensional vector is an eigenvector of this identity matrix. If an eigenvalue has multiplicity (or repeated eigenvalues), some orthonormalization is needed. Interested readers can refer to more advanced literature on linear algebra.

The orthogonal property is useful for structural analysis. In real-world applications, structures may be very complex, but the stiffness matrices for modelling such systems are symmetric and thus the natural frequencies as eigenvalues are real and distinct. In addition, the vibration modes as eigenvectors are orthogonal, which means that any vibrations can be decomposed or represented as a linear combination of the basic mode shapes (eigenvectors). This allows us to simplify structural analysis and gain more insight into structural behaviour.

## 5.2. Definiteness of a Matrix

A square symmetric matrix $\mathbf{A}$ is said to be positive definite if all its eigenvalues are strictly positive ($\lambda_i > 0$ where $i = 1, 2, ..., n$). By multiplying $\mathbf{A}\mathbf{u} = \lambda\mathbf{u}$ by $\mathbf{u}^T$, we have

$$\mathbf{u}^T\mathbf{A}\mathbf{u} = \mathbf{u}^T\lambda\mathbf{u} = \lambda\mathbf{u}^T\mathbf{u}, \tag{3.98}$$

which leads to

$$\lambda = \frac{\mathbf{u}^T\mathbf{A}\mathbf{u}}{\mathbf{u}^T\mathbf{u}}. \tag{3.99}$$

Since $\mathbf{u}^T\mathbf{u} = \mathbf{u} \cdot \mathbf{u} = \|\mathbf{u}\|^2$ is the square of the length $\|\mathbf{u}\|$ of a non-zero vector $\mathbf{u}$, it means that $\mathbf{u}^T\mathbf{u} > 0$. Therefore, we have

$$\mathbf{u}^T\mathbf{A}\mathbf{u} > 0, \qquad \text{if } \lambda > 0. \tag{3.100}$$

In fact, for any non-zero vector $\mathbf{v}$, the following relationship holds

$$\mathbf{v}^T\mathbf{A}\mathbf{v} > 0. \tag{3.101}$$

For $\mathbf{v}$ can be a unit vector, thus all the diagonal elements of $\mathbf{A}$ should be strictly positive as well.

If all the eigenvalues are non-negative or $\lambda_i \geq 0$, then the matrix is called positive semi-definite. On the other hand, if all the eigenvalues are non-positive (i.e., $\lambda_i \leq 0$), the matrix is called negative semi-definite. In general, an indefinite matrix can have both positive and negative eigenvalues. Furthermore, the inverse of a positive definite matrix is also positive definite.

There are many applications of definiteness. For example, for a linear system $\mathbf{Au} = \mathbf{f}$, if $\mathbf{A}$ is positive definite, the system can be solved more efficiently by matrix decomposition methods.

## Example 3.19

In order to determine the definiteness of a $2 \times 2$ symmetric matrix $\mathbf{A}$

$$\mathbf{A} = \begin{pmatrix} \alpha & \beta \\ \beta & \alpha \end{pmatrix},$$

we first have to determine its eigenvalues. From $|\mathbf{A} - \lambda\mathbf{I}| = 0$, we have

$$\det\begin{pmatrix} \alpha - \lambda & \beta \\ \beta & \alpha - \lambda \end{pmatrix} = (\alpha - \lambda)^2 - \beta^2 = 0,$$

or

$$\lambda = \alpha \pm \beta.$$

Their corresponding eigenvectors are

$$\mathbf{v}_1 = \frac{1}{\sqrt{2}}\begin{pmatrix} 1 \\ 1 \end{pmatrix}, \qquad \mathbf{v}_2 = \frac{1}{\sqrt{2}}\begin{pmatrix} 1 \\ -1 \end{pmatrix}.$$

Eigenvectors associated with distinct eigenvalues of a symmetric square matrix are orthogonal. Indeed, they are orthogonal since we have

$$\mathbf{v}_1^T \mathbf{v}_2 = \begin{pmatrix} 1/\sqrt{2} & 1/\sqrt{2} \end{pmatrix}\begin{pmatrix} 1/\sqrt{2} \\ -1/\sqrt{2} \end{pmatrix} = \frac{1}{\sqrt{2}} \times \frac{1}{\sqrt{2}} + \frac{1}{\sqrt{2}} \times (\frac{-1}{\sqrt{2}}) = 0.$$

The matrix $\mathbf{A}$ will be positive definite if

$$\alpha \pm \beta > 0,$$

which means $\alpha > 0$ and $\alpha > \max(+\beta, -\beta)$. The inverse

$$\mathbf{A}^{-1} = \frac{1}{\alpha^2 - \beta^2}\begin{pmatrix} \alpha & -\beta \\ -\beta & \alpha \end{pmatrix},$$

will also be positive definite. For example,

$$A = \begin{pmatrix} 10 & -9 \\ -9 & 10 \end{pmatrix}, \qquad B = \begin{pmatrix} 3 & -2 \\ -2 & 3 \end{pmatrix},$$

are positive definite because all the eigenvalues are positive, so are their inverses

$$A^{-1} = \frac{1}{19} \begin{pmatrix} 10 & 9 \\ 9 & 10 \end{pmatrix}, \qquad B^{-1} = \frac{1}{5} \begin{pmatrix} 3 & 2 \\ 2 & 3 \end{pmatrix}.$$

But

$$D = \begin{pmatrix} 7 & 9 \\ 9 & 7 \end{pmatrix}$$

is indefinite because one eigenvalue 16 is positive and one eigenvalue -2 is negative. On the other hand, the following matrix

$$Q = \begin{pmatrix} 2 & -3 \\ 4 & -5 \end{pmatrix}$$

is negative definite because its eigenvalues are $-1$ and $-2$.

It is worth pointing out that the positiveness of elements or entries in a matrix does not bear any direct relationship to the definiteness, as is clear from the above examples.

Eigenvalues and eigenvectors can have many real-world applications. Apart from the above-mentioned eigenfrequecies and eigenmodes in structures, principal stresses and strains in structures are also closely linked to eigenvalues and eigenvectors. In addition, the energy levels of atoms can be obtained by solving the Schrödinger equation in quantum mechanics in terms of its eigenvalues. Image transformation and image processing are also closely related to eigenvalue problems.

## 6. Tensors

The concept of tensors is more complicated, though the representations of tensors are usually done in terms of matrices for computation and modelling. For example, stresses and strains in civil engineering applications are tensors, and they are mainly represented by two-dimensional arrays.

## 6.1. Summation Notations

In tensor analysis, the summation convention and notations for subscripts are widely used. Any lowercase subscript that appears exactly twice in any term of an expression means that sum is over all the possible values of the subscript. This convention is also called Einstein's summation convention or the index form.

## Example 3.20

For example, in the three-dimensional case, we have

$$\alpha_i x_i \equiv \sum_{i=1}^{3} a_i x_i = \alpha_1 x_1 + \alpha_2 x_2 + \alpha_3 x_3. \tag{3.102}$$

The dot product of two 3D vectors $\mathbf{a}$ and $\mathbf{b}$ can be written as

$$\mathbf{a} \cdot \mathbf{b} = a_i b_i = a_1 b_1 + a_2 b_2 + a_3 b_3. \tag{3.103}$$

In general, the product of two matrices can be written

$$A_{ij} B_{jk} \equiv \sum_{j=1}^{n} A_{ij} B_{jk} = A_{i1} B_{1k} + A_{i2} B_{2k} + \dots + A_{in} B_{nk}, \tag{3.104}$$

where $n$ is the number of columns of matrix $\mathbf{A}$ or the number of rows of matrix $\mathbf{B}$. Furthermore, the trace of an $n \times n$ square matrix $\mathbf{A} = [a_{ij}]$ can be written as

$$\text{tr}(\mathbf{A}) = a_{ii} = a_{11} + a_{22} + \dots + a_{nn}.$$

The Kronecker delta $\delta_{ij}$, which is a unity tensor (like the unity matrix $\mathbf{I}$ in matrix analysis), is defined as

$$\delta_{ij} = \begin{cases} 1 & (\text{if } i = j), \\ 0 & (\text{if } i \neq j). \end{cases} \tag{3.105}$$

Similar to $\delta_{ij}$, the three subscripts Levi-Civita symbol (not a standard tensor) is defined as

$$\epsilon_{ijk} = \begin{cases} +1 & (\text{if } i, j, k \text{ is an even permutation of } 1, 2, 3), \\ -1 & (\text{if } i, j, k \text{ is an odd permutation of } 1, 2, 3), \\ 0 & (\text{otherwise}). \end{cases} \tag{3.106}$$

Both $\delta_{ij}$ and $\epsilon_{ijk}$ are related by

$$\epsilon_{ijk} \epsilon_{kpq} = \delta_{ip} \delta_{jq} - \delta_{iq} \delta_{jp}. \tag{3.107}$$

For example, using the summation conventions, the matrix equation

$$\mathbf{Ax} = \mathbf{b}$$

can alternatively be written as

$$A_{ij} x_j = b_i, \qquad (i = 1, 2, \dots, n). \tag{3.108}$$

## 6.2. Tensors

When changing the basis from the standard Cartesian $e_1 = i$, $e_2 = j$, $e_3 = k$ to a new basis $e'_1, e'_2, e'_3$, a position vector $x = (x_1, x_2, x_3)$ can be written as

$$x = x_1 e_1 + x_2 e_2 + x_3 e_3 = x_i e_i. \tag{3.109}$$

Mathematically speaking, a vector is a tensor of rank 1, which will be defined more formally later.

The vector $x$ in the old bases is related to the new vector $x' = (x'_1, x'_2, x'_3)$ (here $'$ is not the derivative):

$$x' = x'_1 e'_1 + x'_2 e'_2 + x'_3 e'_3 = x'_i e'_i, \tag{3.110}$$

through a special matrix $S = [S_{ij}]$ defined by entries of dot products

$$S_{ij} = e'_i \cdot e_j. \tag{3.111}$$

The new vector $x'$ can be obtained from the old vector $x$ through

$$x' = Sx, \tag{3.112}$$

or in the present notation convention

$$x'_i = S_{ij} x_j. \tag{3.113}$$

## Example 3.21

In a special case when the transformation is a simple rotation with an angle $\theta$ about a fixed axis such as the $z$-axis, we have

$$R = [R_{ij}] = \begin{pmatrix} \cos\theta & \sin\theta & 0 \\ -\sin\theta & \cos\theta & 0 \\ 0 & 0 & 1 \end{pmatrix}. \tag{3.114}$$

The orthogonality of $S_{ij}$ requires that

$$RR^T = R^T R = I,$$

which means that

$$R^{-1} = R^T.$$

The above condition can also be written as

$$R_{ij} R_{jk} = \delta_{ik}, \quad R_{ki} R_{kj} = \delta_{ij}. \tag{3.115}$$

If the components $u_i$ of any variable $u$ are transformed to the components $u'_i$ in a new basis in the same manner as

$$u'_i = S_{ij}u_j, \qquad u_i = S_{ji}u'_j, \qquad (3.116)$$

then $u_i$ are said to form a first-order Cartesian tensor (or vector in this case). If components of a variable $\sigma_{ij}$ are transformed as

$$\sigma'_{ij} = S_{ip}S_{jq}\sigma_{pq}, \qquad \sigma_{ij} = S_{pi}S_{qj}\sigma'_{pq}, \qquad (3.117)$$

we say these components form a second-order tensor such as stresses and strains.

The order of a tensor is also called its rank or degree. Loosely speaking, the rank of a tensor is the dimensionality (or number of indices) of an array needed to represent the tensor uniquely. Therefore, a vector is a rank-1 tensor, while a stress tensor is a tensor of rank 2. Interestingly, a scalar (or a simple number) can be considered as a tensor of rank 0.

For a tensor $T_{ij}$, its contraction (or summation of repetitive indices) $T_{ii}$ is a scalar

$$T_{ij}\delta_{ij} = T_{ii} = T_{jj}, \qquad (3.118)$$

and this scalar value remains the same when changing the coordinates in all frames. For this reason, the contraction here acts like an inner product.

On the other hand, for two vectors $\mathbf{u}$ and $\mathbf{v}$, their outer product gives a tensor

$$T_{ij} = u_i v_j = \mathbf{u} \otimes \mathbf{v}, \qquad (3.119)$$

which is also called a dyad. It is worth pointing out that the inner (dot) product of $\mathbf{u}$ and $\mathbf{v}$ is a scalar (or a zeroth order/rank tensor), while their cross product $\mathbf{u} \times \mathbf{v}$ is a vector (a first-order/rank tensor). The third possibility of vector products is the tensor product or a dyadic product

$$T_{ij} = \mathbf{u} \otimes \mathbf{v} \qquad (3.120)$$

that is a second-order tensor.

## Example 3.22

For two vectors $\mathbf{u} = u_1\mathbf{i} + u_2\mathbf{j} + u_3\mathbf{k}$ and $\mathbf{v} = v_1\mathbf{i} + v_2\mathbf{j} + v_3\mathbf{k}$, their dot product is

$$\mathbf{u} \cdot \mathbf{v} = u_i v_i = u_1 v_1 + u_2 v_2 + u_3 v_3.$$

Their outer or cross product is

$$\mathbf{u} \times \mathbf{v} = \epsilon_{ijk} u_i v_j \mathbf{e}_k,$$

where $\mathbf{e}_k = \mathbf{i}, \mathbf{j}, \mathbf{k}$ for $k = 1, 2, 3$, respectively. With this notation, the above two vectors can be written as

$$\mathbf{u} = u_i \mathbf{e}_i, \quad \mathbf{v} = v_j \mathbf{e}_j.$$

In addition, their tensor product is

$$\mathbf{u} \otimes \mathbf{v} = \mathbf{u}\mathbf{v}^T = \begin{pmatrix} u_1 \\ u_2 \\ u_3 \end{pmatrix} \begin{pmatrix} v_1 & v_2 & v_3 \end{pmatrix} = \begin{pmatrix} u_1 v_1 & u_1 v_2 & u_1 v_3 \\ u_2 v_1 & u_2 v_2 & u_3 v_3 \\ u_3 v_1 & u_3 v_2 & u_3 v_3 \end{pmatrix}.$$

---

For second-order tensors, a tensor $\tau_{ij}$ is said to be symmetric if $\tau_{ij} = \tau_{ji}$, and anti-symmetric if $\tau_{ij} = -\tau_{ji}$. For example, the Kronecker $\delta_{ij}$ is symmetric, while $\epsilon_{ijk}$ is anti-symmetric in any pairs of its three indices.

## 6.3. Elasticity

The basic Hooke's law of elasticity concerns an elastic body such as a spring, and it states that the extension $x$ is proportional to the load $F$, that is

$$F = kx, \tag{3.121}$$

where $k$ the spring constant. It is worth pointing out that because the force is a restoring force, a negative sign is usually incorporated in most formulations. That is, $F = -kx$.

However, the above equation only works for 1-D deformations. For a bar of uniform cross-section with a length $L$ and a cross section area $A$, it is more convenient to use strain $\varepsilon$ and stress $\sigma$. The stress and strain are defined by

$$\sigma = \frac{F}{A}, \quad \varepsilon = \frac{\Delta L}{L}, \tag{3.122}$$

where $\Delta L$ is the extension. The unit of stress is N/m$^2$, while the strain is dimensionless, though it is conventionally expressed in m/m or % (percentage) in engineering. For the elastic bar, the stress-strain relationship is

$$\sigma = E\varepsilon, \tag{3.123}$$

where $E$ is the Young's modulus of elasticity. Written in terms $F$ and $x = \Delta L$, we have

$$F = \frac{EA}{L}\Delta L = kx, \quad k = \frac{EA}{L}, \tag{3.124}$$

where $k$ is the equivalent spring constant for the bar. This equation is valid only for any unidirectional compression or extension. For any 2D and 3D deformation, we need to generalize Hooke's law.

In the formulation of stress-strain relationship, we often use $x_1 = x$, $x_2 = y$, and

$x_3 = z$, and thus write the strain tensor as

$$\varepsilon = \begin{pmatrix} \varepsilon_{11} & \varepsilon_{12} & \varepsilon_{13} \\ \varepsilon_{21} & \varepsilon_{22} & \varepsilon_{23} \\ \varepsilon_{31} & \varepsilon_{32} & \varepsilon_{33} \end{pmatrix} = \begin{pmatrix} \varepsilon_{xx} & \varepsilon_{xy} & \varepsilon_{xz} \\ \varepsilon_{yx} & \varepsilon_{yy} & \varepsilon_{yz} \\ \varepsilon_{zx} & \varepsilon_{zy} & \varepsilon_{zz} \end{pmatrix} = \begin{pmatrix} \varepsilon_{xx} & \gamma_{xy}/2 & \gamma_{xz}/2 \\ \gamma_{yx}/2 & \varepsilon_{yy} & \gamma_{yz}/2 \\ \gamma_{zx}/2 & \gamma_{zy}/2 & \varepsilon_{zz} \end{pmatrix}, \tag{3.125}$$

where $\gamma_{xy}$ and other $\gamma$ components are shear strains.

The general stress tensor (also called Cauchy stress tensor) can be written as

$$\sigma = \begin{pmatrix} \sigma_{11} & \sigma_{12} & \sigma_{13} \\ \sigma_{21} & \sigma_{22} & \sigma_{23} \\ \sigma_{31} & \sigma_{32} & \sigma_{33} \end{pmatrix} = \begin{pmatrix} \sigma_{xx} & \sigma_{xy} & \sigma_{xz} \\ \sigma_{yx} & \sigma_{yy} & \sigma_{yz} \\ \sigma_{zx} & \sigma_{zy} & \sigma_{zz} \end{pmatrix} = \begin{pmatrix} \sigma_{xx} & \tau_{xy} & \tau_{xz} \\ \tau_{yx} & \sigma_{yy} & \tau_{yz} \\ \tau_{zx} & \tau_{zy} & \sigma_{zz} \end{pmatrix}, \tag{3.126}$$

where $\tau_{xy}$ and other $\tau$ components are shear stresses. For isotropic materials, both strain tensor and stress tensor are symmetric; that is $\varepsilon_{ij} = \varepsilon_{ji}$ and $\sigma_{ij} = \sigma_{ji}$. This means there are only six independent components of each tensor.

The generalized Hooke's law can concisely be written as

$$\varepsilon_{ij} = \frac{1+v}{E}\sigma_{ij} - \frac{v}{E}\sigma_{kk}\delta_{ij}, \tag{3.127}$$

where we have used the Einstein's summation convention $\sigma_{kk} = \sigma_{xx} + \sigma_{yy} + \sigma_{zz}$. Here, $v$ is the Poisson's ratio, and it measures the tendency of extension in transverse directions (say, $x$ and $y$) when the elastic body is stretched in one direction (say, $z$). It can be defined as the ratio of the transverse contraction strain (normal to the applied load) to the axial strain in a stretched cylindrical bar in the direction of the applied force. For a perfectly incompressible material, $v = 0.5$, and $v = 0 \sim 0.5$ for most common materials.

In a special case when the stress and strain can be considered as a two-dimensional problem, we have a plane stress or plane strain problem, depending on the conditions. If all stresses act within a plane, we have $\sigma_{zz} = 0$, $\sigma_{xz} = \sigma_{zx} = 0$ and $\sigma_{yz} = \sigma_{zy} = 0$. This becomes a so-called plane stress problem. This applies to the situation where the structure is a laminate plate where one dimension (e.g., $z$) is much smaller than the other two. Thus, the loading can be considered as uniform on a surface or plane. For isotropic materials under plane stress conditions, we have $\tau_{xy} = \tau_{yx}$ and thus we have

$$\sigma = \begin{pmatrix} \sigma_{xx} & \tau_{xy} & 0 \\ \tau_{xy} & \sigma_{yy} & 0 \\ 0 & 0 & 0 \end{pmatrix}. \tag{3.128}$$

Substituting these into Eq.(3.127) and using $\sigma_{kk} = \sigma_{xx} + \sigma_{yy}$, we have

$$\varepsilon_{xx} = \frac{1+v}{E}\sigma_{xx} - \frac{v}{E}(\sigma_{xx} + \sigma_{yy}) = \frac{1}{E}\sigma_{xx} - \frac{v}{E}\sigma_{yy}, \tag{3.129}$$

$$\varepsilon_{yy} = \frac{1+\nu}{E}\sigma_{yy} - \frac{\nu}{E}(\sigma_{xx} + \sigma_{yy}) = -\frac{\nu}{E}\sigma_{xx} + \frac{1}{E}\sigma_{yy}, \tag{3.130}$$

and

$$\varepsilon_{xy} = \frac{1+\nu}{E}\sigma_{xy}, \quad \text{or} \quad \gamma_{xy} = 2\varepsilon_{xy} = \frac{2(1+\nu)}{E}\tau_{xy}, \tag{3.131}$$

where we have used $\delta_{12} = \delta_{21} = 0$ and $\delta_{11} = \delta_{22} = 1$. Other components are $\varepsilon_{xy} = \varepsilon_{yx} = 0$.

In engineering applications, we often use a vectorized representation of such components by rewriting the above equations as a compact form as

$$\left\{\begin{array}{c} \varepsilon_{xx} \\ \varepsilon_{yy} \\ \gamma_{xy} \end{array}\right\} = \frac{1}{E}\begin{bmatrix} 1 & -\nu & 0 \\ -\nu & 1 & 0 \\ 0 & 0 & 2(1+\nu) \end{bmatrix}\left\{\begin{array}{c} \sigma_{xx} \\ \sigma_{yy} \\ \tau_{xy} \end{array}\right\}. \tag{3.132}$$

Here, the slightly different notations from the vectors and matrices are used to highlight that this is not a standard vector. It is worth pointing out that the strain $\varepsilon_{zz}$

$$\varepsilon_{zz} = \frac{1+\nu}{E}\sigma_{zz} - \frac{\nu}{E}(\sigma_{xx} + \sigma_{yy}) = -\frac{\nu}{E}(\sigma_{xx} + \sigma_{yy}), \tag{3.133}$$

is not zero in general. If it requires that $\varepsilon_{zz} = 0$, we need the plane strain conditions.

An important concept in civil engineering for two-dimensional (2D) stress analysis is Mohr's circle. In this case, we have a simple 2D stress tensor

$$\sigma = \begin{pmatrix} \sigma_{xx} & \sigma_{xy} \\ \sigma_{xy} & \sigma_{yy} \end{pmatrix} = \begin{pmatrix} \sigma_x & \tau_{xy} \\ \tau_{xy} & \sigma_y \end{pmatrix},$$

where we used the simpler notations $\sigma_x = \sigma_{xx}$ and $\sigma_y = \sigma_{yy}$ commonly used in engineering textbooks. The aim is to find the maximum and minimum stresses, called principal stresses, and their directions. To do this, let us define an average normal stress

$$\bar{\sigma} = \frac{1}{2}(\sigma_x + \sigma_y), \tag{3.134}$$

and stress radius

$$R_\sigma = \sqrt{\tau_{xy}^2 + [(\sigma_x - \sigma_y)/2]^2}. \tag{3.135}$$

Then, the maximum stress and the minimum stress, respectively, are

$$\sigma_{max} = \bar{\sigma} + R_\sigma, \quad \sigma_{min} = \bar{\sigma} - R_\sigma. \tag{3.136}$$

Obviously, the directions of the minimum stress and the maximum are perpendicular to each other. This is essentially an eigenvalue and eigenvector problem.

The principal angle $\theta$ of the maximum stress to the $x$-axis is determined by

$$\tan(2\theta) = \frac{2\tau_{xy}}{(\sigma_x - \sigma_y)}. \tag{3.137}$$

## Example 3.23

For example, for a stress tensor of $\sigma = \begin{pmatrix} 200 & -50 \\ -50 & 100 \end{pmatrix}$, we have $\sigma_x = 200$ MPa, $\sigma_y = 100$ MPa and $\tau_{xy} = -50$ MPa. Thus, the average normal stress is

$$\bar{\sigma} = \frac{1}{2}(\sigma_x + \sigma_y) = \frac{1}{2}(200 + 100) = 150, \quad R_\sigma = \sqrt{(-50)^2 + [(200 - 100)/2]^2} \approx 70.7.$$

The maximum stress is

$$\sigma_{max} = \bar{\sigma} - R_\sigma = 150 + 70.7 \approx 220.7,$$

whose principal angle is

$$\tan(2\theta) = \frac{2\tau_{xy}}{(\sigma_x - \sigma_y)} = \frac{2 \times (-50)}{(200 - 100)} = -1,$$

which means that $2\theta = -45°$ or $\theta = -22.5°$. It is easy to obtain the minimum stress is $\sigma_{min} = \bar{\sigma} - R_\sigma = 150 - 70.7 = 79.3$ MPa.

From Eq.(3.127), we can do a contraction by setting $i = j$ and thus we have

$$\varepsilon_{ii} = \frac{1 + \nu}{E}\sigma_{ii} - \frac{\nu}{E}\sigma_{kk}\delta_{ii}. \tag{3.138}$$

Since $\delta_{ii} = 1 + 1 + 1 = 3$ and $\sigma_{ii} = \sigma_{kk}$, we have

$$\varepsilon_{ii} = \frac{1 + \nu}{E}\sigma_{ii} - \frac{3\nu}{E}\sigma_{kk} = \left(\frac{1 + \nu}{E} - \frac{3\nu}{E}\right)\sigma_{ii} = \frac{(1 - 2\nu)}{E}\sigma_{ii}, \tag{3.139}$$

which gives

$$\sigma_{ii} = \frac{E}{(1 - 2\nu)}\varepsilon_{ii}. \tag{3.140}$$

Since $\sigma_{ii} = \sigma_{kk}$ and $\varepsilon_{ii} = \varepsilon_{kk}$ because contractions are just sums, we can write the above relationship as

$$\sigma_{kk} = \frac{E}{(1 - 2\nu)}\varepsilon_{kk}. \tag{3.141}$$

Now let us invert Eq.(3.127) by rewriting it as

$$\frac{1 + \nu}{E}\sigma_{ij} = \varepsilon_{ij} + \frac{\nu}{E}\sigma_{kk}\delta_{ij}. \tag{3.142}$$

Multiplying both sides by $E/(1 + v)$, we obtain

$$\sigma_{ij} = \frac{E}{(1 + v)}\varepsilon_{ij} + \frac{v}{(1 + v)}\sigma_{kk}\delta_{ij}. \tag{3.143}$$

Substituting Eq.(3.141), we have

$$\sigma_{ij} = \frac{E}{(1 + v)}\varepsilon_{ij} + \frac{vE}{(1 + v)(1 - 2v)}\varepsilon_{kk}\delta_{ij} = \frac{E}{(1 + v)}\varepsilon_{ij} + \lambda\varepsilon_{kk}\delta_{ij}, \tag{3.144}$$

where $\lambda = vE/(1 + v)(1 - 2v)$ is the Lamé constant.

We can see that the tensor notations and representations make it easier to do inverse of complex tensor relationships. In fact, many quantities and relationships are expressed in tensor forms, including stresses, strains, non-isotropic conductivity and diffusivity and moment of inertia.

## Exercises

**3.1.** For three vectors: $\mathbf{u} = \begin{pmatrix} 1 \\ 2 \\ 0 \end{pmatrix}$, $\mathbf{v} = \begin{pmatrix} 3 \\ -4 \\ 0 \end{pmatrix}$, and $\mathbf{w} = \begin{pmatrix} 2 \\ 3 \\ 1 \end{pmatrix}$, show that

$$|\mathbf{u} + \mathbf{v} + \mathbf{w}| \le |\mathbf{u}| + |\mathbf{v}| + |\mathbf{w}|.$$

**3.2.** For the same vectors as before, calculate $\mathbf{u} \cdot \mathbf{v}$, $\mathbf{u} \times \mathbf{v}$ and $\mathbf{w} \cdot (\mathbf{u} \times \mathbf{v})$.

**3.3.** The rest of this exercise uses the following matrices:

$$\mathbf{A} = \begin{pmatrix} 2 & 3 \\ 3 & 2 \end{pmatrix}, \ \mathbf{B} = \begin{pmatrix} 1 & -2 \\ -3 & 4 \end{pmatrix}.$$

- Compute $\mathbf{AB}$ and $\mathbf{BA}$. What is $\mathbf{AB} - \mathbf{BA}$?
- Compute $\det(\mathbf{A})$, $\det(\mathbf{B})$, $\mathbf{A}^{-1}$ and $\mathbf{B}^{-1}$.

**3.4.** Find the eigenvalues $\lambda_1$ and $\lambda_2$ of $\mathbf{A}$ and their corresponding eigenvectors. Show that $\det(\mathbf{A}) = \lambda_1 \cdot \lambda_2$.

# CHAPTER 4

# Calculus I: Differentiation

Contents

## Key Points

- Introduce gradients, derivatives and higher-order derivatives as well as differentiation rules.
- Introduce approximations to functions by series expansions.
- Introduce partial derivatives and their applications.

Calculus is an important part of the skills set and toolboxes for any engineer, and a vast majority of computation in engineering requires advanced calculus, including differentiation, partial derivatives, integration, multiple integrals, and differential equations as well as integral transforms. This chapter introduces all the fundamentals of differentiation.

## 1. Gradient and Derivative

The calculations of gradients of a curve can be useful in many applications. For example, let us imagine a car travelling on a straight road whose speed is measured at different times, as shown in Fig. 4.1. In the figure, point $B(10, 10)$ means that the speed is 10 m/s at $t = 10$ seconds at Point $B$. That is, $t_B = 10$ seconds and $V_B = 10$ m/s.

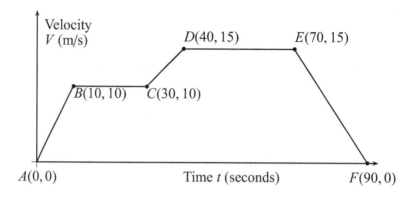

Figure 4.1: The speed of a car at different times.

Though the speed of a car is usually measured in miles per hour (mph), we use meters per second for ease of our calculations here.

From point $A$ to $B$, the car speeds up; what is its acceleration? The acceleration is the gradient of the velocity graph, which is

$$a_{AB} = \frac{V_B - V_A}{t_B - t_A} = \frac{10 - 0}{10 - 0} = 1 \text{ m/s}^2. \tag{4.1}$$

From $B$ to $C$, the speed is constant, so the acceleration is zero. From $C$ to $D$, the car speeds up again, so the acceleration is

$$a_{CD} = \frac{15 - 10}{40 - 30} = 0.5 \text{ m/s}^2. \tag{4.2}$$

From $D$ to $E$, its acceleration is zero as the speed is constant. From $E$ to $F$, the car decelerates to stop. Thus, the acceleration is

$$a_{EF} = \frac{0 - 15}{90 - 70} = -0.75 \text{ m/s}^2. \tag{4.3}$$

The distance travelled by the car is the area under the curve. Thus, from $A$ to $B$, the distance is $d_{AB} = \frac{1}{2} \times 10 \times 10 = 50$ m, and the distance $d_{BC}$ from $B$ to $C$ is $10 \times 20 = 200$ m. Similarly, $d_{CD} = \frac{1}{2}(10 + 15)(40 - 30) = 125$ m, $d_{DE} = 15 \times 30 = 450$ m, and $d_{EF} = \frac{1}{2} \times 15 \times 20 = 150$ m. So the total distance travelled in the 90 seconds is $d = 50 + 200 + 125 + 450 + 150 = 975$ m.

For the above graph, the curve is formed by different line segments. However, curves can be highly complicated in practice. In this case, we need to calculate the gradients using a unified approach, which is differentiation. We know that the gradient of a curve at any point $P$ is the rate of change at that point, and it is the gradient of the tangent to the curve at that point. We also know geometrically what a tangent means, but it is not easy to draw it accurately without calculations.

In order to find the true gradient and the tangent, we normally use some adjustment points and try to use the ratio of the small change in $y$, $\delta y = y_Q - y_P$ to the change in $x$, $\delta x = x_Q - x_P$ to approximate the gradient. That is to say, we use

$$\frac{\delta y}{\delta x} = \frac{y_Q - y_P}{x_Q - x_P}, \tag{4.4}$$

to estimate the gradient. As $Q$ is closer than $R$ to $P$, thus the gradient estimated using $P$ and $Q$ is more accurate than that using $P$ and $R$ (see Fig. 4.2). We hope that the estimate will become the true gradient as the point $Q$ becomes very, very close to $P$. How do we describe such closeness? A simple way is to use the distance $h = \delta x = PS = x_Q - x_P$ and let $h$ tend to zero. That is to say, $h \to 0$.

Since we know that the point $x_P = a$, its coordinates for the curve $x^3$ are $(a, a^3)$. The adjustment point $Q$ now has coordinates $(a + h, (a + h)^3)$. Therefore, the gradient at point $P$ is approximately

$$\frac{\delta y}{\delta x} = \frac{(a + h)^3 - a^3}{h}. \tag{4.5}$$

Since $(a + h)^3 = (a^3 + 3a^2h + 3ah^2 + h^3)$, we now have

$$\frac{\delta y}{\delta x} = \frac{(a^3 + 3a^2h + 3ah^2 + h^3) - a^3}{h} = 3a^2 + 3ah + h^2. \tag{4.6}$$

Since $h \to 0$, and also $h^2 \to 0$, both terms $3ah$ and $h^2$ will tend to zero. This means that the true gradient of $x^3$ at $x = a$ is $3a^2$. Since $x = a$ is any point, so in general the gradient of $x^3$ is $3x^2$. That is

$$(x^3)' = 3x^2, \tag{4.7}$$

where we use the $'$ notation to denote the gradient.

Now let us introduce a more formal notation for the gradient of any curve $y = f(x)$. We define the gradient as

$$f'(x) \equiv \frac{dy}{dx} \equiv \frac{df(x)}{dx} = \lim_{h \to 0} \frac{f(x + h) - f(x)}{h}. \tag{4.8}$$

The gradient is also called the first derivative. The three notations $f'(x)$, $dy/dx$ and $df(x)/dx$ are interchangeable.

Conventionally, the notation $dy/dx$ is called Leibnitz's notation, while the prime notation $'$ is called Lagrange's notation. Newton's dot notation $\dot{y} = dy/dt$ is now exclusively used for time derivatives. The choice of such notations is purely for clarity, convention and/or personal preference.

In addition, here the notation 'lim' is used, as it is associated with the fact that

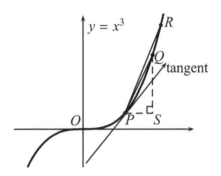

Figure 4.2: The tangent and gradient of a curve with $h=PS=x_Q - x_P$.

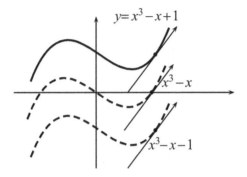

Figure 4.3: The gradients of a family of curves $y = x^3 - x + k$ (where $k = 0, \pm1$) at $x = 1$ are the same $3x^2 - 1$.

$h \to 0$. The limit can be understood that for any $0 < h \ll 1$ where $h$ is much less than 1 and infinitesimally small, then there always exists another even smaller number $0 < \epsilon < h$ so that we can find a better estimate for the gradient using $\epsilon$.

Using this standard notation, our earlier example becomes

$$(x^3)' \equiv \frac{dx^3}{dx} = 3x^2. \tag{4.9}$$

From the basic definition of the derivative, we can verify that such differentiation manipulation is a linear operator. That is to say that for any two functions $f(x)$, $g(x)$ and two constants $\alpha$ and $\beta$, the derivative or gradient of a linear combination of the two functions can be obtained by differentiating the combination term by term. We have

$$[\alpha f(x) + \beta g(x)]' = \alpha f'(x) + \beta g'(x), \tag{4.10}$$

which can easily be extended to multiple terms. Furthermore, if $k$ is a real constant, then its derivative is zero. For example, if we have $f(x) = x^3 - x + k$, we have

$$\frac{df(x)}{dx} = \frac{d(x^3 - x + k)}{dx}$$

$$= \frac{d(x^3)}{dx} - \frac{dx}{dx} + \frac{dk}{dx} = 3x^2 - 1 + 0 = 3x^2 - 1, \tag{4.11}$$

which means that for a family of curves shifted by a constant $k$, the gradients at any same point (say $x = 1$) are all the same, and the tangent lines are parallel to each other (see Fig. 4.3).

## Example 4.1

To find the gradient $f(x) = x^n$ where $n = 1, 2, 3, 4, \ldots$ is a positive integer, we use

$$f'(x) = \lim_{h \to 0} \frac{f(x + h) - f(x)}{h} = \lim_{h \to 0} \frac{(x + h)^n - x^n}{h}.$$

Using the binomial theorem

$$(x + h)^n = \binom{n}{0} x^n + \binom{n}{1} x^{n-1} h + \binom{n}{2} x^{n-2} h^2 + \ldots + \binom{n}{n} h^n,$$

and the fact that

$$\binom{n}{0} = 1, \quad \binom{n}{1} = n, \quad \binom{n}{2} = \frac{n(n-1)}{2!}, \quad \binom{n}{k} = \frac{n!}{k!(n-k)!}, \quad \binom{n}{n} = 1,$$

we have

$$(x + h)^n - x^n = (x^n + nx^{n-1} h + \frac{n(n-1)}{2!} x^{n-2} h^2 + \ldots + h^n) - x^n$$

$$= nx^{n-1} h + \frac{n(n-1)}{2!} x^{n-2} h^2 + \ldots + h^n.$$

Therefore, we get

$$f'(x) = \lim_{h \to 0} \frac{nx^{n-1} h + \frac{n(n-1)}{2!} x^{n-2} h^2 + \ldots + h^n}{h}$$

$$= \lim_{h \to 0} \left[ nx^{n-1} + \frac{n(n-1)}{2!} x^{n-2} h + \ldots + h^{n-1} \right].$$

Since $h \to 0$, all the terms except the first $(nx^{n-1})$ will tend to zero. We finally have

$$f'(x) = nx^{n-1}.$$

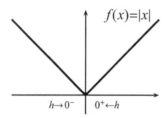

Figure 4.4: Limit and thus the derivative at $x = 0$ is not defined.

In fact, this formula is valid for any integer $(n = 0, \pm 1, \pm 2, ...)$.

It is worth pointing out that the limit $\lim_{h \to 0}$ must exist. If such a limit does not exist, then the derivative or the gradient is not defined. For example, the gradient of a seemingly rather simple function $f(x) = |x|$ where $|x|$ means the absolute value of $f'(x)$ does not exist at $x = 0$ (see Fig. 4.4). This is because we will get different values depending on how we approach $x = 0$. If we approach $h \to 0$ from the right $(h > 0)$ and use the notation $h \to 0^+$, we have

$$f'(0^+) = \lim_{h \to 0^+} \frac{|h| - |0|}{h} = \lim_{h \to 0^+} \frac{h}{h} = \lim_{h \to 0^+} 1 = 1. \tag{4.12}$$

However, if we approach from the left $(h < 0)$, we have

$$f'(0^-) = \lim_{h \to 0^-} \frac{|h| - |0|}{h} = \lim_{h \to 0^-} \frac{-h}{h} = \lim_{h \to 0^-} -1 = -1. \tag{4.13}$$

The correct definition of the derivative should be independent of the way we approach the point using $h \to 0$. In this case, we say, the gradient or derivative at $x = 0$ does not exist.

We can obtain the derivative of any function from first principles. For example, we now try to derive $d(\sin x)/dx$.

In order to obtain the derivative of $f(x) = \sin(x)$, we now use the definition

$$f'(x) = \lim_{h \to 0} \frac{\sin(x + h) - \sin x}{h}.$$

From trigonometry, we know that

$$\sin(x + h) - \sin x = 2 \cos \frac{[(x + h) + x]}{2} \sin \frac{[(x + h) - x]}{2}$$

$$= 2\cos(x + \frac{h}{2})\sin\frac{h}{2}.$$

We have

$$f'(x) = \lim_{h \to 0} \frac{2\cos(x + \frac{h}{2})\sin\frac{h}{2}}{h} = \lim_{h \to 0} \cos(x + \frac{h}{2})\frac{\sin\frac{h}{2}}{\frac{h}{2}}.$$

Using

$$\lim_{h \to 0} \frac{\sin(h/2)}{h/2} = 1, \qquad \lim_{h \to 0} \cos(x + \frac{h}{2}) = \cos x,$$

we finally have

$$f'(x) = \frac{d(\sin x)}{dx} = \cos x.$$

Following the same procedure, we can also derive that

$$\frac{d(\cos x)}{dx} = -\sin x,$$

which is left as an exercise.

This example demonstrates that even if we are able to calculate the derivative using first principles, it is usually not the best way to do so in practice. The good thing is that we only have to do it once to understand how it works. For more complicated functions, we should use certain rules such as the chain rule and product rule. In addition, we can sometimes use tables or mathematical software packages. Now let us introduce some differentiation rules.

## 2. Differentiation Rules

We know how to differentiate $x^2$ and $\sin(x)$, and a natural question is how we can differentiate $\sin(x^2)$? Here we need the chain rule. The trick for deriving the chain rule is to consider $dx$ and $dy$ (and other similar quantities) as infinitesimal but non-zero quantities or increments $\delta x$ and $\delta y$ so that we can divide and multiply them as necessary. In the limit of $\delta x \to 0$, we have $\delta x \to dx$ and $\delta y/\delta x \to dy/dx$. If $f(g)$ is a function of $g$, and $g$ is in turn a function of $x$ (thus a composite function $(f \circ g)(x)$ as discussed in Chapter 1), we want to calculate $dy/dx$. We then have

$$\frac{dy}{dx} = \frac{dy}{dg} \cdot \frac{dg}{dx}, \tag{4.14}$$

or

$$\frac{df(g(x))}{dx} = \frac{df(g)}{dg} \cdot \frac{dg(x)}{dx}. \tag{4.15}$$

This is the well-known chain rule.

## Example 4.2

Now we come back to our original problem of differentiating $\sin(x^2)$. First we let $g = x^2$, we then

$$\frac{d(\sin(x^2))}{dx} = \frac{d(\sin g)}{dg} \cdot \frac{dx^2}{dx} = \cos g \cdot 2x^1 = 2x \cos(x^2).$$

As we practise more, we can write the derivatives more compactly and quickly. For example,

$$\frac{d(\sin^n(x))}{dx} = n \sin^{n-1}(x) \cdot \cos x,$$

where we consider $\sin^n(x)$ to be in the form of $g^n$ and $g = \sin(x)$.

A further question is if we can differentiate $x^5 \sin(x)$ easily, since we already know how to differentiate $x^5$ and $\sin x$ separately? For this, we need the differentiation rule for products. Let $y = uv$ where $u(x)$ and $v(x)$ are (often simpler) functions of $x$. From the definition, we have

$$\frac{dy}{dx} = \frac{d(uv)}{dx} = \lim_{h \to 0} \frac{u(x+h)v(x+h) - u(x)v(x)}{h}. \tag{4.16}$$

Since adding a term and deducting the same term does not change an expression, we have

$$\frac{u(x+h)v(x+h) - u(x)v(x)}{h}$$

$$= \frac{u(x+h)v(x+h) - \overbrace{u(x+h)v(x) + u(x+h)v(x)}^{=0} - u(x)v(x)}{h}$$

$$= u(x+h)\frac{[v(x+h) - v(x)]}{h} + v(x)\frac{[u(x+h) - u(x)]}{h}. \tag{4.17}$$

In addition, $u(x+h) \to u(x)$ as $h \to 0$, we finally have

$$\frac{d(uv)}{dx} = \lim_{h \to 0} \left\{ u(x+h)\frac{[v(x+h) - v(x)]}{h} + v(x)\frac{[u(x+h) - u(x)]}{h} \right\}$$

$$= u\frac{dv}{dx} + v\frac{du}{dx}, \tag{4.18}$$

or simply

$$(uv)' = uv' + vu'. \tag{4.19}$$

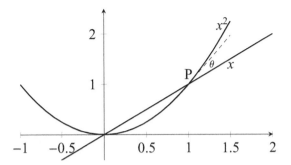

Figure 4.5: The angle of intersection of two curves $f(x) = x$ and $g(x) = x^2$.

This is the formula for products or product rule. If we replace $v$ by $1/v = v^{-1}$ and apply the chain rule

$$\frac{d(v^{-1})}{dx} = -1 \times v^{-1-1} \times \frac{dv}{dx} = -\frac{1}{v^2}\frac{dv}{dx}, \tag{4.20}$$

we have the formula for quotients or the quotient rule

$$\frac{d(\frac{u}{v})}{dx} = \frac{d(uv^{-1})}{dx} = u(\frac{-1}{v^2})\frac{dv}{dx} + v^{-1}\frac{du}{dx} = \frac{v\frac{du}{dx} - u\frac{dv}{dx}}{v^2}. \tag{4.21}$$

Now let us apply these rules to differentiating $\tan x$.

## Example 4.3

Since $\tan x = \sin x / \cos x$, we have $u = \sin x$ and $v = \cos x$. Therefore, we get

$$\frac{d(\tan x)}{dx} = \frac{\cos x \frac{d \sin x}{dx} - \sin x \frac{d \cos x}{dx}}{\cos^2 x} = \frac{\cos x \cos x - \sin x \times (-\sin x)}{\cos^2 x}$$

$$= \frac{\cos^2 x + \sin^2 x}{\cos^2 x} = \frac{1}{\cos^2 x}.$$

One of the applications of differentiation in engineering is to find the angle $\theta$ of intersection of two curves $f(x)$ and $g(x)$, which is given by

$$\tan \theta = \frac{g'(x) - f'(x)}{1 + f'(x)g'(x)}, \tag{4.22}$$

which requires that both derivatives $f'(x)$ and $g'(x)$ must exist (see Fig. 4.5). If $\tan(\theta) > 0$, the angle is acute; otherwise the angle is obtuse for $\tan(\theta) < 0$.

## Example 4.4

For example, for two curves $f(x) = x$ and $g(x) = x^2$ as shown in Fig. 4.5, we have $f'(x) = 1$ and $g'(x) = 2x$. The intersection occurs at two points $(0,0)$ (the origin) and $P(1,1)$.

At $P(1,1)$, we have $f'(x) = 1$ and $g'(1) = 2$. Thus, we have

$$\tan(\theta) = \frac{g'(x) - f'(x)}{1 + f'(x)g'(x)} = \frac{2 - 1}{1 + 1 \times 2} = \frac{1}{3},$$

which means that it is an acute angle $\theta = \tan^{-1}(1/3) \approx 18.4°$. However, at the origin, we have $f'(x) = 1$ and $g'(0) = 0$, so that

$$\tan(\theta) = \frac{0 - 1}{1 + 1 \times 0} = -1,$$

which means the angle is obtuse. That is $\theta = \tan^{-1} = 135°$.

Sometimes, it is not easy to find the derivative $dy/dx$ directly. In this case, it is usually a good idea to try to find $dx/dy$ since

$$\frac{dy}{dx} = 1 / \frac{dx}{dy}, \tag{4.23}$$

or carry out the derivatives term-by-term and then find $dy/dx$. This is especially the case for implicit functions. Let us demonstrate this with an example.

## Example 4.5

To find the gradient of the curve $\sin^2 x + y^2 - 3y = e^{4x}$, at the point $(0, 1)$, we first differentiate each term with respect to $x$, and we have

$$\frac{d\sin^2 x}{dx} + \frac{dy^2}{dx} - \frac{d(3y)}{dx} = \frac{de^{4x}}{dx},$$

or

$$2 \sin x \cos(x) + 2y\frac{dy}{dx} - 3\frac{dy}{dx} = 4e^{4x},$$

where we have used the chain rule

$$\frac{du}{dx} = \frac{du}{dy} \cdot \frac{dy}{dx},$$

so that $u = y^2$ gives

$$\frac{dy^2}{dx} = 2y\frac{dy}{dx}.$$

After some rearrangement, we have

$$(2y - 3)\frac{dy}{dx} = 4e^{4x} - 2\sin x \cos x,$$

or

$$\frac{dy}{dx} = \frac{4e^{4x} - 2\sin x \cos x}{(2y - 3)}.$$

Therefore, the gradient at $x = 0$ and $y = 1$ becomes

$$\frac{dy}{dx} = \frac{4e^{4\times 0} - 2 \times \sin 0 \times \cos 0}{2 \times 1 - 3} = \frac{4}{-1} = -4.$$

The first derivatives of commonly used mathematical functions are listed in Table 4.1, and the derivatives of other functions can be obtained using the differentiation rules in combination with this table.

Table 4.1: First Derivatives of common functions.

| $f(x)$ | $f'(x)$ | $f(x)$ | $f'(x)$ |
|--------|---------|--------|---------|
| $x^n$ | $nx^{n-1}$ | $e^x$ | $e^x$ |
| $a^x$ | $a^x \ln a \; (a > 0)$ | $\ln x$ | $\frac{1}{x} \; (x > 0)$ |
| $\sin x$ | $\cos x$ | $\cos x$ | $-\sin x$ |
| $\tan x$ | $1 + \tan^2 x$ | $\sec x$ | $\sec x \tan x$ |
| $\log_a x$ | $\frac{1}{x \ln a}$ | $\tanh x$ | $\operatorname{sech}^2 x$ |
| $\sinh x$ | $\cosh x$ | $\cosh x$ | $\sinh x$ |
| $\sin^{-1} x$ | $\frac{1}{\sqrt{1-x^2}}$ | $\cos^{-1} x$ | $\frac{-1}{\sqrt{1-x^2}}$ |
| $\tan^{-1} x$ | $\frac{1}{1+x^2}$ | $\sinh^{-1} x$ | $\frac{1}{\sqrt{x^2+1}}$ |
| $\tanh^{-1} x$ | $\frac{1}{1-x^2}$ | $\cosh^{-1} x$ | $\frac{1}{\sqrt{x^2-1}}$ |

The derivative we discussed so far is the gradient or the first derivative of a function $f(x)$. If we want to see how the gradient itself varies, we may need to take the gradient of the gradient of a function. In this case, we are in fact calculating the second derivative. We write

$$\frac{d^2 f(x)}{dx^2} \equiv f''(x) \equiv \frac{d}{dx}\left(\frac{df(x)}{dx}\right). \tag{4.24}$$

Following a similar procedure, we can write any higher-order derivative as

$$\frac{d^3 f(x)}{dx^3} = f'''(x) = \frac{d(f''(x))}{dx}, \quad \ldots, \quad \frac{d^n f(x)}{dx^n} = f^{(n)}(x). \tag{4.25}$$

Let us look some examples.

## Example 4.6

For example, $f(x) = x^4$, we have

$$f'(x) = 4x^3, \quad f''(x) = d(4x^3)/dx = 4 \times 3x^2 = 12x^2, \quad f'''(x) = 24x$$

and

$$f''''(x) = 24, \quad f'''''(x) = 0.$$

For $g(x) = e^x \sin(x^2)$, we have

$$g' = [e^x \sin(x^2)]' = (e^x)' \sin(x^2) + e^x(\sin(x^2))' = e^x \sin(x^2) + e^x \cos(x^2)2x.$$

$$g'' = [e^x \sin(x^2)]' + [e^x \cos(x^2)2x]' = e^x[\sin(x^2) + 4x \cos(x^2) - 4x^2 \sin(x^2) + 2 \cos(x^2)].$$

The expressions can get quite lengthy for higher-order derivatives.

It is worth pointing out that notation for the second derivative is $\frac{d^2 y}{dx^2}$, not $\frac{d^2 y}{d^2 x}$ or $\frac{dy^2}{dx^2}$ (which is wrong).

## 3. Maximum, Minimum and Radius of Curvature

In many engineering applications, designs have to optimize something (such as to maximize loads and performance and to minimize the costs). This becomes optimization and the function to be maximized or minimized is called a cost or objective function.

For a given curve or a univariate function $f(x)$, the maximum or minimum usually occurs when the rate of change is zero (see Fig. 4.6). That is at zero gradient, which means that

$$f'(x) = 0. \tag{4.26}$$

For example, the minimum of $f(x) = x^2$ occurs at $x = 0$ when $f'(x) = 2x = 0$, and its minimum is $f_{min} = 0^2 = 0$. Whether it is a maximum or minimum is determined by its second derivatives $f''(x)$. If $f''(x) > 0$, it is a minimum. If $f''(x) < 0$, it corresponds to a maximum. For $f(x) = x^2$, we have $f''(x) = 2 > 0$, thus $x = 0$ corresponds a minimum. If a maximum is the highest function value in the whole domain, it is also called the global maximum. Similarly, the lowest function value in the whole domain

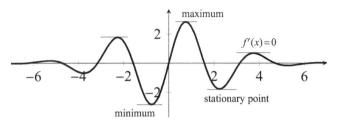

Figure 4.6: Stationary points ($f'(x) = 0$), maximum and minimum of a univariate function.

is called the global minimum. Clearly, the global minimum of $f(x) = x^2$ is zero, while its global maximum is $+\infty$.

For example, for $f(x) = x^3 - 9x^2 + 15x$, we have its first derivative

$$f'(x) = 3x^2 - 18x + 15 = 3(x - 1)(x - 5) = 0, \qquad (4.27)$$

which gives two stationary points: $x = 1$ and $x = 5$. The second derivative $f''(x) = 6x - 18$ at $x = 1$ is $f''(1) = 6 \times 1 - 18 = -12 < 0$, thus $x = 1$ corresponds to a maximum $f(1) = 7$. However, the stationary point at $x = 5$ gives $f''(5) = 12 > 0$, which corresponds a minimum point $f(5) = -25$. These two points are neither global maximum nor minimum.

As we mentioned earlier, $f'(x)$ determines the locations of stationary points, while the sign of $f''(x)$ determines if it is a maximum or minimum. There is a special case when $f'(x) = 0$ and $f''(x) = 0$, care should be taken because it either needs higher-order derivatives to determine or it is a saddle point. For example, for $f(x) = x^3$, $f'(x) = 3x^2 = 0$ also means that $f''(x) = 6x = 0$. In fact, $x = 0$ of $x^3$ is a saddle point. However, for $g(x) = x^4$, we have $g'(x) = 4x^3 = 0$, which gives $x = 0$ and also $f''(x) = 12x^2 = 0$. But $f''''(x) = 24$. In fact, $g(x) = x^4$ has a global minimum at $x = 0$.

## Example 4.7

A beam of length $L$ with both fixed ends is deflected by a uniformly distributed load $W > 0$. The moment along the beam is given by

$$M(x) = \frac{W}{2L}\left(Lx - x^2 - \frac{L^2}{6}\right),$$

while the deflection is given by

$$D(x) = \frac{Wx^2}{24EIL}(L^2 - 2Lx + x^2),$$

where $E$ is Young's modulus, $I$ is the area moment of inertia and $x$ is the distance

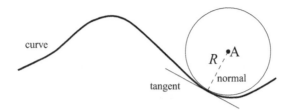

Figure 4.7: Radius of curvature ($R$), centre of curvature ($A$), tangent and normal.

from one end.

What is the location with the maximum moment? Where does the maximum deflection occur?

From $M(x)$, we know that

$$M'(x) = \frac{W}{2L}(L - 2x) = 0,$$

which gives $x = L/2$. Thus, the maximum moment at $x = L/2$ is

$$M_{max} = \frac{W}{2L}\left(L\frac{L}{2} - \frac{L^4}{2} - \frac{L^2}{6}\right) = \frac{WL}{24}.$$

It is easy to check that $M''(x) = -W/L < 0$, thus this is indeed a maximum.

In addition, from $D(x)$, we know that

$$D'(x) = \frac{W}{24EIL}\left(2L^2x - 6Lx^2 + 4x^3\right) = \frac{W}{12EIL}\left[x(2x - L)(x - L)\right] = 0,$$

which gives three stationary points $x = 0$, $x = L/2$ and $x = L$. The two points $x = 0$ and $x = L$ correspond to two fixed ends. The maximum deflection occurs at $x = L/2$ (centre) with

$$D_{max} = \frac{W}{24EIL}\left(\frac{L}{2}\right)^2\left[L^2 - 2L\frac{L}{2} - \left(\frac{L}{2}\right)^2\right] = \frac{W}{24EIL}\frac{L^4}{16} = \frac{WL^3}{384EI}.$$

Optimization such as finding the maximum or minimum of a function can be extended to functions with multiple variables. In this case, we have to deal with partial derivatives, gradient vectors and Hessian matrix, to be introduced later in this chapter.

In describing particle motion and paths in engineering, we often have to deal with the radius of curvature (see Fig. 4.7). For a given curve $y(x)$, its curvature $\kappa = 1/R$ is linked to the radius of curvature $R$. In Cartesian coordinates, we have

$$R = \frac{\left\{1 + [y'(x)]^2\right\}^{3/2}}{y''(x)}, \tag{4.28}$$

which is the signed radius of curvature. In some textbooks, the radius of curvature is defined as the absolute value of this. That is

$$R = \left| \frac{\left\{ 1 + [y'(x)]^2 \right\}^{3/2}}{y''(x)} \right|. \tag{4.29}$$

It is worth pointing out that the special case of $y''(x) = 0$ corresponds to an inflection point whose radius of curvature is infinite. The center of curvature (A) is a point along the normal with a distance $R$ from point $(x, y)$, and this normal line is perpendicular to the tangent (see Fig. 4.7).

## Example 4.8

For a curve $y(x) = xe^{-x}$, what is the radius of curvature of this curve at $x = 0$? Where is its centre of curvature?

From $y(x) = xe^{-x}$, we have

$$y'(x) = (1 - x)e^{-x}, \quad y''(x) = (x - 2)e^{-x}.$$

Thus, the radius of curve at $x$ is

$$R = \left| \frac{\left\{ 1 + [y'(x)]^2 \right\}^{3/2}}{y''(x)} \right| = \left| \frac{[1 + (1 - x)^2 e^{-2x}]^{3/2}}{(x - 2)e^{-x}} \right|.$$

Clearly, the inflection point is $x = 2$ from $y'' = 0$. At $x = 0$, we have

$$R = \left| \frac{[1 + (1 - 0)^2 e^{-2 \times 0}]^{3/2}}{-2e^{-0}} \right| = \frac{\sqrt{2^{3/2}}}{2} = \sqrt{2}.$$

Since $y'(0) = 1$ at $x = 0$, the tangent is $45°$ from the $x$-axis. Thus, the normal must be $-45°$ from the $x$-axis. The distance from the center of curvature to the origin $(0, 0)$ is $\sqrt{2}$, which means that the centre of curvature is at $(1, -1)$.

---

Bending of a beam is governed by the fact that the curvature $\kappa$ of a slim beam is proportional to the applied bending moment $M$. That is

$$\kappa = -\frac{M}{EI}, \quad \text{or} \quad M = -EI\kappa, \tag{4.30}$$

where $E$ is the modulus of elasticity or Young's modulus and $I$ is the area moment of inertia. The product $EI$ is often called flexural rigidity. Here, the minus sign comes from the sign convention used in structural analysis.

Let $u(x)$ be the deflection $u(x)$ (here upwards is positive) at a distance $x$ from a support. For a slim beam, $u(x)$ is much smaller than the length $L$ of the beam and we can usually assume $u'(x) \approx 0$. Thus, the curvature $\kappa$ can be approximated by $u''(x)$

(i.e., $\kappa \approx u''(x)$). This leads to the Euler-Bernoulli beam equation

$$M = -EI\frac{d^2u}{dx^2}. \tag{4.31}$$

This is a differential equation to be introduced later in this book. With appropriate boundary conditions, we can find the deflection by solving this differential equation.

## 4. Series Expansions and Taylor Series

In numerical methods and some mathematical analysis, series expansions make some calculations easier. For example, we can write the exponential function $e^x$ as a series about $x_0 = 0$

$$e^x = \alpha_0 + \alpha_1 x + \alpha_2 x^2 + \alpha_3 x^3 + ... + \alpha_n x^n. \tag{4.32}$$

Now let us try to determine these coefficients. At $x = 0$, we have

$$e^0 = 1 = \alpha_0 + \alpha_1 \times 0 + \alpha_2 \times 0^2 + ...\alpha_n \times 0^n = \alpha_0, \tag{4.33}$$

which gives $\alpha_0 = 1$. In order to reduce the power or order of the expansion so that we can determine the next coefficient, we first differentiate both sides of (4.32) once; we have

$$e^x = \alpha_1 + 2\alpha_2 x + 3\alpha_3 x^2 + ... + n\alpha_n x^{n-1}. \tag{4.34}$$

By setting again $x = 0$, we have

$$e^0 = 1 = \alpha_1 + 2\alpha_2 \times 0 + ... + n\alpha_n \times 0^{n-1} = \alpha_1, \tag{4.35}$$

which gives $\alpha_1 = 1$. Similarly, differentiating it again, we have

$$e^x = (2 \times 1) \times \alpha_2 + 3 \times 2\alpha_3 x + ... + n(n-1)\alpha_n x^{n-2}. \tag{4.36}$$

At $x = 0$, we get

$$e^0 = (2 \times 1) \times \alpha_2 + 3 \times 2\alpha_3 \times 0 + ... + n(n-1)\alpha_n \times 0^{n-2} = 2\alpha_2, \tag{4.37}$$

or $\alpha_2 = 1/(2 \times 1) = 1/2!$. Following the same procedure and differentiating it $n$ times, we have

$$e^x = n!\alpha_n, \tag{4.38}$$

and $x = 0$ leads to $\alpha_n = 1/n!$. Therefore, the final series expansion can be written as

$$e^x = 1 + x + \frac{1}{2!}x^2 + \frac{1}{3!}x^3 + ... + \frac{1}{n!}x^n. \tag{4.39}$$

Obviously, we can follow a similar process to expand other functions. We have seen here the importance of differentiation and derivatives.

If we know the value of $f(x)$ at $x_0$, we can use some approximations in a small

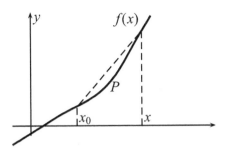

Figure 4.8: Expansion and approximations for $f(x) = f(x_0 + h)$ where $h = x - x_0$.

interval $h = x - x_0$ (see Fig. 4.8). Following the similar idea as Eq.(4.32), we can first write the approximation in the following general form:

$$f(x) = a_0 + a_1(x - x_0) + a_2(x - x_0)^2 + ... + a_n(x - x_0)^n, \qquad (4.40)$$

and then try to figure out the unknown coefficients $a_i(i = 0, 1, 2, ...)$. For the above approximation to be valid at $x = x_0$, we have

$$f(x_0) = a_0 + 0 \quad \text{(all the other terms are zeros)}, \qquad (4.41)$$

so that $a_0 = f(x_0)$.

Now let us first take the first derivative of (4.40),

$$f'(x) = 0 + a_1 + 2a_2(x - x_0) + ... + na_n(x - x_0)^{n-1}. \qquad (4.42)$$

By setting $x = x_0$, we have

$$f'(x_0) = 0 + a_1 + 0 + ... + na_n \times 0, \qquad (4.43)$$

which gives

$$a_1 = f'(x_0). \qquad (4.44)$$

Similarly, we differentiate Eq.(4.40) twice with respect to $x$ and we have

$$f''(x) = 0 + 0 + a_2 \times 2 \times 1 + ... + n(n - 1)a_n(x - x_0)^2. \qquad (4.45)$$

Setting $x = x_0$, we have

$$f''(x_0) = 2!a_2, \quad \text{or} \quad a_2 = \frac{f''(x_0)}{2!}. \qquad (4.46)$$

Following the same procedure, we have

$$a_3 = \frac{f'''(x_0)}{3!}, \quad a_4 = \frac{f''''(x_0)}{4!}, \quad ..., \quad a_n = \frac{f^{(n)}(x_0)}{n!}. \tag{4.47}$$

Thus, we finally obtain

$$f(x) = f(x_0) + f'(x_0)(x - x_0) + \frac{f''(x_0)}{2!}(x - x_0)^2$$

$$+\frac{f'''(a)}{3!}(x - x_0)^3 + ... + \frac{f^{(n)}(x_0)}{n!}(x - x_0)^n, \tag{4.48}$$

which is the well-known Taylor series.

In a special case when $x_0 = 0$ and $h = x - x_0 = x$, the above Taylor series becomes zero centred, and such expansions are traditionally called Maclaurin series

$$f(x) = f(0) + f'(0)x + \frac{f''(0)}{2!}x^2 + \frac{f'''(0)}{3!}x^3 + ... + \frac{f^{(n)}}{n!}x^n + ..., \tag{4.49}$$

named after mathematician Colin Maclaurin.

In theory, we can use as many terms as possible, but in practice, the series converges very quickly and only a few terms are sufficient. It is straightforward to verify that the exponential series for $e^x$ is identical to the results given earlier. Now let us look at other examples.

## Example 4.9

Let us expand $f(x) = \sin x$ about $x_0 = 0$. We know that

$$f'(x) = \cos x, \quad f''(x) = -\sin x, \quad f'''(x) = -\cos x, \quad ...,$$

or $f'(0) = 1$, $f''(0) = 0$, $f'''(0) = -1$, $f''''(0) = 0, ...$, which means that

$$\sin x = \sin 0 + x f'(0) + \frac{f''(0)}{2!}x^2 + \frac{f'''(0)}{3!}x^3 + ... = x - \frac{x^3}{3!} + \frac{x^5}{5!} + ...,$$

where the angle $x$ is in radians.

For example, we know that $\sin 30° = \sin \frac{\pi}{6} = 1/2$. We now use the expansion to estimate it for $x = \pi/3 = 0.523598$

$$\sin \frac{\pi}{6} \approx \frac{\pi}{6} - \frac{(\pi/6)^3}{3!} + \frac{(\pi/6)^5}{5!}$$

$$= 0.523599 - 0.02392 + 0.0000328 \approx 0.5000021326,$$

which is very close to the true value $1/2$.

If we continue the process to infinity, we then reach the infinite power series and the error $f^{(n)}(0)x^n/n!$ becomes negligibly small if the series converges. For example, some common series are

$$\frac{1}{1-x} = 1 + x + x^2 + x^3 + \dots + x^n + \dots, \quad [x \in (-1, 1)], \tag{4.50}$$

$$\sin x = x - \frac{x^3}{3!} + \frac{x^5}{5!} - \dots, \quad \cos x = 1 - \frac{x^2}{2!} + \frac{x^4}{4!} - \dots, \quad (x \in \mathcal{R}), \tag{4.51}$$

$$\tan(x) = x + \frac{x^3}{3} + \frac{2x^5}{15} + \frac{17x^7}{315} + \dots, \quad [x \in (-\pi/2, \pi/2)], \tag{4.52}$$

and

$$\ln(1 + x) = x - \frac{x^2}{2} + \frac{x^3}{3} - \frac{x^4}{4} + \frac{x^5}{5} - \dots, \quad x \in (-1, 1]. \tag{4.53}$$

As an exercise, we leave the reader to prove the above series.

## 5. Partial Derivatives

The derivative defined earlier is for function $f(x)$ which has only one independent variable $x$, and the gradient will generally depend on the location $x$. For functions $f(x, y)$ of two variables $x$ and $y$, their gradient will depend on both $x$ and $y$ in general. In addition, the gradient or rate of change will also depend on the direction (along $x$-axis or $y$-axis or any other directions). For example, the function $f(x, y) = x(y - 1)$ has different gradients at $(0, 0)$ along $x$-axis and $y$-axis. The gradients along the positive $x$- and $y$- directions are called the partial derivatives with respect to $x$ and $y$, respectively. They are denoted as $\frac{\partial f}{\partial x}$ and $\frac{\partial f}{\partial y}$, respectively.

The partial derivative of $f(x, y)$ with respect to $x$ can be calculated assuming that $y$ =constant. Thus, we have

$$\frac{\partial f(x, y)}{\partial x} \equiv f_x \equiv \frac{\partial f}{\partial x}\bigg|_y = \lim_{\Delta x \to 0, y = \text{const}} \frac{f(x + \Delta x, y) - f(x, y)}{\Delta x}. \tag{4.54}$$

Similarly, we have

$$\frac{\partial f(x, y)}{\partial y} \equiv f_y \equiv \frac{\partial f}{\partial y}\bigg|_x = \lim_{\Delta y \to 0, x = \text{const}} \frac{f(x, y + \Delta y) - f(x, y)}{\Delta y}. \tag{4.55}$$

The notation $\frac{\partial}{\partial x}\big|_y$ emphasizes the fact that $y$ is held constant. The subscript notation $f_x$ (or $f_y$) emphasizes that the derivative is carried out with respect to $x$ (or $y$).

As the partial derivatives treating all the other variables as constants (except for the one variable of interest), the usual properties and rules for differentiation still apply. That is, for multiple additive terms, you can differentiate each term individually. In addition, both chain rules and product rules also apply here.

## Example 4.10

For a function

$$f(x,y) = x^3 + y^2 + 7xy,$$

its partial derivatives are

$$f_x = \frac{\partial f}{\partial x} = \frac{\partial x^3}{\partial x} + \frac{\partial y^2}{\partial x} + \frac{\partial(7xy)}{\partial x} = 3x^2 + 0 + 7y = 3x^2 + 7y,$$

$$f_y = \frac{\partial f}{\partial y} = \frac{\partial x^3}{\partial y} + \frac{\partial y^2}{\partial y} + \frac{\partial(7xy)}{\partial y} = 0 + 2y + 7x = 2y + 7x.$$

Mathematicians like to use the subscript forms, as they are simpler notations and can be easily generalized. For example,

$$f_{xx} = \frac{\partial^2 f}{\partial x^2}, \qquad f_{xy} = \frac{\partial^2 f}{\partial x \partial y}. \tag{4.56}$$

Since $\Delta x \Delta y = \Delta y \Delta x$, we have $f_{xy} = f_{yx}$. For any small change $\Delta f = f(x + \Delta x, y + \Delta y) - f(x, y)$ due to $\Delta x$ and $\Delta y$, the total infinitesimal change $df$ can be written as

$$df = \frac{\partial f}{\partial x} dx + \frac{\partial f}{\partial y} dy. \tag{4.57}$$

If $x$ and $y$ are functions of another independent variable $\xi$, then the above equation leads to the following chain rule

$$\frac{df}{d\xi} = \frac{\partial f}{\partial x}\frac{dx}{d\xi} + \frac{\partial f}{\partial y}\frac{dy}{d\xi}, \tag{4.58}$$

which is very useful in calculating the derivatives in a parametric form or for change of variables.

## Example 4.11

For the same function $f = x^3 + y^2 + 7xy$ as in the previous example, we have

$$f_{xx} = \frac{\partial f_x}{\partial x} = \frac{\partial(3x^2 + 7y)}{\partial x} = 6x + 0 = 6x. \quad f_{yy} = \frac{\partial f_y}{\partial y} = \frac{\partial(2y + 7x)}{\partial y} = 2 + 0 = 2.$$

In addition, we have

$$f_{xy} = \frac{\partial f_y}{\partial x} = \frac{\partial(2y + 7x)}{\partial x} = 0 + 7 = 7. \quad f_{yx} = \frac{\partial f_x}{\partial y} = \frac{\partial(3x^2 + 7y)}{\partial y} = 0 + 7 = 7,$$

which indeed shows that

$$f_{xy} = f_{yx}.$$

Partial derivatives are very useful in computing many quantities in engineering, especially for quantities involving field quantities such as displacement, potential, flow, stress and strain.

## Example 4.12

The strain tensor $s_{ij}$ can be defined by

$$\varepsilon_{ij} = \frac{1}{2}\left(\frac{\partial u_i}{\partial x_j} + \frac{\partial u_j}{\partial x_i}\right), \quad (i, j = 1, 2, 3),$$

where $\mathbf{u} = (u_1, u_2, u_3)$ are displacements. This tensor is is clearly symmetric because

$$\varepsilon_{ji} = \frac{1}{2}\left(\frac{\partial u_j}{\partial x_i} + \frac{\partial u_i}{\partial x_j}\right) = \frac{1}{2}\left(\frac{\partial u_i}{\partial x_j} + \frac{\partial u_j}{\partial x_i}\right) = \varepsilon_{ij}.$$

If a complicated function $f(x)$ can be written in terms of simpler functions $u$ and $v$ so that $f(x) = g(x, u, v)$ where $u(x)$ and $v(x)$ are known functions of $x$, then we have the generalized chain rule for the total derivative

$$\frac{dg}{dx} = \frac{\partial g}{\partial x} + \frac{\partial g}{\partial u}\frac{du}{dx} + \frac{\partial g}{\partial v}\frac{dv}{dx}. \tag{4.59}$$

## 6. Differentiation of Vectors

The differentiation of a vector is carried out over each component, treating each component as the usual differentiation of a scalar. For example, a position vector of a particle moving in the three-dimensional (3D) space can be written

$$\mathbf{P}(t) = x(t)\mathbf{i} + y(t)\mathbf{j} + z(t)\mathbf{k}, \tag{4.60}$$

where $(x(t), y(t), z(t))$ are its three coordinates and the parameter $t$ is time, which shows that the position is changing with time $t$.

The velocity of the particle is its vector derivative and we can write its velocity as

$$\mathbf{v} = \frac{d\mathbf{P}}{dt} = \dot{x}(t)\mathbf{i} + \dot{y}(t)\mathbf{j} + \dot{z}(t)\mathbf{k}, \tag{4.61}$$

and acceleration as

$$\mathbf{a} = \frac{d^2\mathbf{P}}{dt^2} = \ddot{x}(t)\mathbf{i} + \ddot{y}(t)\mathbf{j} + \ddot{z}(t)\mathbf{k}, \tag{4.62}$$

where $\dot{()} = d()/dt$.

From the basic definition of differentiation, it is easy to check that the differentiation of vectors has the following properties:

$$\frac{d(\alpha\mathbf{a})}{dt} = \alpha\frac{d\mathbf{a}}{dt}, \quad \frac{d(\mathbf{a}\cdot\mathbf{b})}{dt} = \frac{d\mathbf{a}}{dt}\cdot\mathbf{b} + \mathbf{a}\cdot\frac{d\mathbf{b}}{dt}, \tag{4.63}$$

and

$$\frac{d(\mathbf{a}\times\mathbf{b})}{dt} = \frac{d\mathbf{a}}{dt}\times\mathbf{b} + \mathbf{a}\times\frac{d\mathbf{b}}{dt}. \tag{4.64}$$

## Example 4.13

For a particle with its position at

$$x(t) = \cos(\omega t), \quad y(t) = \sin(\omega t), \quad z(t) = \frac{1}{2}gt^2,$$

where $\omega$ is a constant, what are its velocity and acceleration vectors?
    The velocity vector is given by

$$\mathbf{v} = \dot{x}(t)\mathbf{i} + \dot{y}(t)\mathbf{j} + \dot{z}\mathbf{k} = -\omega\sin(\omega t)\mathbf{i} + \omega\cos(\omega t)\mathbf{j} + gt\mathbf{k}.$$

Its acceleration is

$$\mathbf{a} = \dot{\mathbf{v}} = \ddot{x}(t)\mathbf{i} + \ddot{y}(t)\mathbf{j} + \ddot{z}(t)\mathbf{k} = -\omega^2\cos(\omega t)\mathbf{i} - \omega^2\sin(\omega t)\mathbf{j} + g\mathbf{k}.$$

The orbit is a spiral with the free fall along the $z$-axis.

## 6.1. Polar Coordinates

A point $P$ on a plane can be represented uniquely by their Cartesian coordinates $(x, y)$. The same point can also be represented using a different coordinate system. For example, in the polar coordinate system, the same point $P$ can be represented by a distance $(r)$ from a reference point (the origin $O$, often referred to as the pole) and the angle $(\theta)$ from the reference direction (the $x$−axis in this case) as shown in Fig. 4.9.

The distance is also referred to as the radial coordinate, while the angle is also called the polar angle (or azimuth). To ensure the uniqueness of polar coordinates, it

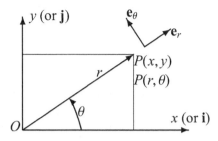

Figure 4.9: Polar coordinates and their relationship to Cartesian coordinates.

is necessary for $r \geq 0$ and $0 \leq \theta < 2\pi$ (or $\theta \in [0, 360°)$. In some textbooks, the angle interval can also be $-\pi < \theta \leq \pi$ (or $\theta \in (-180°, 180°]$). In addition, the pole can be represented by the point $(0, \theta)$ where $\theta$ can be any value.

In addition, in the Cartesian coordinate system, the unit vectors are simply $\mathbf{i}$ and $\mathbf{j}$ along $x$-axis and $y$-axis, respectively. In the polar coordinate system, the unit vectors are $\mathbf{e}_r$ and $\mathbf{e}_\theta$ along the $r$ and $\theta$ directions, respectively. The angle $\theta$ is considered as positive if it is anticlockwise (see Fig. 4.9). The relationships between the unit vectors ($\mathbf{i}$ and $\mathbf{j}$) in the Cartesian system and the unit vectors $\mathbf{e}_r$ and $\mathbf{e}_\theta$ in the polar system are as follows:

$$\mathbf{e}_r = \cos(\theta)\mathbf{i} + \sin(\theta)\mathbf{j}, \tag{4.65}$$

and

$$\mathbf{e}_\theta = -\sin(\theta)\mathbf{i} + \cos(\theta)\mathbf{j}. \tag{4.66}$$

From Fig. 4.9 and trigonometrical functions, we can easily write down the relationship that links the polar coordinate system to its Cartesian coordinate system

$$x = r\cos\theta, \quad y = r\sin\theta, \tag{4.67}$$

and

$$r = \sqrt{x^2 + y^2}, \quad \tan\theta = \frac{y}{x} \quad [\text{or } \theta = \tan^{-1}(\frac{y}{x})]. \tag{4.68}$$

Thus, the same point $P(x, y)$ or $P(r, \theta)$ can be written as

$$\overrightarrow{OP} = x\mathbf{i} + y\mathbf{j} = r\mathbf{e}_r + \theta\mathbf{e}_\theta. \tag{4.69}$$

Here, the angle can also be represented by other trigonometrical functions and we have

$$\cos\theta = \frac{x}{\sqrt{x^2 + y^2}}, \quad \sin\theta = \frac{y}{\sqrt{x^2 + y^2}}. \tag{4.70}$$

Sometimes, there are significant advantages to using the polar coordinate system. For example, a circle of radius $a$ can be represented in the polar coordinate system by a simple equation

$$r = a, \tag{4.71}$$

which is equivalent to the standard equation $x^2 + y^2 = a^2$ in the Cartesian system. On the other hand, a line in the $(x, y)$ system is in general represented by $y = kx + b$ as we have seen in Chapter 1. In the $(r, \theta)$ system, a line can simply be represented by

$$\theta = \theta_0, \tag{4.72}$$

where $\tan \theta_0$ is the slope or gradient of the line.

## 6.2. Three Basic Operators

There are three important operators commonly used in vector analysis. These operators are: gradient operator (grad or $\nabla$), the divergence operator (div or $\nabla \cdot$) and the curl operator (curl or $\nabla \times$). They often appear in science and engineering, especially in fluid dynamics, electromagnetic dynamics and solid mechanics.

Sometimes, it is useful to calculate the directional derivative of a function $\psi$ at the point $(x, y, z)$ in the direction of $\mathbf{n}$

$$\frac{\partial \psi}{\partial \mathbf{n}} = \mathbf{n} \cdot \nabla \psi = \frac{\partial \psi}{\partial x} \cos(\alpha) + \frac{\partial \psi}{\partial y} \cos(\beta) + \frac{\partial \psi}{\partial z} \cos(\gamma), \tag{4.73}$$

where $\mathbf{n} = (\cos \alpha, \cos \beta, \cos \gamma)$ is a unit vector and $\alpha, \beta, \gamma$ are the directional angles. Generally speaking, the gradient of any scalar function $\psi$ of $x, y, z$ can be written in a similar way

$$\text{grad } \psi = \nabla \psi = \frac{\partial \psi}{\partial x} \mathbf{i} + \frac{\partial \psi}{\partial y} \mathbf{j} + \frac{\partial \psi}{\partial z} \mathbf{k}. \tag{4.74}$$

This is equivalent to applying the del operator $\nabla$ to the scalar function $\psi$

$$\nabla = \frac{\partial}{\partial x} \mathbf{i} + \frac{\partial}{\partial y} \mathbf{j} + \frac{\partial}{\partial z} \mathbf{k}. \tag{4.75}$$

Let us look at an example.

## Example 4.14

For a function $\psi$ that describes a field, we have

$$\psi(x) = r \sin(2z), \quad r = \sqrt{x^2 + y^2}.$$

We have

$$\frac{\partial \psi}{\partial x} = \frac{x}{r} \sin(2z), \quad \frac{\partial \psi}{\partial y} = \frac{y}{r} \sin(2z), \quad \frac{\partial \psi}{\partial z} = 2r \cos(2z).$$

The grad of $\psi$ can be obtained by

$$\text{grad } \psi = \frac{\partial \psi}{\partial x} \mathbf{i} + \frac{\partial \psi}{\partial y} \mathbf{j} + \frac{\partial \psi}{\partial z} \mathbf{k}.$$

$$= \frac{x}{\sqrt{x^2 + y^2}} \sin(2z)\mathbf{i} + \frac{y}{\sqrt{x^2 + y^2}} \sin(2z)\mathbf{j} + 2\sqrt{x^2 + y^2} \cos(2z)\mathbf{k}.$$

Gradient operators are useful in expressing many physical relationships in engineering applications.

## Example 4.15

The strain tensor mentioned earlier can be defined in terms of the displacement $\mathbf{u}^T = (u_1, u_2, u_3)$

$$\varepsilon_{ij} = \frac{1}{2}\left(\frac{\partial u_i}{\partial x_j} + \frac{\partial u_j}{\partial x_i}\right), \tag{4.76}$$

where $x_1 = x$, $x_2 = y$, and $x_3 = z$. Sometimes, it is useful to write

$$\varepsilon = \frac{1}{2}\left(\nabla \mathbf{u} + (\nabla \mathbf{u})^T\right), \tag{4.77}$$

where $T$ means transpose.

In addition, Darcy's law in hydrology links the flow flux $q$ to the pressure gradient $\nabla p$ in the following form:

$$q = -\frac{\kappa}{\mu}\nabla p,$$

where $\kappa$ is the permeability of the medium such as soil and $\mu$ is the viscosity of the fluid such as water.

One of the most common operators in engineering and science, the Laplacian operator, is

$$\nabla^2 = \nabla \cdot \nabla = \frac{\partial^2}{\partial x^2} + \frac{\partial^2}{\partial y^2} + \frac{\partial^2}{\partial z^2}. \tag{4.78}$$

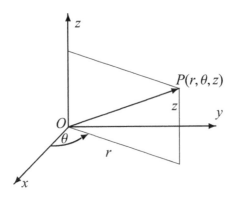

Figure 4.10: Cylindrical coordinates.

Thus, for a function $\Phi$, its Laplace equation can be written as

$$\Delta\Psi = \nabla^2\Psi = \frac{\partial^2\Psi}{\partial x^2} + \frac{\partial^2\Psi}{\partial y^2} + \frac{\partial^2\Psi}{\partial z^2} = 0. \tag{4.79}$$

A solution $\Phi$ satisfies the above Laplace equation is called a harmonic function. This equation is widely used for the studies of potential theory concerning electrical fields, gravitational fields, and steady state heat transfer as well as fluid dynamics.

## 6.3. Cylindrical Coordinates

Another commonly used coordinate system is the cylindrical polar coordinate system or simply cylindrical coordinate system as shown in Fig. 4.10 where the base for $(x, y)$ plan uses the standard polar coordinate system $(r, \theta)$, while the $z$-axis remains the same as the Cartesian $z$-axis.

Let $e_r$ and $e_\theta$ be the unit vectors pointing towards $\mathbf{r}$ and $\theta$ direction (anti-clockwise as positive in the range $0 \le \theta \le 2\pi$), respectively. The direction of the third (or $z$) coordinate $e_z$ is the same as the Cartesian $\mathbf{k}$.

Similar to the relationship between unit vectors in the Cartesian system and the unit vectors in the polar system, the links of unit vectors between the cylindrical and Cartesian systems are as follows:

$$\mathbf{e}_r = \cos(\theta)\mathbf{i} + \sin(\theta)\mathbf{j}, \quad \mathbf{e}_\theta = -\sin(\theta)\mathbf{i} + \cos(\theta)\mathbf{j}, \quad \mathbf{e}_z = \mathbf{k}, \tag{4.80}$$

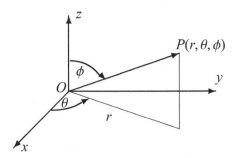

Figure 4.11: Spherical polar coordinates.

which can be written as rotation matrix $\mathbf{R}_1$

$$\begin{pmatrix} \mathbf{e}_r \\ \mathbf{e}_\theta \\ \mathbf{e}_z \end{pmatrix} = \mathbf{R}_1 \begin{pmatrix} \mathbf{i} \\ \mathbf{j} \\ \mathbf{k} \end{pmatrix}, \quad \mathbf{R}_1 = \begin{pmatrix} \cos\theta & \sin\theta & 0 \\ -\sin\theta & \cos\theta & 0 \\ 0 & 0 & 1 \end{pmatrix}. \tag{4.81}$$

The gradient vector of a function (called a field) $\psi(x,y,z)$ in the Cartesian

$$\nabla\psi = \frac{\partial\psi}{\partial x}\mathbf{i} + \frac{\partial\psi}{\partial y}\mathbf{j} + \frac{\partial\psi}{\partial z}\mathbf{k}, \tag{4.82}$$

can be written in the cylindrical coordinates as

$$\nabla\psi = \frac{\partial\psi}{\partial r}\mathbf{e}_r + \frac{1}{r}\frac{\partial\psi}{\partial\theta}\mathbf{e}_\theta + \frac{\partial\psi}{\partial z}\mathbf{e}_z. \tag{4.83}$$

Similarly, the divergence for a vector $\mathbf{u} = (u_r, u_\theta, u_z)$ can be written as

$$\nabla \cdot \mathbf{u} = \frac{1}{r}\frac{\partial(r u_r)}{\partial r} + \frac{1}{r}\frac{\partial u_\theta}{\partial\theta} + \frac{\partial u_z}{\partial z}. \tag{4.84}$$

## 6.4. Spherical Coordinates

Another important coordinate system is the spherical polar coordinate system (simply the spherical coordinates), which uses two angles (see Fig. 4.11).

For spherical polar coordinates $(r, \theta, \phi)$ as shown in Fig. 4.11, where $\phi$ is the zenithal angle between the $z$-axis and the position vector $\mathbf{r}$, and $\phi$ is the azimuthal angle, we have

$$x = r\cos\theta\sin\phi, \quad y = r\sin\theta\sin\phi, \quad z = r\cos\phi, \tag{4.85}$$

where $0 \le r < \infty$, $0 \le \theta \le 2\pi$ and $0 \le \phi \le \pi$.

Its three unit vectors are $\mathbf{e}_r$, $\mathbf{e}_\theta$ and $\mathbf{e}_\phi$, along the directions of $r$, $\theta$ and $\phi$, respectively. Their links to the Cartesian unit vectors are as follows:

$$\mathbf{e}_r = \cos\theta\sin\phi\mathbf{i} + \sin\theta\sin\phi\mathbf{j} + \cos\phi\mathbf{k}, \quad \mathbf{e}_\theta = -\sin\theta\mathbf{i} + \sin\theta\mathbf{j} + 0\mathbf{k}, \tag{4.86}$$

$$\mathbf{e}_\phi = \cos\theta\cos\phi\mathbf{i} + \sin\theta\cos\phi\mathbf{j} - \sin\phi\mathbf{k}. \tag{4.87}$$

Its gradient operator becomes

$$\nabla = \mathbf{e}_r\frac{\partial}{\partial r} + \mathbf{e}_\theta\frac{1}{r\sin\phi}\frac{\partial}{\partial\theta} + \mathbf{e}_\phi\frac{1}{r}\frac{\partial}{\partial\phi}. \tag{4.88}$$

## 7. Jacobian and Hessian Matrices

Jacobian matrix is an important concept for multiple integrals and vector calculus. For a vector function $f(x_1, x_2, ..., x_n)$ with $m$ components:

$$\mathbf{f}(x_1, x_2, ..., x_n) = (f_1, f_2, ..., f_m), \tag{4.89}$$

its Jacobian matrix can be defined as

$$\mathbf{J} \equiv [J_{ij}] \equiv \frac{\partial(f_1, ..., f_m)}{\partial(x_1, ..., x_n)} \equiv \frac{\partial f_i}{\partial x_j}, \quad (i = 1, ..., m; j = 1, ..., n),$$

$$= \begin{pmatrix} \frac{\partial f_1}{\partial x_1} & \frac{\partial f_1}{\partial x_2} & \cdots & \frac{\partial f_1}{\partial x_n} \\ \frac{\partial f_2}{\partial x_1} & \frac{\partial f_2}{\partial x_2} & \cdots & \frac{\partial f_2}{\partial x_n} \\ \vdots & \vdots & \ddots & \vdots \\ \frac{\partial f_m}{\partial x_1} & \frac{\partial f_m}{\partial x_2} & \cdots & \frac{\partial f_m}{\partial x_n} \end{pmatrix}, \tag{4.90}$$

where $\equiv$ means 'exactly equal to'. In general, the Jacobian matrix is an $m \times n$ matrix.

In the case of $m = 1$, $\mathbf{f}$ becomes a simple scalar function $f(x_1, ..., x_n)$ and its Jacobian becomes its gradient

$$J = \nabla f = \left( \frac{\partial f}{\partial x_1} \quad \frac{\partial f}{\partial x_2} \quad \cdots \quad \frac{\partial f}{\partial x_n} \right), \tag{4.91}$$

where $\frac{\partial f}{\partial x_j}$ is the partial gradient along the $x_j$-axis in the $n$-dimensional space.

For $\mathbf{f} = (f_1, f_2, f_3) = (xy, \sin(xy), x + y^2)$, its Jacobian matrix is

$$\mathbf{J} = \frac{\partial(f_1, f_2, f_3)}{\partial(x, y)} = \begin{pmatrix} \frac{\partial f_1}{\partial x} & \frac{\partial f_1}{\partial y} \\ \frac{\partial f_2}{\partial x} & \frac{\partial f_2}{\partial y} \\ \frac{\partial f_3}{\partial x} & \frac{\partial f_3}{\partial y} \end{pmatrix} = \begin{pmatrix} y & x \\ y\cos(xy) & x\cos(xy) \\ 1 & 2y \end{pmatrix}.$$

A special case of Jacobian is when $m = n$, the matrix becomes a square matrix. In this case, the determinant of the Jacobian matrix (or simply Jacobian determinant) has a special meaning in multiple integral as it links to the change of area or volume elements when changing coordinate systems.

Let us consider a small element area. The area element in the $(x, y)$ system is simply $dA = dxdy$, but the area element is spanned by an arc element with the arc length of $rd\theta$ and the thickness of $dr$. Thus, the area element should be $dA = rdrd\theta$ in the polar coordinate system. There is an extra factor $r$ here. This concept becomes clearer when we discuss the multiple integrals in the next chapter.

The Jacobian for transforming from one coordinate system (e.g., $(x, y)$) to another (e.g, $(r, \theta)$) is the determinant in the following form:

$$J = \frac{\partial(x, y)}{\partial(r, \theta)} = \begin{pmatrix} \frac{\partial x}{\partial r} & \frac{\partial x}{\partial \theta} \\ \frac{\partial y}{\partial r} & \frac{\partial y}{\partial \theta} \end{pmatrix}, \tag{4.92}$$

its corresponding Jacobian determinant is

$$\det(J) = \frac{\partial x}{\partial r}\frac{\partial y}{\partial \theta} - \frac{\partial y}{\partial r}\frac{\partial x}{\partial \theta}. \tag{4.93}$$

It is worth pointing out that both the Jacobian matrix and Jacobian determinant are referred to as simply, Jacobian, in the literature.

## Example 4.16

For the polar coordinate system $(r, \theta)$, we know (from the earlier results)

$$\frac{\partial x}{\partial r} = \cos\theta, \quad \frac{\partial y}{\partial r} = \sin\theta,$$

and

$$\frac{\partial x}{\partial \theta} = -r\sin\theta, \quad \frac{\partial y}{\partial \theta} = r\cos\theta.$$

So the Jacobian determinant becomes

$$J = \begin{vmatrix} \cos\theta & -r\sin\theta \\ \sin\theta & r\cos\theta \end{vmatrix} = \cos\theta \cdot (r\cos\theta) - \sin\theta(-r\sin\theta)$$

$$= r^2(\cos^2\theta + \sin^2\theta) = r.$$

Indeed, the Jacobian gives the factor $r$ as discussed in the above regarding the area element.

Jacobian matrices concern the first-order partial derivative. The Hessian matrix of a multivariate function is a square matrix of a scalar function $f(x_1, ..., x_n)$ and can be calculated by

$$
\mathbf{H} \equiv \frac{\partial^2 f}{\partial x_i \partial x_j} =
\begin{pmatrix}
\frac{\partial^2 f}{\partial x_1^2} & \frac{\partial^2 f}{\partial x_1 \partial x_2} & \cdots & \frac{\partial^2 f}{\partial x_1 \partial x_n} \\
\frac{\partial^2 f}{\partial x_2 \partial x_1} & \frac{\partial^2 f}{\partial x_2^2} & \cdots & \frac{\partial^2 f}{\partial x_2 \partial x_n} \\
\vdots & \vdots & \ddots & \vdots \\
\frac{\partial^2 f}{\partial x_n \partial x_1} & \frac{\partial^2 f}{\partial x_n \partial x_2} & \cdots & \frac{\partial^2 f}{\partial x_n^2}
\end{pmatrix},
\tag{4.94}
$$

which is also symmetric due to the fact that $\partial^2 f/\partial x_i \partial x_j = \partial^2 f/\partial x_j \partial x_i$ for all $i, j = 1, .., n$.

### Example 4.17

As a simple example, the Hessian matrix of $f = x^2 + y^2 + xy$ is

$$
\mathbf{H} = \begin{pmatrix} 2 & 1 \\ 1 & 2 \end{pmatrix}.
$$

Similarly, the Hessian matrix of $g = xy + e^{-x-y}$ is

$$
\mathbf{H} = \begin{pmatrix} e^{-x-y} & 1 + e^{-x-y} \\ 1 + e^{-x-y} & e^{-x-y} \end{pmatrix}.
$$

Both Hessian matrices are indeed symmetric.

Differentiation is just one important part of calculus, and another important part is integration, which will be introduced in the next chapter.

## Exercises

**4.1.** Find the first and second derivatives of the following expressions:

$$
f(x) = x^4 + 2x^2 + 3x, \quad g(x) = xe^{-x}, \quad h(x) = x^4 + \sin(x^2) + x \ln x.
$$

**4.2.** Find the first derivative of $f(x) = x^x + x$ for $x > 0$. Also calculate $f''$ and $f'''(x)$.

**4.3.** Find $y'(x)$ from the following equation:

$$
y(x) - \sin(x)e^x + x^2 y(x) = e^x.
$$

**4.4.** Find the derivatives (up to the fifth derivative) of $e^{-x} + x \sin(x)$.

**4.5.** Expand $\exp(-x^2)$ into a series.

**4.6.** Find the partial derivatives of the following functions:

$$
f(x, y) = x^2 + y^2 + xy, \quad g(x, y) = \sin(xy) + (x^2 + y^2)\exp(-x^2 - y^2).
$$

**4.7.** For function $\phi(r) = 1/r$ where $r = \sqrt{x^2 + y^2 + z^2}$, calculate $\nabla\phi$, $\nabla \cdot \phi$ and $\nabla^2\phi$.

# CHAPTER 5

# Calculus II: Integration

Contents

---

## Key Points

- Integration is explained in great detail, including both infinite integral, definite integrals and their basic rules.
- Techniques are explained and demonstrated to obtain the integrals of basic functions, including integration by substitution and integration by parts.
- Multiple integrals are also introduced with examples.

---

Integration is another important part of calculus. This chapter introduces the fundamentals of integration and methods of calculating integrals.

## 1. Integration

Differentiation is used to find the gradient for a given function. Now a natural question is how to find the original function for a given gradient. This is the integration process, which can be considered as the reverse of the differentiation process. Since we know that

$$\frac{d(\sin x)}{dx} = \cos x, \tag{5.1}$$

that is the gradient of $\sin x$ is $\cos x$, we can easily say that the original function is $\sin x$ if we know its gradient is $\cos x$. We can write

$$\int \cos x \, dx = \sin x + C, \tag{5.2}$$

where $C$ is the constant of integration. Here $\int dx$ is the standard notation showing the integration is with respect to $x$, and we usually call this the indefinite integral. The function $\cos x$ is called the integrand.

The integration constant comes from the fact that a family of curves shifted by a constant will have the same gradient at their corresponding points (see Fig. 4.3). This means that the integration can be determined up to an arbitrary constant. For this reason, we call it the indefinite integral.

Integration is more complicated than differentiation in general. Even when we know the derivative of a function, we have to be careful. For example, we know that $(x^{n+1})' = (n+1)x^n$ or

$$\left(\frac{1}{n+1}x^{n+1}\right)' = x^n$$

for any $n$ integers, so we can write

$$\int x^n dx = \frac{1}{n+1}x^{n+1} + C. \tag{5.3}$$

However, there is a possible problem when $n = -1$ because $1/(n+1)$ will become $1/0$. In fact, the above integral is valid for any $n$ except for $n = -1$. When $n = -1$, we have

$$\int \frac{1}{x}dx = \ln x + C. \tag{5.4}$$

If we know that the gradient of a function $F(x)$ is $f(x)$ or $F'(x) = f(x)$, it is possible and sometimes useful to express where the integration starts and ends, and we often write

$$\int_a^b f(x)dx = \left[F(x)\right]_a^b = F(x)\Big|_a^b = F(b) - F(a). \tag{5.5}$$

Here $a$ is called the lower limit of the integration, while $b$ is the upper limit of the integration. In this case, the constant of integration has dropped out because the integral can be determined accurately. The integral becomes a definite integral and it corresponds to the area under a curve $f(x)$ from $a$ to $b \geq a$ (see Fig. 5.1).

It is worth pointing out that the area in Fig. 5.1 may have different signs depending on whether the curve $f(x)$ is above or below the $x$-axis. If $f(x) > 0$, then that section of area is positive; otherwise, it is negative.

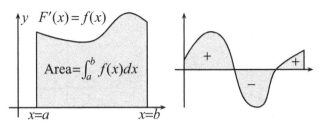

Figure 5.1: Integration and geometrical interpretation.

## Example 5.1

To calculate the integral $\int_0^1 \sqrt[3]{x^2}\,dx$, we know that $n = 2/3$, so we have

$$\int_0^1 \sqrt[3]{x^2}\,dx = \left[\frac{1}{\frac{2}{3}+1}x^{\frac{2}{3}+1}\right]_0^1 = \left[\frac{3}{5}x^{\frac{5}{3}}\right]_0^1 = \frac{3}{5}1^{5/3} - \frac{3}{5}0^{5/3} = \frac{3}{5}.$$

From (5.5), if we interchange the limits of the integral, we have

$$\int_b^a f(x)dx = \left[F(x)\right]_b^a = F(a) - F(b) = -\left(F(b) - F(a)\right) = -\int_a^b f(x)dx. \qquad (5.6)$$

This means the interchange of the integration limits is equivalent to multiplying the integral by $-1$.

## Example 5.2

Since we have already obtained that

$$\int_0^1 \sqrt[3]{x^2}\,dx = \frac{3}{5},$$

we can easily conclude that

$$\int_1^0 \sqrt[3]{x^2}\,dx = -\frac{3}{5}.$$

In addition, since differentiation is linear, that is $[\alpha f(x) + \beta g(x)]' = \alpha f'(x) + \beta g'(x)$ where $\alpha$ and $\beta$ are two real constants, it is easy to understand that integration is also a

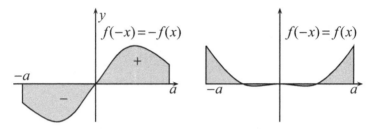

Figure 5.2: Integrals of odd and even functions.

linear operator. We have

$$\int_a^b [\alpha f(x) + \beta g(x)]dx = \alpha \int_a^b f(x)dx + \beta \int_a^b g(x)dx. \tag{5.7}$$

Also if $a \le c \le b$, we have

$$\int_a^b f(x)dx = \int_a^c f(x)dx + \int_c^b f(x)dx. \tag{5.8}$$

This may become useful when dealing with certain integrals, especially if we do not want the areas (positive and negative) to simply add together.

In the special case of an odd function $f(-x) = -f(x)$, we have

$$\int_{-a}^a f(x)dx = \int_{-a}^0 f(x)dx + \int_0^a f(x)dx = -\int_0^{-a} f(x)dx + \int_0^a f(x)dx$$

$$= \int_0^a f(-x)dx + \int_0^a f(x)dx = \int_0^a [-f(x) + f(x)]dx = 0. \tag{5.9}$$

The other way of understanding this is that the area for $x > 0$ is opposite in sign to the area for $x < 0$. So the total area or the integral is zero.

Similarly, for an even function $f(-x) = f(x)$, the areas for both $x > 0$ and $x < 0$ are the same, so we have

$$\int_{-a}^a f(x)dx = 2 \int_0^a f(x)dx. \tag{5.10}$$

The odd and even properties can become useful in simplifying some complex integrals, especially when the integrand is odd and the integration limits are symmetric.

## Example 5.3

For example, to evaluate the integral

$$I = \int_{-\pi}^{\pi} \sin(x)e^{-x^{100}}dx,$$

may be difficult because there may not be any explicit form for the integral. However, we know that $x^{100}$ is an even function and $\exp(-x^{100})$ is also an even function. We also know that $\sin(x)$ is an odd function, thus the integrand $f(x) = \sin(x)e^{-x^{100}}$ is an odd function. Alternatively, we know that

$$f(-x) = \sin(-x)e^{-(-x)^{100}} = -\sin(x)e^{-x^{100}} = -f(x),$$

which also confirms that $f(x)$ is an odd function.

Therefore, we have

$$\int_{-\pi}^{0} \sin(x)e^{-x^{100}}dx = -\int_{0}^{\pi} \sin(x)e^{-x^{100}}dx,$$

which means that

$$\int_{-\pi}^{\pi} \sin(x)e^{-x^{100}}dx = 0.$$

On the other hand, an even integrand can still make the calculation slightly easier. To calculate $\int_{-\pi/2}^{\pi/2} \cos(x)dx$, we know that $\cos(x)$ is an even function, and the integration limits are symmetric, so we have

$$\int_{-\pi/2}^{0} \cos(x)dx = \int_{0}^{\pi/2} \cos(x)dx.$$

Thus, we have

$$\int_{-\pi/2}^{\pi/2} \cos(x)dx = 2\int_{0}^{\pi/2} \cos(x)dx = 2\sin(x)\Big|_{0}^{\pi/2} = 2[\sin(\frac{\pi}{2}) - \sin(0)] = 2[1 - 0] = 2.$$

Sometimes, the integral that we evaluate can have precise physical meaning. For example, the energy stored in a simple linear spring with a spring constant $k$ and displacement $x$ is $\frac{1}{2}kx^2$. For a nonlinear spring, the force can be expressed by

$$F(x) = kx + \alpha x^3, \quad k > 0. \tag{5.11}$$

If $\alpha > 0$, the spring is hardening, while $\alpha < 0$ means the spring is softening. The potential energy stored in a nonlinear spring with an extension of $L$ can be obtained by

$$E = \int_{0}^{L} F(x)dx = \int_{0}^{L} [kx + \alpha x^3]dx = \left[\frac{kx^2}{2} + \frac{\alpha x^4}{4}\right]_{0}^{L} = \frac{kL^2}{2} + \frac{\alpha L^4}{4}. \tag{5.12}$$

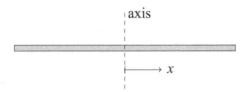

Figure 5.3: Moment of inertia of a rod of length $L$ and mass $m$.

Let us look at an example about the moment of inertia in mechanics.

## Example 5.4

For a thin rod of length $L$ with a total mass $m$ as shown in Fig. 5.3, the density $\rho$ per unit length is

$$\rho = \frac{m}{L}.$$

The moment of inertia rotating about the axis through the centre of gravity is defined by

$$I = \int x^2 dm = \int x^2 \rho dx,$$

where we have used $dm = \rho dx$. However, we have to be careful about the integration limits. Since $x$ starts at the centre, we have

$$I = 2 \int_0^{L/2} x^2 \rho dx = 2\rho \int_0^{L/2} x^2 dx,$$

where the factor 2 is due to symmetry. Thus, we have

$$I = 2\rho \Big[\frac{x^3}{3}\Big]_0^{L/2} = 2\rho\Big[\frac{(L/2)^3}{3} - \frac{0^3}{3}\Big] = \rho\frac{L^3}{12}.$$

From $\rho = m/L$, we have

$$I = \frac{m}{L} \cdot \frac{L^3}{12} = \frac{mL^2}{12}.$$

The integrals of commonly used functions are listed in Table 5.1. For many simple functions, we can easily write their integrals using the table of differentiation. For more complicated functions, what are the general techniques to obtain integrals? There are many techniques such as integration by parts, method of substitution, complex integrals and others. In the next section, we will briefly introduce integration by parts.

Table 5.1: Integrals of common functions.

| $f(x)$ | $\int f(x)dx$ | $f(x)$ | $\int f(x)dx$ |
|---|---|---|---|
| $e^x$ | $e^x$ | $\frac{1}{x}$ | $\ln x$ |
| $x^n$ | $\frac{1}{n+1}x^{n+1}$ $(n\neq-1)$ | $a^u$ | $\frac{a^u}{\ln a}$ $(a>0)$ |
| $\ln x$ | $x\ln x - x$ | $\log_a x$ | $x\log_a x - \frac{x}{\ln a}$ |
| $\frac{1}{a^2+x^2}$ | $\frac{1}{a}\tan^{-1}\frac{x}{a}$ | $\frac{1}{a^2-x^2}$ | $\frac{1}{2a}\ln\frac{a+x}{a-x}$ |
| $\frac{1}{x^2-a^2}$ | $\frac{1}{2a}\ln\frac{x-a}{x+a}$ | $\frac{1}{\sqrt{a^2-x^2}}$ | $\sin^{-1}\frac{x}{a}$ |
| $\frac{1}{\sqrt{x^2+a^2}}$ | $\sinh^{-1}\frac{x}{a}$ | $\frac{1}{\sqrt{x^2-a^2}}$ | $\cosh^{-1}\frac{x}{a}$ |
| $\sin x$ | $-\cos x$ | $\cos x$ | $\sin x$ |
| $\sec^2 x$ | $\tan x$ | $\csc^2 x$ | $-\cot x$ |
| $\sinh x$ | $\cosh x$ | $\cosh x$ | $\sinh x$ |
| $\tanh x$ | $\ln\cosh x$ | $\coth x$ | $\ln\sinh x$ |

It is worth pointing out that some integrals with seemingly simple integrands may not be well defined and can be divergent. For example, we know

$$\int \frac{1}{x}dx = \ln(x) + C, \qquad (5.13)$$

so

$$\int_0^1 = \ln(x)\Big|_0^1 = \ln(1) - \ln(0), \qquad (5.14)$$

but $\ln(0)$ is not well defined since $\ln(0) \to -\infty$, which means that above definite integral is divergent (or does not exist).

When the integral limits involve an infinity, some care should be taken when evaluating the definite integral.

## Example 5.5

A very useful definite integral is

$$I = \int_0^\infty e^{-x}dx.$$

We know that $[e^{-x}]' = -e^{-x}$ or $[-e^{-x}]' = e^{-x}$. So we have

$$I = \int_0^\infty e^{-x}dx = \Big[-e^{-x}\Big]_0^\infty = -e^{-\infty} - (-e^{-0}) = 0 - (-1) = 1.$$

Here, we have used $\exp[-\infty] = 0$. However, the integral $\int_0^\infty e^x dx$ does not exist because it diverges (involving the value of infinity).

In fact, many definite integrals with one infinite integration limits may diverge. For example, $\int_{-\infty}^{0} e^{-x}dx$, $\int_{0}^{\infty} xdx$ and $\int_{0}^{1} \frac{1}{x^2}dx$ will all diverge.

## 2. Integration by Parts

In differentiation, we can easily find the gradient of $x \sin x$ if we know the derivatives of $x$ and $\sin x$ by using the rule of products. In integration, can we find the integral of $x \sin x$ if we know the integrals of $x$ and $\sin x$? The answer is yes, this is the integration by parts.

From the differentiation rule

$$\frac{d(uv)}{dx} = u\frac{dv}{dx} + v\frac{du}{dx}, \tag{5.15}$$

we integrate it with respect to $x$, we have

$$uv = \int u\frac{dv}{dx}dx + \int v\frac{du}{dx}dx. \tag{5.16}$$

By rearranging, we have

$$\int u\frac{dv}{dx}dx = uv - \int v\frac{du}{dx}dx, \tag{5.17}$$

or

$$\int uv'dx = uv - \int vu'dx. \tag{5.18}$$

This is the well-known formula for the technique known as integration by parts.

Since $dx$ appears in both $\frac{du}{dx}$ and $dx$, it makes sense to write the above as

$$\int udv = uv - \int vdu, \tag{5.19}$$

and this notation is also widely used in many textbooks.

You may wonder where the constant of integration is? You are right, there is a constant of integration for $uv$, but as we know they exist for indefinite integrals, we simply omit to write it out in the formula. Obviously, we have to remember to put it back in the end when we write the final formula. Now let us look at an example to calculate $\int x \sin(x)dx$.

## Example 5.6

We know that $\int x dx = \frac{x^2}{2} + C$, and $\int \sin x dx = -\cos x + C$, we now try to find the integral

$$I = \int x \sin x dx. \tag{5.20}$$

Let $u = x$ and $dv/dx = \sin x$, we have

$$\frac{du}{dx} = 1, \qquad v = \int \sin x dx = -\cos x. \tag{5.21}$$

Therefore, we have, after substituting into the formula and using integration by parts

$$I = \int x \sin x dx = x(-\cos x) - \int (-\cos x)\frac{du}{dx} dx$$

$$= -x \cos x + \int \cos x dx = -x \cos x + \sin x + C. \tag{5.22}$$

Obviously, the way of treating definite integrals using integration by parts is identical to that for indefinite integrals. The only difference is that we write out the integration limits explicitly. Thus, the formula for integration by parts becomes

$$\int_a^b uv' dx = [uv]\Big|_a^b - \int_a^b vu' dx. \tag{5.23}$$

## Example 5.7

Let us evaluate the integral

$$I = \int_0^\infty xe^{-x}dx.$$

We use

$$u = x, \quad v' = e^{-x},$$

which gives

$$du = dx, \quad u' = 1, \quad v = -e^{-x}.$$

Thus we have

$$I = \int_0^\infty xe^{-x}dx = \left[x(-e^{-x})\right]_0^\infty - \int_0^\infty 1 \times (-e^{-x})dx = \left[-xe^{-x}\right]_0^\infty + \int_0^\infty e^{-x}dx.$$

From an earlier example in the previous section, we know that

$$\int_0^\infty e^{-x}dx = 1.$$

How do we calculate the term of $Q = xe^{-x}$ as $x$ becomes infinity? One trick as used in many textbooks is that we first let $x = L$ and then take a limit of $L$. That is

$$Q = \lim_{L \to \infty} Le^{-L}.$$

Since for any value of $L$, $\exp[-L]$ decreases much faster than the increase rate of $L$, and their production is very small. Thus, as $L \to \infty$, $Q \to 0$, using the above results, we finally get

$$I = \lim_{L \to \infty} Le^{-L} - 0e^{-0} + \int_0^\infty e^{-x}dx = 0 - 0 + 1 = 1.$$

Here, we have used the fact that $x$ increases much more slowly than the decrease of $e^{-x}$ as $x \to \infty$. That is $xe^{-x} \to 0$ as $x \to \infty$. In fact, we have

$$x^n e^{-x} \to 0, \quad x \to \infty, \quad \text{for all } n > 0.$$

As a useful exercise, you can use integration by parts many times to show that

$$\int_0^\infty x^n e^{-x}dx = n!, \tag{5.24}$$

for $n = 1, 2, ...$, (all positive integer values). This is in fact a special case of the Gamma integral $\Gamma(z) = \int_0^\infty x^{z-1}e^{-x}dx$.

## 3. Integration by Substitution

Sometimes, it is not easy to carry out the integration directly. However, it might become easier if we used the change of variables or integration by substitution. For example, we want to calculate the integral

$$I = \int f(x)dx. \tag{5.25}$$

We can change the variable $x$ into another variable $u = g(x)$ where $g(x)$ is a known function of $x$. This means

$$\frac{du}{dx} = g'(x), \quad \text{or} \quad du = g'(x)dx, \quad dx = \frac{1}{g'(x)}du. \tag{5.26}$$

This means that

$$I = \int f(x)dx = \int f[g^{-1}(u)]\frac{1}{g'}du, \tag{5.27}$$

where it is usually not necessary to calculate $g^{-1}(u)$ as it is relatively obvious.

## Example 5.8

For example, in order to do the integration

$$I = \int x^2 e^{x^3+7} dx, \tag{5.28}$$

we let $u = x^3 + 7$ and we have

$$\frac{du}{dx} = (x^3 + 7)' = 3x^2, \tag{5.29}$$

or

$$du = 3x^2 dx, \tag{5.30}$$

which means $dx = \frac{1}{3x^2} du$. Therefore, we have

$$I = \int x^2 e^{x^3+7} dx = \int x^2 e^u \frac{1}{3x^2} du = \frac{1}{3} \int e^u du = \frac{1}{3} e^u + A = \frac{1}{3} e^{x^3+7} + A,$$

where $A$ is the constant of integration. Here we have substituted $u = x^3 + 7$ back in the last step.

Sometimes, the integrals involve trigonometrical functions and in many cases a combination of techniques may be required.

Now let us try to calculate the area of an ellipse. We know the equation of an ellipse in the Cartesian coordinates is

$$\frac{x^2}{a^2} + \frac{y^2}{b^2} = 1, \quad a, b > 0. \tag{5.31}$$

## Example 5.9

Using symmetry, we only need to calculate the area in the first quadrant. Thus, we can solve $y$ in terms of $x$ and we have

$$y = \frac{b}{a} \sqrt{a^2 - x^2}, \quad 0 \le x \le a. \quad A = 4 \int_0^a y(x) dx = \frac{4b}{a} \int_0^a \sqrt{a^2 - x^2} dx,$$

We can set $x = a \sin(t)$ for $t \in [0, \pi/2]$, so $dx = a \cos(t) dt$. Thus, we have (using $\sin^2(t) + \cos^2(t) = 1$)

$$A = \frac{4b}{a} \int_0^{\pi/2} \sqrt{a^2 - a^2 \sin^2(t)} (a \cos(t)) dt = \frac{4b}{a} \int_0^{\pi/2} a^2 \cos^2(t) dt = 4ab \int_0^{\pi/2} \cos^2(t) dt.$$

Using $\cos^2(t) = [1 + \cos(2t)]/2$, we have

$$A = 2ab \int_0^{\pi/2} [1 + \cos(2t)]dt = 2ab\left[ \int_0^{\pi/2} 1dt + \int_0^{\pi} \cos(2t)dt \right]$$

$$= 2ab\left[\frac{\pi}{2} + \frac{\sin(2t)}{2}\Big|_0^{\pi/2}\right] = 2ab[\frac{\pi}{2} + 0] = \pi ab.$$

There are other techniques for integration. However, an interesting fact about calculus is that it is always possible to find the derivatives of complex functions and express them in explicit forms, but it is usually difficult and sometimes impossible to find the explicit form of integrals. Some seemingly simple integrals with simple integrands can be very hard to evaluate. For example, the following integral

$$I = \int_0^x e^{-u^2} du,$$

is not easy to calculate and it is related to the so-called error function:

$$\text{erf}(x) = \frac{2}{\sqrt{\pi}} \int_0^x e^{-u^2} dt. \tag{5.32}$$

We will introduce this function later in this chapter.

## 4. Double Integrals and Multiple Integrals

The integrals we have discussed so far involve only one variable and such integration is a reverse operation of differentiation. As we know, differentiation of functions with more than one variable leads to partial derivatives, the integration can also extend to multiple variables, leading to multiple integrals.

Let us start with a simple function with two variables $f(x,y) = xy$. When we calculate the partial derivative with respect to $x$, we can essentially consider $y$ as a constant parameter. Similarly, we take the partial derivative with respect to $y$, we can consider $x$ as a constant parameter. Thus, we have

$$\frac{\partial f}{\partial x} = y, \quad \frac{\partial f}{\partial y} = x. \tag{5.33}$$

Now let us evaluate the integral of $f(x,y) = xy$ with respect to both $x$ and $y$. Since this integral has two variables, it becomes a double integral, and we can write

$$I = \iint f(x,y)dxdy, \tag{5.34}$$

which is an indefinite double integral. However, in engineering, we are more con-

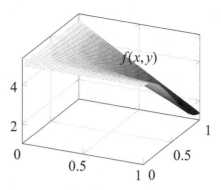

Figure 5.4: The meaning of a double integral.

cerned with definite integrals. With proper integration limits, we have the definite double integral or simple double integral as

$$I = \int_c^d \int_a^b f(x,y)dxdy, \tag{5.35}$$

for a rectangular domain $x \in [a, b]$ and $y \in [c, d]$. Here, the domain can be written as $D = [a, b] \times [c, d]$. In this case, we can simply write the above integral as

$$I = \iint_D f(x,y)dxdy. \tag{5.36}$$

The definite integral of a univariate $f(x)$ corresponds to the area enclosed by the curve $f(x)$ and the $x$-axis, treating the area above the axis as positive and the area under the axis as negative. Similarly, the definite double integral of a bivariate function $f(x,y)$ over the domain $D$ corresponds to the volume enclosed by the surface $f(x,y)$ and the domain $D$ on the plane $(x,y)$ (or $z = 0$ in the 3D Cartesian coordinates) as shown in Fig. 5.4.

The volume above the plane $(x,y)$ is considered as positive, while the volume under the plane is considered as negative.

One way of obtaining a double integral is the direct integration with respect to each variable. Let us look at an example.

## Example 5.10

To evaluate

$$V = \int_0^1 \int_0^1 xydxdy,$$

we have the integrand $f(x,y) = xy$. We can first integrate it with respect to $x$, considering $y$ as a constant. We have

$$V = \int_0^1 \int_0^1 xy\,dx\,dy = \int_0^1 \left[ \int_0^1 xy\,dx \right] dy = \int_0^1 \left[ y \int_0^1 x\,dx \right] dy.$$

Since $\int_0^1 x\,dx = 1/2$, we have

$$V = \int_0^1 \frac{y}{2}\,dy = \frac{1}{2} \int_0^1 y\,dy = \frac{1}{4}.$$

You may wonder what happens if we integrate $y$ first? We have

$$V = \int_0^1 \int_0^x y\,dx\,dy = \int_0^1 \left[ \int_0^1 xy\,dy \right] dx = \int_0^1 \left[ x \int_0^1 y\,dy \right] dx = \int_0^1 \frac{x}{2}\,dx = \frac{1}{4},$$

which is the same answer.

The same answer for the above integral is independent of the order of the integration as guaranteed by Fubini's theorem, which essentially states that a double integral can be considered as an iterated integral. That is

$$\iint_D f(x,y)\,dx\,dy = \int_a^b \left[ \int_c^d f(x,y)\,dy \right] dx = \int_c^d \left[ \int_a^b f(x,y)\,dx \right] dy. \qquad (5.37)$$

This is true as long as the integral for the absolute value of the integrand is finite, which means

$$\iint_D |f(x,y)|\,dx\,dy < \infty. \qquad (5.38)$$

If we take the absolute value of the integrand, it means the volume under the plane enclosed by the surface $f(x,y)$ is also considered as positive. In this case, the above condition means the total sum of the volume must be finite. Thus, as long as the total volume is finite, the double integral can be calculated in any order as iterated integrals.

One application of calculus is to find the centre of mass or gravity for a laminate shape in two-dimensional Cartesian coordinates $(x, y)$, defined by

$$\bar{x} = \frac{1}{A_0} \iint x\,dx\,dy, \quad \bar{y} = \frac{1}{A_0} \iint y\,dx\,dy, \qquad (5.39)$$

where $A_0$ is the area of the domain and $dA = dx\,dy$ is the area element. Many rules such as linearity that are valid for univariate integrals are also valid for multiple integrals. In addition, the domain $D$ can in general be any shape and in this case some of the integration limit may not be fixed. Let us show this using an example.

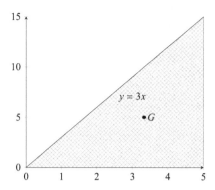

Figure 5.5: Centre of mass of a triangular domain $D = [0, 5] \times [0, 3x]$.

## Example 5.11

Now let us calculate the centre of mass of the triangular domain (see Fig. 5.5)

$$\bar{x} = \frac{1}{A_0} \iint_D x\,dx\,dy,$$

where $A_0 = 5 \times 15/2 = 75/2$ (the area of the triangular region). The triangular domain $D$ is enclosed by $0 \geq x \leq 5$ and $y = 3x$, as shown in Fig. 5.5. The integrand is $f(x,y) = x$, so we have

$$I = \iint_D x\,dx\,dy = \int_0^5 x\Big[\int_0^{3x} dy\Big]dx = \int_0^5 x\Big[y\Big|_{y=0}^{y=3x}\Big]dx$$

$$= \int_0^5 x(3x - 0)dx = 3\int_0^5 x^2 dx = 3\Big[\frac{x^3}{3}\Big]\Big|_0^5 = 125.$$

Thus, the centre of mass (G) lies at $\bar{x} = \frac{125}{A_0} = \frac{10}{3}$. Similarly, we have

$$\bar{y} = \frac{1}{A_0} \iint_D y\,dx\,dy = \frac{1}{A_0} \int_0^5 \Big[\int_0^{3x} y\,dy\Big]dx = \frac{1}{A_0} \int_0^5 [\frac{(3x)^2}{2} - 0]dx$$

$$= \frac{9}{2A_0} \int_0^5 x^2 dx = \frac{9}{2A_0} \cdot [\frac{5^5}{3} - \frac{0^3}{3}] = \frac{375}{2 \times 75/2} = 5.$$

Double integrals are important in a diverse range of engineering applications. For example, in order to estimate the behaviour of a structure, we often have to calculate the bending moment and second moment of area.

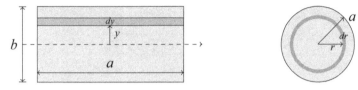

Figure 5.6: Area moment of inertia of a rectangular plate (left) and polar moment of inertia of a disc (right).

## Example 5.12

The area moment of inertia (or second moment of area) of a thin rectangular plate, with the width $a$ and the depth $b$ (see Fig. 5.6), is defined by

$$I = \iint_\Omega y^2 dS = \iint_\Omega y^2 dxdy.$$

The plate can be divided into four equal parts, and we have

$$I = 4\int_0^{a/2} \left[ \int_0^{b/2} y^2 dy \right] dx = 4\int_0^{a/2} \frac{1}{3}\left(\frac{b}{2}\right)^3 dx = \frac{b^3}{6}\int_0^{a/2} dx = \frac{ab^3}{12}.$$

In describing the behaviour of rotation and rigid-body motion, we have to calculate the moment of inertia

$$I = \int_\Omega r^2 dm = \int_\Omega \rho r^2 dV, \tag{5.40}$$

where $dm = \rho dV$ is the mass of the volume element $dV$ and $\rho$ is the density. The integration must be over the whole body $\Omega$.

## Example 5.13

For a thin disc of radius $a$ with a uniform thickness $t$ and density $\rho$ (see Fig. 5.6), we have $dV = tdA = trdrd\theta$ and we only need to integrate in two dimensions. Thus, we have

$$I = \int_\Omega \rho r^2 dV = \int_0^{2\pi} \int_0^a (\rho tr^2) rdrd\theta = t\rho \int_0^{2\pi} d\theta \int_0^a r^3 dr$$

$$= t\rho(2\pi)\int_0^a r^3 dr = 2\pi\rho t\left(\frac{r^4}{4}\Big|_0^a\right) = \pi\rho t\frac{a^4}{2}.$$

Let $m$ be the total mass of the thin disc, we have its volume is $V = \pi a^2 t$ and its density is $\rho = m/V = m/(\pi a^2 t)$. Finally, we have

$$I = \pi t \cdot \frac{m}{\pi a^2 t} \cdot \frac{a^4}{2} = \frac{ma^2}{2}.$$

This is the polar moment of inertia for the thin disc with respect to the axis through the centre but perpendicular to the disc surface (see Fig. 5.6).

It is worth pointing out that the polar moment of inertia has been calculated using the double integral. In fact, it is possible to use rotational symmetry to do the calculation using a single integral. From Fig. 5.6, we know that $dm = \rho dA$ where $\rho = m/(\pi a^2)$ is the density per unit area. Since $dA = 2\pi r dr$, we have

$$I = \int_0^a r^2 dm = \int_0^a r^2 \rho(2\pi r)dr = 2\pi\rho \int_0^a r^3 dr = 2\pi\rho \left[\frac{r^4}{4}\right]\Big|_0^a = \frac{\pi\rho a^4}{2}. \tag{5.41}$$

Using $\rho = m/(\pi a^2)$, we have

$$I = \frac{\pi a^4}{2} \cdot \frac{m}{\pi a^2} = \frac{ma^2}{2}, \tag{5.42}$$

which is exactly the same result we obtained in the above example. Therefore, the use of symmetry can simplify calculations.

In the case of a thick-walled cylindrical tube with an inner radius $a$ and an outer radius $b$, the integral (5.41) becomes

$$I = \int_a^b r^2 dm = 2\pi\rho \left[\frac{r^4}{4}\right]\Big|_a^b = \frac{\pi\rho}{2}(b^4 - a^4). \tag{5.43}$$

However, even with the same total mass $m$, the density will be different because the area of the cross section is $A = \pi(b^2 - a^2)$. So the density $\rho = m/A = \frac{m}{\pi(b^2-a^2)}$, and we have

$$I = \frac{\pi\rho}{2}(b^4 - a^4) = \frac{\pi}{2} \cdot \frac{m}{\pi(b^2 - a^2)} \cdot (b^4 - a^4) = \frac{m}{2}(a^2 + b^2). \tag{5.44}$$

In general, multiple integrals with $n$ variables can be written as

$$I_n = \int \int \cdots \int_\Omega f(x_1, x_2, ..., x_n)dx_1 dx_2 ... dx_n, \tag{5.45}$$

where $\Omega$ the domain in the $n$-dimensional space.

## 5. Jacobian Determinant

Sometimes it is necessary to change variables when evaluating an integral. For a simple one-dimensional integral, the change of variables from $x$ to a new variable $v$

(say) leads to $x = x(v)$. This is relatively simple as $dx = \frac{dx}{dv}dv$, and we have

$$\int_{x_a}^{x_b} f(x)dx = \int_a^b f(x(v))\frac{dx}{dv}dv, \tag{5.46}$$

where the integration limits change so that $x(a) = x_a$ and $x(b) = x_b$. Here the extra factor $dx/dv$ in the integrand is referred to as the Jacobian. For a double integral, it is more complicated. Assuming $x = x(\xi, \eta)$, $y = y(\xi, \eta)$, we have

$$\iint f(x,y)dxdy = \iint f(\xi\eta)|J|d\xi d\eta, \tag{5.47}$$

where $J$ is the Jacobian. That is

$$J \equiv \frac{\partial(x,y)}{\partial(\xi,\eta)} = \begin{vmatrix} \frac{\partial x}{\partial \xi} & \frac{\partial x}{\partial \eta} \\ \frac{\partial y}{\partial \xi} & \frac{\partial y}{\partial \eta} \end{vmatrix} = \begin{vmatrix} \frac{\partial x}{\partial \xi} & \frac{\partial y}{\partial \xi} \\ \frac{\partial x}{\partial \eta} & \frac{\partial y}{\partial \eta} \end{vmatrix}. \tag{5.48}$$

The notation $\partial(x,y)/\partial(\xi,\eta)$ is just a useful shorthand. This is equivalent to saying that the change of the infinitesimal area $dA = dxdy$ becomes

$$dxdy = \left|\frac{\partial(x,y)}{\partial(\xi,\eta)}\right|d\xi d\eta = \left|\frac{\partial x}{\partial \xi}\frac{\partial y}{\partial \eta} - \frac{\partial x}{\partial \eta}\frac{\partial y}{\partial \xi}\right|d\xi d\eta. \tag{5.49}$$

## Example 5.14

When transforming from $(x, y)$ to polar coordinates $(r, \theta)$, we have the following relationships

$$x = r\cos\theta, \qquad y = r\sin\theta.$$

Thus, we have

$$\frac{\partial x}{\partial r} = \cos\theta, \quad \frac{\partial x}{\partial \theta} = -r\sin\theta, \quad \frac{\partial y}{\partial r} = \sin\theta, \quad \frac{\partial y}{\partial \theta} = r\cos\theta,$$

so that the Jacobian is

$$J = \frac{\partial(x,y)}{\partial(r,\theta)} = \frac{\partial x}{\partial r}\frac{\partial y}{\partial \theta} - \frac{\partial x}{\partial \theta}\frac{\partial y}{\partial r} = \cos\theta \times r\cos\theta - (-r\sin\theta) \times \sin\theta$$

$$= r[\cos^2\theta + \sin^2\theta] = r.$$

Thus, an integral in $(x, y)$ will be transformed into

$$\iint \phi(x,y)dxdy = \iint \phi(r\cos\theta, r\sin\theta)rdrd\theta.$$

It is worth pointing out that in the change from the Cartesian system to the cylindrical coordinate system, the Jacobian determinant is the same (i.e., $\det(J) = r$) and as an exercise, we leave the reader to show that this is the case.

In a similar fashion, the change of variables in triple integrals gives

$$V = \iiint_\Omega \phi(x,y,z)dxdydz = \iiint_\omega \psi(\xi,\eta,\zeta)|J|d\xi d\eta d\zeta, \tag{5.50}$$

and

$$J \equiv \frac{\partial(x,y,z)}{\partial(\xi,\eta,\zeta)} = \begin{vmatrix} \frac{\partial x}{\partial \xi} & \frac{\partial y}{\partial \xi} & \frac{\partial z}{\partial \xi} \\ \frac{\partial x}{\partial \eta} & \frac{\partial y}{\partial \eta} & \frac{\partial z}{\partial \eta} \\ \frac{\partial x}{\partial \zeta} & \frac{\partial y}{\partial \zeta} & \frac{\partial z}{\partial \zeta} \end{vmatrix} = \begin{vmatrix} \frac{\partial x}{\partial \xi} & \frac{\partial x}{\partial \eta} & \frac{\partial x}{\partial \zeta} \\ \frac{\partial y}{\partial \xi} & \frac{\partial y}{\partial \eta} & \frac{\partial y}{\partial \zeta} \\ \frac{\partial z}{\partial \xi} & \frac{\partial z}{\partial \eta} & \frac{\partial z}{\partial \zeta} \end{vmatrix}. \tag{5.51}$$

For spherical polar coordinates $(r, \theta, \phi)$ as shown in Fig. 4.11, we know that

$$x = r\cos\theta\sin\phi, \qquad y = r\sin\theta\sin\phi, \qquad z = r\cos\phi. \tag{5.52}$$

Therefore, the Jacobian determinant is

$$J = \frac{\partial(x,y,z)}{\partial(r,\theta,\phi)} = \begin{vmatrix} \cos\theta\sin\phi & -r\sin\theta\sin\phi & r\cos\theta\cos\phi \\ \sin\theta\sin\phi & r\cos\theta\sin\phi & r\sin\theta\cos\phi \\ \cos\phi & 0 & -r\sin\phi \end{vmatrix}$$

$$= \cos\phi \begin{vmatrix} -r\sin\theta\sin\phi & r\cos\theta\cos\phi \\ r\cos\theta\sin\phi & r\sin\theta\cos\phi \end{vmatrix} + 0 - r\sin\phi \begin{vmatrix} \cos\theta\sin\phi & -r\sin\theta\sin\phi \\ \sin\theta\sin\phi & r\cos\theta\sin\phi \end{vmatrix}$$

$$= r^2 \sin\phi. \tag{5.53}$$

Thus, the volume element change in the spherical system is

$$dV = dxdydz = r^2 \sin\phi \, drd\theta d\phi. \tag{5.54}$$

## 6. Special Integrals

Some integrals appear so frequently in engineering mathematics that they deserve special attention. Most of these special integrals are also called special functions as they have certain varying parameters or integral limits. We discuss only Gaussian integral and the error function.

### 6.1. Line Integral

An important class of integrals in this context is the line integral, which integrates along a curve $\mathbf{r}(x,y,z) = x\mathbf{i} + y\mathbf{j} + z\mathbf{k}$. For example, in order to calculate the arc length

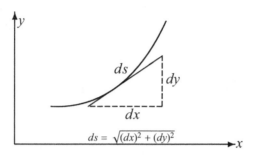

Figure 5.7: Arc length along a curve.

$L$ of curve $\mathbf{r}$ as shown in Fig. 5.7, we have to use the line integral.

$$L = \int_{s_0}^{s} ds = \int_{s_0}^{s} \sqrt{(dx)^2 + (dy)^2} = \int_{x_0}^{x} \left[\sqrt{\frac{(dx)^2}{(dx)^2} + \frac{(dy)^2}{(dx)^2}}\right] dx = \int_{x_0}^{x} \sqrt{1 + (\frac{dy}{dx})^2}\ dx.$$

Thus, for a curve described by a function $y = f(x)$, its arc length from $x = a$ and $x = b$ can be calculated by

$$L = \int_{a}^{b} \sqrt{1 + [f'(x)]^2}\ dx. \tag{5.55}$$

For example, the arc length of the parabola $y(x) = \frac{1}{2}x^2$ from $x = -1$ to $x = 1$ is given by

$$L = \int_{0}^{1} \sqrt{1 + y'^2}\,dx = \int_{0}^{1} \sqrt{1 + x^2}\,dx = \frac{1}{2}\left[x\sqrt{1 + x^2} + \ln(x + \sqrt{1 + x^2})\right]\Big|_{-1}^{1}$$

$$= \frac{1}{2}\left[\left(\sqrt{2} + \ln(\sqrt{2} + 1)\right) - \left(-\sqrt{2} + \ln(-1 + \sqrt{2})\right)\right] \approx 2.295587. \tag{5.56}$$

## 6.2. Gaussian Integrals

The Gaussian integral appears in many situations in engineering mathematics and statistics. It can be defined by

$$I(\alpha) = \int_{-\infty}^{\infty} e^{-\alpha x^2}\,dx. \tag{5.57}$$

In order to evaluate the integral, let us first evaluate $I^2$. Since the Gaussian integral is a definite integral and must give a constant value, we can change the dummy variable

$x$ to anything appropriate ($y$) as we wish. Thus we have

$$I^2 = [\int_{-\infty}^{\infty} e^{-\alpha x^2} dx]^2 = \int_{-\infty}^{\infty} e^{-\alpha x^2} dx \int_{-\infty}^{\infty} e^{-\alpha y^2} dy = \int_{-\infty}^{\infty} \int_{-\infty}^{\infty} e^{-\alpha(x^2+y^2)} dx dy. \quad (5.58)$$

Changing into the polar coordinates $(r, \theta)$ and noticing that $r^2 = x^2 + y^2$ and $dx dy = r dr d\theta$, we have

$$I^2 = \int_0^{\infty} dr \int_0^{2\pi} r e^{-\alpha r^2} d\theta = 2\pi \int_0^{\infty} e^{-\alpha r^2} r dr$$

$$= 2\pi \int_0^{\infty} \frac{1}{2\alpha} e^{-\alpha r^2} d(\alpha r^2) = \frac{\pi}{\alpha} \int_0^{\infty} e^{-\alpha r^2} d(\alpha r^2) = \frac{\pi}{\alpha}, \quad (5.59)$$

where we have used

$$\int_0^{\infty} e^{-\alpha r^2} d(\alpha r^2) = \int_0^{\infty} e^{-u} du = 1, \quad u = \alpha r^2. \quad (5.60)$$

Therefore, the integral becomes

$$I(\alpha) = \int_{-\infty}^{\infty} e^{-\alpha x^2} dx = \sqrt{\frac{\pi}{\alpha}}. \quad (5.61)$$

Since $\alpha$ is a parameter, we can differentiate both sides of this equation with respect to $\alpha$, and we have

$$\int_{-\infty}^{\infty} x^2 e^{-\alpha x^2} dx = \frac{1}{2\alpha} \sqrt{\frac{\pi}{\alpha}}. \quad (5.62)$$

## 6.3. Error Functions

The error function, which appears frequently in heat conduction and diffusion problems, is defined by

$$\text{erf}(x) = \frac{2}{\sqrt{\pi}} \int_0^x e^{-\eta^2} d\eta. \quad (5.63)$$

Its complementary error function is defined by

$$\text{erfc}(x) = 1 - \text{erf}(x) = \frac{2}{\sqrt{\pi}} \int_x^{\infty} e^{-t^2} dt. \quad (5.64)$$

The error function is an odd function: $\text{erf}(-x) = -\text{erf}(x)$. Using the results from the Gaussian integral $\int_{-\infty}^{\infty} e^{-\eta^2} d\eta = \sqrt{\pi}$, together with the basic definition, we have $\text{erf}(0) = 0$, and $\text{erf}(\infty) = 1$.

The error function cannot be easily evaluated in a closed form. Using Taylor series for

$$e^x = 1 + x + \frac{x^2}{2!} + \frac{x^3}{3!} + \ldots + \frac{x^n}{n!} + \ldots, \tag{5.65}$$

and setting $x = -\eta^2$, we have the integrand

$$e^{-\eta^2} = 1 - \eta^2 + \frac{1}{2}\eta^4 - \frac{1}{6}\eta^6 + \ldots + \frac{(-1)^n \eta^{2n}}{n!} + \ldots, \tag{5.66}$$

and integrating term by term, we have

$$\text{erf}(x) = \frac{2}{\sqrt{\pi}}\left[x - \frac{x^3}{3} + \frac{x^5}{10} + \ldots + \frac{(-1)^n \eta^{2n+1}}{n!(2n+1)} + \ldots\right] = \frac{2}{\sqrt{\pi}}\sum_{n=0}^{\infty}\frac{(-1)^n}{2n+1}\frac{x^{2n+1}}{n!}. \tag{5.67}$$

The integrals of the complementary error function are defined by

$$\text{ierfc}(x) = \int_x^{\infty} \text{erfc}(\eta)d\eta = \int_x^{\infty} [1 - \text{erf}(\eta)]\,d\eta = \frac{e^{-x^2}}{\sqrt{\pi}} - x\,\text{erfc}(x). \tag{5.68}$$

In fact, the evaluations of complex integrals require numerical methods such as the Guassian integration method and Monte Carlo methods.

## Exercises

**5.1.** Integrate the following integrals:
- $\int(x - \frac{1}{x})dx$
- $\int(x^2 + \cos(x) + e^{-x})dx$
- $\int \ln x\,dx$
- $\int x^2 \cos(x^3)dx$

**5.2.** Find the values of the following definite integrals:
- $\int_0^{\pi/2} x\sin(2x)dx$
- $\int_0^{\infty} xe^{-x}dx$
- $\int_0^{\infty} xe^{-x^2}dx$

**5.3.** Calculate the area under the curve $|x|e^{-x^2}$.

**5.4.** Evaluate the following double integrals:
- $\int_0^1 \int_0^2 xy^2 dxdy$
- $\iint_D(x^4 + y^4 + 2x^2y^2)dxdy$ with $D$ as a circular domain $x^2 + y^2 \leq 1$.

**5.5.** Show that the volume $V = \frac{4\pi}{3}abc$ of an ellipsoid given by

$$\frac{x^2}{a^2} + \frac{y^2}{b^2} + \frac{z^2}{c^2} \leq 1, \quad a, b, c > 0.$$

# CHAPTER 6

# Complex Numbers

Contents

## Key Points

- Basic concepts of complex numbers are introduced, and complex algebraic operations are also explained, including modulus, Argand diagram and their links to vectors.
- The well-known Euler's formula is explained and examples are given to show how to use this formula in its proper context.
- Hyperbolic functions are introduced, and their links with trigonometrical functions are also discussed, including some identities.
- Complex integrals are introduced briefly with worked examples.

## 1. Complex Numbers

You may wonder why complex numbers are necessary. There are many reasons and situations where complex numbers are needed. From the mathematical point of view, complex numbers make the number system complete. For example, when we solve quadratic equations, we say the square root of a negative number, for example $\sqrt{-1}$, does not exist because there is no number whose square is negative (for all real numbers $x$, $x^2 \geq 0$). This is only true in the context of real numbers. Such limitations mean that the system of real numbers is incomplete, as the mathematical operations of numbers could lead to something which does not belong to the number system.

On the other hand, from the engineering point of view, it is much easier to study the behaviour of certain systems (such as control systems and vibrations) using Laplace transforms, which require complex numbers. For example, circuit theory and the modelling of power engineering can rely on the complex models, and complex numbers can make such models simpler.

A significant extension is to introduce imaginary numbers by defining an imaginary unit

$$i = \sqrt{-1}, \qquad i^2 = (\sqrt{-1})^2 = -1. \tag{6.1}$$

This is a seemingly simple step but it may have many profound consequences. It is worth pointing out that $i$ is a special notation, but you cannot use $i^2 = (\sqrt{-1})^2 = \sqrt{(-1)^2} = \sqrt{1} = \pm 1$ because this may lead to some confusion. To avoid such possible confusion, it is better to think of $i$ as $\sqrt{-}$ or the imaginary unit, so for any real number $a > 0$

$$\sqrt{-a} = \sqrt{(-1) \times a} = \sqrt{-1}\sqrt{a} = i\sqrt{a}. \tag{6.2}$$

For example, $\sqrt{-2} = i\sqrt{2}$ and $\sqrt{-25} = i5 = 5i$ (we prefer to write numbers first followed by $i$).

## Example 6.1

The imaginary number $i$ follows the same rules for mathematical functions defined in the real-number system. For example, we can calculate the following

$$i^3 = i^2 i = -1i = -i, \quad i^4 = (i^2)^2 = (-1)^2 = 1, \quad i^5 = i^4 i = i, \quad i^6 = i^4 \cdot i^2 = -1,$$

$$i^7 = i^4 \cdot i^2 \cdot i = -i, \quad i^8 = (i^4)^2 = 1, \quad i^9 = i^8 \cdot i = i, \quad i^{10} = i^{4 \times 2 + 2} = (i^4)^2 \cdot i^2 = -1.$$

Thus

$$i^{50} = i^{4 \times 12 + 2} = (i^4)^{12} \times i^2 = -1, \quad i^{101} = i^{4 \times 25 + 1} = i,$$

and

$$i^{1001} = i^{4 \times 250 + 1} = i, \quad i^{12345} = i^{12344 + 1} = i^{4 \times 3086 + 1} = i.$$

So the best way to estimate such an expression is to try to write exponents in terms of the multiples of 4 (as close as possible). To simply use odd or even exponents is not enough, and often leads to incorrect results.

It is worth pointing out that the notation $j$ (instead of $i$) is often used in engineering textbooks. One of the reasons is that $i$ is often associated with a current in circuit theory. Whatever the notation may be, it is just $j \equiv i = \sqrt{-1}$.

You might ask what $\sqrt{i}$ or $i^{1/2}$ is? Obviously, we can calculate it, but we first have to define the algebraic rule for complex numbers. Before we proceed to do this, let us

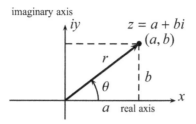

Figure 6.1: Complex plane, angle and modulus.

try to represent complex numbers geometrically.

Another way of thinking of a complex number is to represent it as a pair of two numbers $x$ and $y$ on a complex plane (see Fig. 6.1), similar to the Cartesian coordinates $(x, y)$ where each point corresponds uniquely to an ordered pair of two coordinates (real numbers), but with a significant difference. Here the vertical axis is no longer a real axis; it becomes the imaginary axis $iy$-axis. Similarly, a point on the complex plane corresponds also uniquely to a complex number

$$z = a + bi, \tag{6.3}$$

which consists of a real part $a$ and an imaginary part $bi$. This also means that it corresponds to an ordered pair of real numbers $a$ and $b$. For a given $z = a + bi$, the real part is denoted by $a = \Re(z)$ and imaginary part is denoted by $b = \Im(z)$. The length or modulus of a complex number $z$ is defined as

$$|z| = \sqrt{a^2 + b^2}, \tag{6.4}$$

which is always non-negative (i.e., $|z| \geq 0$).

## 2. Complex Algebra

Now we can define the mathematical operations of any two complex numbers $z_1 = a + bi$ and $z_2 = c + di$. The addition of two complex numbers is carried out by adding their real parts and imaginary parts, respectively. That is

$$(a + bi) + (c + di) = (a + c) + (b + d)i. \tag{6.5}$$

Similarly, the subtraction of two complex numbers is defined by

$$(a + bi) - (c + di) = (a - c) + (b - d)i. \tag{6.6}$$

Two complex numbers are equal if, and only if, their real parts and imaginary parts are equal, respectively. That is, $z_1 = z_2$ if, and only if, $a = c$ and $b = d$.

The multiplication of two complex numbers is carried out similarly to expanding an expression, using $i^2 = -1$ when necessary

$$(a + bi) \cdot (c + di) = a \cdot (c + di) + bi \cdot (c + di)$$

$$= ac + adi + bci + bdi^2 = (ac - bd) + (bc + ad)i. \tag{6.7}$$

The division of two complex numbers

$$\frac{a + bi}{c + di} = \frac{(a + bi) \cdot (c - di)}{(c + di) \cdot (c - di)}$$

$$= \frac{ac - adi + bci - bdi^2}{c^2 - d^2 i^2} = \frac{(ac + bd)}{(c^2 + d^2)} + \frac{(bc - ad)}{(c^2 + d^2)}i, \tag{6.8}$$

where we have used the equality $(x + y)(x - y) = x^2 - y^2$.

## Example 6.2

Now let us try to find $\sqrt{i}$. Let $z = a + bi$ and $z^2 = i$, we have

$$(a + bi)^2 = (a^2 - b^2) + 2abi = i = 0 + i.$$

As two complex numbers, here $(a^2 - b^2) + 2abi$ and $0 + i$, must be equal on both sides, we have

$$a^2 - b^2 = 0, \qquad 2ab = 1.$$

The first condition gives $a = \pm b$. In case of $a = b$, the second condition gives

$$a^2 = b^2 = \frac{1}{2}, \quad \text{or} \quad a = b = \pm\frac{1}{\sqrt{2}} = \pm\frac{\sqrt{2}}{2}.$$

In case of $a = -b$, the second condition leads to $-2a^2 = 1$, which does not have any solution for any real number $a$.

Therefore, the only possibility is that $a = b$ and thus $\sqrt{i}$ has two distinct roots

$$\sqrt{i} = \frac{\sqrt{2}}{2} + \frac{\sqrt{2}}{2}i, \quad \text{and} \quad \sqrt{i} = -\frac{\sqrt{2}}{2} - \frac{\sqrt{2}}{2}i.$$

The complex conjugate $z^*$ of a complex number $z = a + bi$ is defined by changing the sign of the imaginary part

$$z^* = a - bi, \tag{6.9}$$

which is the reflection in the real axis of the original $z$ (see Fig. 6.2). The definition

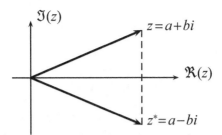

Figure 6.2: Complex conjugate and reflection.

also leads to

$$(z^*)^* = (a - bi)^* = a + bi = z. \tag{6.10}$$

It is worth pointing out that the complex conjugate is also denoted as $\bar{z} = z^*$.

## Example 6.3

For any number $z = a + bi$, we can prove that $zz^* = |z|^2 = a^2 + b^2$ because

$$zz^* = (a + bi)(a - bi) = a^2 + \underline{abi} - \underline{abi} - b^2 i^2 = a^2 - b^2(-1) = a^2 + b^2.$$

It is also straightforward to show that

$$|z|^2 = |z^2|.$$

From $|z| = \sqrt{a^2 + b^2}$, we get $|z|^2 = a^2 + b^2$.
    Since $z^2 = (a + bi)^2 = a^2 + 2abi + b^2 i^2 = (a^2 - b^2) + 2abi$, we have

$$|z^2| = \sqrt{(a^2 - b^2)^2 + (2ab)^2} = \sqrt{a^4 - 2a^2 b^2 + b^4 + 4a^2 b^2}$$

$$= \sqrt{a^4 + 2a^2 b^2 + b^4} = \sqrt{(a^2 + b^2)^2} = (a^2 + b^2),$$

which means that $|z^2| = |z^2|$.

From Fig. 6.1, we can also represent a complex number in the polar form in terms of an ordered pair $(r, \theta)$. From basic trigonometry, we know that

$$\sin(\theta) = \frac{b}{r}, \qquad \cos(\theta) = \frac{a}{r}, \qquad \text{or} \quad a = r\cos(\theta), \quad b = r\sin(\theta), \tag{6.11}$$

where $\theta$ is the argument or the angle of $z$, and $r$ is the modulus or magnitude of the complex number $z = a + bi$, which can be obtained using Pythagoras' theorem

$$r = |z| = \sqrt{a^2 + b^2}. \tag{6.12}$$

For any given $z = a + bi$, the angle $\theta$ is given by

$$\theta = \arg(z) = \tan^{-1}(\frac{b}{a}), \tag{6.13}$$

where we only use the range $-\pi < \theta \le \pi$, called the principal values. In some text-books, the principal values are defined as $0 \le \theta < 2\pi$, i.e., $[0, 2\pi)$.

Thus, the same complex number can be expressed as the polar form

$$z = a + bi = r\cos(\theta) + ir\sin(\theta). \tag{6.14}$$

Sometimes, it is also conveniently written as the exponential form

$$z = re^{i\theta} = r[\cos(\theta) + i\sin(\theta)], \tag{6.15}$$

which requires Euler's formula

$$e^{i\theta} = \cos(\theta) + i\sin(\theta). \tag{6.16}$$

The proof of this formula usually involves the expansion of infinite power series.

From the power series $e^x$ in Chapter 4, we know that

$$e^x = 1 + x + \frac{x^2}{2!} + \frac{x^3}{3!} + \dots + \frac{x^n}{n!} + \dots, \tag{6.17}$$

and we can set $x = i\theta$ so that

$$e^{i\theta} = 1 + i\theta + \frac{(i\theta)^2}{2!} - \frac{(i\theta)^3}{3!} + \frac{(i\theta)^4}{4!} + \frac{(i\theta)^5}{5!} + \dots = 1 + i\theta - \frac{\theta^2}{2!} + i\frac{\theta^3}{3!} + \frac{\theta^4}{4!} + i\frac{\theta^5}{5!} + \dots$$

$$= \left(1 - \frac{\theta^2}{2!} + \frac{\theta^4}{4!} + \dots\right) + i\left(\theta - \frac{\theta^3}{3!} + \frac{\theta^5}{5!} + \dots\right). \tag{6.18}$$

Using the power expansions

$$\sin\theta = \theta - \frac{\theta^3}{3!} + \frac{\theta^5}{5!} + \dots, \quad \cos\theta = 1 - \frac{\theta^2}{2!} + \frac{\theta^4}{4!} + \dots, \tag{6.19}$$

and substituting into the above expression of $\exp(i\theta)$, we have

$$e^{i\theta} = \cos\theta + i\sin\theta. \tag{6.20}$$

The Euler formula provides a way of analyzing harmonic motion in civil engineering using complex forms or representations because any sinusoidal oscillations $\cos(\omega t)$ can be represented by the real part

$$\cos(\omega t) = \Re\{e^{i\omega t}\} \text{ (real part)} = \Re\{\cos(\omega t) + i\sin(\omega t)\}. \tag{6.21}$$

This becomes convenient for solving differential equations using the form $Ae^{i\omega t}$ as we will see in the next chapter. If we treat $A$ as a vectorized amplitude $\tilde{A}$, we can include a phase angle $\phi$ in the harmonic motion $\cos(\omega t + \phi)$ because

$$e^{i(\omega t + \phi)} = \cos(\omega t + \phi) + i\sin(\omega t + \phi) = Ae^{i\phi}e^{i\omega t} = \tilde{A}e^{i\omega t}, \tag{6.22}$$

or

$$\cos(\omega t + \phi) = \mathcal{R}\{\cos(\omega t + \phi) + i\sin(\omega t + \phi)\} = \mathcal{R}\{\tilde{A}e^{i\omega t}\} = \mathcal{R}\{\tilde{A}\cos\omega t\}, \tag{6.23}$$

where $\tilde{A} = Ae^{i\phi}$.

An interesting extension is to replace $\theta$ by $-\theta$ in Euler's formula; we have

$$e^{-i\theta} = \cos(-\theta) + i\sin(-\theta) = \cos(\theta) - i\sin(\theta). \tag{6.24}$$

Adding this to the original formula (6.16), we have

$$e^{i\theta} + e^{-i\theta} = [\cos(\theta) + \cos(\theta)] + i[\sin(\theta) - \sin(\theta)] = 2\cos(\theta), \tag{6.25}$$

which gives

$$\cos(\theta) = \frac{e^{i\theta} + e^{-i\theta}}{2}. \tag{6.26}$$

If we follow the same procedure, but subtract these two formulas, we have

$$e^{i\theta} - e^{-i\theta} = [\cos(\theta) - \cos(\theta)] + 2\sin(\theta)i, \quad \text{or} \quad \sin(\theta) = \frac{e^{i\theta} - e^{-i\theta}}{2i}. \tag{6.27}$$

The polar form is especially convenient for multiplication, division, exponential manipulations and other mathematical manipulations. For example, the complex conjugate $z = re^{i\theta}$ is simply $z^* = re^{-i\theta}$.

For example, for two complex numbers $z_1 = r_1 e^{i\theta_1}$ and $z_2 = r_2 e^{i\theta_2}$, their product is simply

$$z_1 z_2 = r_1 e^{i\theta_1} \times r_2 e^{i\theta_2} = r_1 r_2 e^{i(\theta_1 + \theta_2)}.$$

Their ratio is

$$\frac{z_1}{z_2} = \frac{r_1 e^{i\theta_1}}{r_2 e^{i\theta_2}} = \frac{r_1}{r_2} e^{i(\theta_1 - \theta_2)}.$$

Furthermore, for $z = re^{i\theta}$, we have

$$z^n = (re^{i\theta})^n = r^n(e^{i\theta})^n = r^n[\cos(\theta) + i\sin(\theta)]^n.$$

Also using Euler's formula, we have

$$z^n = r^n(e^{i\theta})^n = r^n e^{in\theta} = r^n[\cos(n\theta) + i\sin(n\theta)].$$

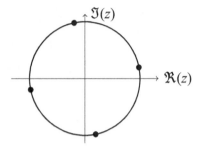

Figure 6.3: Four complex roots of $1 + i$ are uniformly distributed on the unit circle.

Combining the above two equations, we have

$$[\cos(\theta) + i\sin(\theta)]^n = \cos(n\theta) + i\sin(n\theta),$$

which is the famous de Moivre's formula. This formula can also extended to fractions.

This formula is useful for finding the $n$th root of a complex number $z$. For $z = re^{i\theta}$, we have

$$z^{1/n} = r^{1/n}e^{i\theta/n} = r^{1/n}\left[\cos\left(\frac{\theta + 2\pi k}{n}\right) + i\sin\left(\frac{\theta + 2\pi k}{n}\right)\right], \qquad (6.28)$$

where $k = 0, 1, 2, ..., n - 1$. The reason for including $k$ is based on the fact that there are $n$ distinct roots. In fact, when $r = 1$, all the points lie on a unit circle and these points are equally spaced.

## Example 6.4

In case of $z = 1 + i = \sqrt{2}e^{\pi/4}$, we have $n = 4$, $r = \sqrt{2}$ and $\theta = \pi/4$. Thus, we get

$$(1 + i)^{1/4} = 2^{1/8}e^{i\theta/4} = 2^{1/8}\left[\cos\left(\frac{\frac{\pi}{4} + 2\pi k}{4}\right) + i\sin\left(\frac{\frac{\pi}{4} + 2\pi k}{4}\right)\right], \quad k = 0, 1, 2, 3.$$

For $k = 0$, we have

$$2^{1/8}[\sin\frac{\pi}{16} + i\sin\frac{\pi}{16}] \approx 1.06955 + 0.21274i.$$

Similarly, for $k = 1, 2, 3$, we have

$$2^{1/8}[\cos\frac{9\pi}{16} + i\sin\frac{9\pi}{16}], \quad 2^{1/8}[\cos\frac{17\pi}{16} + i\sin\frac{17\pi}{16}], \quad 2^{1/8}[\cos\frac{25\pi}{16} + i\sin\frac{25\pi}{16}].$$

These four roots are show in Fig. 6.3.

One of the applications of complex numbers is the analysis of alternating current

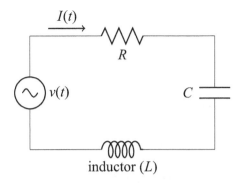

Figure 6.4: Complex impedance of an AC circuit.

(AC) circuits using the complex impedance method. In this method, a resistor is represented by $Z_R = R$, while an inductor $L$ is represented by a complex impedance $Z_L = i\omega L$. In addition, a capacitor $C$ is represented by its complex impedance $Z_C = 1/(i\omega C)$. A simple RLC circuit is shown in Fig. 6.4 under the AC source $v(t) = v_0 \cos(\omega t)$. What is the current $I(t)$ through this circuit?

As the circuit is in series, the total complex impedance is

$$Z = Z_R + Z_C + Z_L = R + \frac{1}{i\omega C} + i\omega L. \tag{6.29}$$

Since $v(t) = v_0 \cos(\omega t) = \Re\{v_0 e^{i\omega t}\}$ (real part). We can also write $I(t) = \Re\{I_Z e^{i\omega t}\}$ (real part) where $I_Z$ can be determined by

$$I_Z = \frac{v_0}{Z} = \frac{v_0}{R + i\omega L + \frac{1}{i\omega C}} = \frac{v_0}{R + i(\omega L - \frac{1}{\omega C})} = \frac{v_0}{R^2 + (\omega L - \frac{1}{\omega C})^2}\left[R - i(\omega L - \frac{1}{\omega C})\right].$$

From the above discussion in Eq. (6.23), we can also write $I(t) = A \cos(\omega t + \phi)$ with the amplitude $|I_Z|$ as

$$A = |I_Z| = \frac{v_0}{\sqrt{R^2 + (\omega L - \frac{1}{\omega C})^2}}, \tag{6.30}$$

and the phase angle

$$\phi = -\tan^{-1}\left(\frac{\omega L - \frac{1}{\omega C}}{R}\right). \tag{6.31}$$

If parameters $L$ and $C$ are adjustable, it is obvious that the maximum occurs when

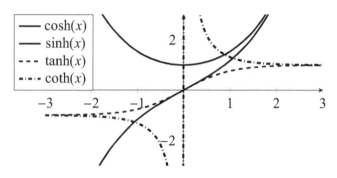

Figure 6.5: Hyperbolic functions.

$\omega L - 1/(\omega C) = 0$, which gives a resonance freqency

$$\omega_0 = \frac{1}{\sqrt{LC}}. \tag{6.32}$$

Real functions such as logarithms can be extended directly to the functions of complex numbers. For example, we know from Euler's formula when $\theta = \pi$

$$e^{i\pi} = \cos(\pi) + i\sin(\pi) = -1 + 0i = -1, \quad \text{or} \quad e^{i\pi} + 1 = 0, \tag{6.33}$$

which links all five important quantities $e$, $\pi$, $i$, 0 and 1 in a single equation.

If we use the logarithm definition, we should have

$$i\pi = \log_e(-1) = \ln(-1). \tag{6.34}$$

In the standard $\log(x)$, we usually assume that $x > 0$. Now we can see that $x < 0$ is also valid, though its logarithm is a complex number.

## 3. Hyperbolic Functions

Hyperbolic functions have many similarities to trigonometrical functions, and are commonly used in engineering and science.

The hyperbolic sine and cosine functions are defined as

$$\sinh x = \frac{e^x - e^{-x}}{2}, \quad \cosh x = \frac{e^x + e^{-x}}{2}, \tag{6.35}$$

where $x$ is a real number. If we replace $x$ by $-x$, and substitute it into the above definitions, we have

$$\sinh(-x) = \frac{e^{-x} - e^{-(-x)}}{2} = \frac{-[e^x - e^{-x}]}{2} = -\sinh x, \tag{6.36}$$

and

$$\cosh(-x) = \frac{e^{-x} + e^{-(-x)}}{2} = \frac{e^{-x} + e^x}{2} = \cosh x. \tag{6.37}$$

This suggests that $\sinh x$ is an odd function while $\cosh x$ is an even function. Their graphs are shown in Fig. 6.5. It is worth pointing out that the domain for both sinh and cosh are all the real numbers $(-\infty < x < +\infty)$.

It is easy to check that

$$\sinh(0) = \frac{e^0 - e^{-0}}{2} = \frac{1-1}{2} = 0, \quad \cosh(0) = \frac{e^0 + e^{-0}}{2} = \frac{1+1}{2} = 1. \tag{6.38}$$

Since both $\exp[x]$ and $\exp[-x]$ are always positive for any real number $x$, we can conclude that $\cosh(x)$ is always positive. In fact, $\cosh(x) \geq 1$. Therefore, the range of $\cosh(x)$ is $[1, +\infty)$, while the range of $\sinh(x)$ is the whole range of real numbers.

For a suspension cable, hung at both ends at equal heights, with a horizontal span $s$, the catenary arc length $\Lambda$, measured from the lowest point of the cable, can be calculated by

$$\Lambda = \frac{H}{q_0} \sinh\left(\frac{q_0 x}{H}\right), \tag{6.39}$$

where $q_0$ is the load per unit length horizontally and $x$ is measured horizontally from the sag (the lowest point of the cable). Here, $H$ is the horizontal component of cable tension, which is equivalent to the midspan tension

$$H = \frac{q_0 s^2}{8d}, \tag{6.40}$$

where $d$ is the sag (vertical distance from the supports). When $x = s/2$ (at supports), $\Lambda$ should be half of the total length $L$ of the cable (i.e., $\Lambda = L/2$). Thus, we have

$$\frac{L}{2} = \frac{H}{q_0} \sinh\left(\frac{q_0 s}{2H}\right) = \frac{s^2}{8d} \sinh\left(\frac{4d}{s}\right), \tag{6.41}$$

where we have used

$$\frac{s q_0}{(2H)} = \frac{q_0 s}{2[q_0 s^2/(8d)]} = \frac{4d}{s}.$$

From Taylor series of $\exp(x) = 1 + x + x^2/2! + x^3/3! + ...$, it is straightforward to verify that

$$\sinh(x) = \frac{e^x - e^{-x}}{2} = \frac{1}{2}\left[\left(1 + x + \frac{x^2}{2!} + \frac{x^3}{3!} + ...\right) - \left(1 - x + \frac{(-x)^2}{2!} + \frac{(-x)^3}{3!} + ...\right)\right]$$

$$= x + \frac{x^3}{3!} + \frac{x^5}{5!} + \frac{x^7}{7!} + \frac{x^9}{9!} + ..., \quad (x \in \mathbb{R}). \tag{6.42}$$

When $x$ is small, we can use the first two terms to approximate $\sinh(x) = x + x^3/6$. In case when the sag $d$ is very small compared to $s$ (i.e., $d/s \ll 1$), we can obtain

$$\frac{L}{2} = \frac{s^2}{8d} \sin\left(\frac{4d}{s}\right) = \frac{s^2}{8d}\left[\frac{4d}{s} + \frac{(4d)^3}{6s^3} + ...\right], \tag{6.43}$$

which gives

$$\frac{L}{2} = \frac{s}{2} + \frac{8d^2}{6s}, \quad \text{or} \quad L = s + \frac{8d^2}{3s}, \tag{6.44}$$

which is a parabolic approximation as the formula given by Hicks.[1]

Other hyperbolic functions are defined in a similar manner to the ratio of the basic hyperbolic sine and cosine functions. For example, we have the hyperbolic tangent

$$\tanh x = \frac{\sinh x}{\cosh x} = \frac{e^x - e^{-x}}{e^x + e^{-x}}, \tag{6.45}$$

and $\coth x = 1/\tanh x$, $\operatorname{sech} x = 1/\cosh x$, and $\operatorname{cosech} x = 1/\sinh x$.

When $x$ is large, these functions can have some asymptotic values. For example, for $\tanh(x)$ for any $x > 0$, we have

$$\tanh(x) = \frac{e^x - e^{-x}}{e^x + e^{-x}} = \frac{(e^x - e^{-x})}{(e^x + e^{-x})} \cdot \frac{e^{-x}}{e^{-x}} = \frac{1 - e^{-2x}}{1 + e^{-2x}}. \tag{6.46}$$

If $x$ is very large and positive, we have $e^{-x} \to 0$. So we have $\tanh(x) \approx \frac{1-0}{1+0} \approx 1$, as $x \to +\infty$. Similarly, we have $\tanh(x) \approx -1$, as $x \to -\infty$.

For example, the speed ($v$) of water waves in the ocean can be estimated by

$$v = \sqrt{\frac{g\lambda}{2\pi} \tanh\left(\frac{2\pi h}{\lambda}\right)}, \tag{6.47}$$

where $h$ is the depth of the water and $\lambda$ is the wavelength. Here $g = 9.8$ m/s$^2$ is the acceleration due to gravity.

This formula indicates that waves with longer wavelengths travel faster than waves with shorter wavelengths. Waves travel faster in deeper waters than in shallow waters. When we say deep waters, it means that the wavelength $\lambda$ is much smaller than the depth $h$ and we write this as $\lambda \ll h$. In this case, we have $2\pi h/\lambda$ becomes very large (so we write $2\pi h/\lambda \to \infty$).

[1]TG Hicks, Civil Engineering Formulas, McGraw-Hill, 2010.

Setting $x = 2\pi h/\lambda$ and using $\tanh(x) \approx 1$ if $x$ is sufficiently large, we have

$$v = \sqrt{\frac{g\lambda}{2\pi}} \tanh\left(\frac{2\pi h}{\lambda}\right) \approx \sqrt{\frac{g\lambda}{2\pi}}.$$

## Example 6.5

A tsunami is a giant water wave whose wavelength and speed are constantly changing as it travels towards the shore. In deep ocean waters, its wavelength is about 25 km to 50 km. So for $\lambda = 25$ km $= 25 \times 1000 = 25000$ m, we have

$$v = \sqrt{\frac{g\lambda}{2\pi}} = \sqrt{\frac{9.8 \times 25000}{2\pi}} \approx 197 \text{ m/s},$$

which is about 440 mph. For longer wavelength $\lambda = 50$ km, its speed is about 630 mph. Obviously, as they travel towards shores, their wavelengths will gradually reduce to the range of about 1.5 km to 5 km and their speed can also reduce to the range of about 200 mph to 25 mph, depending on the structures of the shores. However, such huge waves can reach up to 30 metres, which can cause significant damage onshore.

If we replace $x$ by $ix$ in the hyperbolic cosine function, we know from (6.26) in our earlier discussion that

$$\cosh(ix) = \frac{e^{ix} + e^{-ix}}{2} = \cos x, \quad \sinh(ix) = \frac{e^{ix} - e^{-ix}}{2} = i \sin x, \qquad (6.48)$$

where we have used $\sin x = (e^{ix} - e^{-ix})/2$ from (6.27).

The hyperbolic functions also have similar identities to their trigonometrical counterparts. For example, to prove $\cosh^2 x - \sinh^2 x = 1$, we start from

$$\cosh^2 x - \sinh^2(x) = (\cosh x + \sinh x)(\cosh x - \sinh x)$$

$$= \left(\frac{e^x + e^{-x}}{2} + \frac{e^x - e^{-x}}{2}\right)\left(\frac{e^x + e^{-x}}{2} - \frac{e^x - e^{-x}}{2}\right) = \left(\frac{2e^x}{2}\right)\left(\frac{2e^{-x}}{2}\right) = 1. \qquad (6.49)$$

This is similar to $\cos^2 x + \sin^2 x = 1$.

There is a quick way to obtain the corresponding identities from those identities for trigonometrical functions. Taking the squares of both sides of (6.48), we have

$$\cos^2 x = \cosh^2(ix), \qquad \sin^2 x = \frac{1}{i^2} \sinh^2(ix) = -\sinh^2(ix). \qquad (6.50)$$

This implies that we can replace $\cos^2 x$ by $\cosh^2 x$ and $\sin^2 x$ by $-\sinh^2 x$ in the identity $\cos^2 x + \sin^2 x = 1$; we will get the identity $\cosh^2 x - \sinh^2 x = 1$ for hyperbolic functions. That is Osborn's rule for converting identities where we only need to change the sign of any terms containing the squares of a sine function including tangent (tan) and

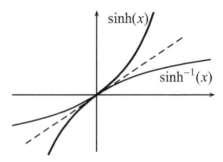

Figure 6.6: Graphs of $\sinh(x)$ and $\sinh^{-1}(x)$.

cotangent (cot). For example, we can change

$$\tan 2\theta = \frac{2\tan\theta}{1 - \tan^2\theta}, \quad \text{into} \quad \tanh 2x = \frac{2\tanh x}{1 + \tanh^2 x}. \tag{6.51}$$

In fact, we can even extend Osborn's rule further to include $\cos \to \cosh$ and $\sin \to i\sinh$. For example, from the identity $\cos(A + B) = \cos A \cos B - \sin A \sin B$, we have

$$\cosh(A + B) = \cosh A \cosh B - i\sinh A \cdot i\sinh B$$

$$= \cosh A \cosh B + \sinh A \sinh B, \tag{6.52}$$

where we have used $i^2 = -1$.

The inverse of $\sinh x$ can easily be obtained (graphically shown in Fig. 6.6) by simply reflecting the graph of $\sinh x$ in the line of $y = x$ (dashed line). Mathematically speaking, we want to find $y = \sinh^{-1} x$ such that $\sinh y = x$. From the identity $\cosh^2 y - \sinh^2 y = 1$, we have

$$\cosh^2 y = 1 + \sinh^2 y = 1 + x^2. \tag{6.53}$$

Since $\cosh y = (e^y + e^{-y})/2 \geq 1$, the above equation becomes

$$\cosh y = \sqrt{1 + x^2}. \tag{6.54}$$

In addition, from the previous example, we know that

$$\cosh y + \sinh y = \frac{e^y + e^{-y}}{2} + \frac{e^y - e^{-y}}{2} = e^y.$$

After combining the above two equations, we now have

$$\sqrt{1 + x^2} + \sinh y = e^y. \tag{6.55}$$

Using $\sinh y = x$, we have $\sqrt{1 + x^2} + x = e^y$. Taking the logarithm of both sides, we finally obtain

$$y = \sinh^{-1} x = \ln[x + \sqrt{1 + x^2}].$$    (6.56)

The inverse of other functions can be obtained in a similar manner.

In order to get $y = \cosh^{-1} x$ or $\cosh y = x$, we use

$$\cosh^2 y - \sinh^2 y = 1,$$

so that $\sinh^2 y = \cosh^2 y - 1 = x^2 - 1$, which gives $\sinh y = \sqrt{x^2 - 1}$ where $(x \geq 1)$. Again using the identity $\cosh y + \sinh y = e^y$, we have

$$x + \sinh y = x + \sqrt{x^2 - 1} = e^y.$$

Taking the logarithms, we now have

$$y = \cosh^{-1}(x) = \ln[x + \sqrt{x^2 - 1}],$$    (6.57)

where $x \geq 1$. It is worth pointing out that there are two branches of $\cosh^{-1}(x)$, and we have assumed here that $\cosh^{-1}(x) > 0$.

## 4. Analytical Functions

Any function of real variables can be extended to the function of complex variables in the same form while treating the real numbers $x$ as $x + i0$. For example, $f(x) = x^2, x \in \mathbb{R}$ becomes $f(z) = z^2, z \in \mathbb{C}$. Any complex function $f(z)$ can be written as a real part $u = \mathfrak{R}(f(z))$ and an imaginary part $v = \mathfrak{I}(f(z))$. That is

$$f(z) = f(x + iy) = \mathfrak{R}(f(z)) + i\mathfrak{I}(f(z)) = u(x, y) + iv(x, y),$$    (6.58)

where $u(x, y)$ and $v(x, y)$ are real-valued functions of two real variables: $x$ and $y$.

In some textbooks, the domain of $f(z)$ is called $z$-plane, while the mapping or a function of a point $z = x + iy$ to a point $\omega = u + iv$, and all the points $\omega = f(z) = u + iv$ form a so-called $\omega$-plane.

In many cases, the extension of real functions to complex-valued functions is relatively straightforward. Sometimes, however, care should be taken because a function of complex variables may become a multiple-valued function.

For a simple function $f(z) = z^2$, we have

$$f(z) = z^2 = (x + iy)^2 = x^2 + 2xyi + (iy)^2 = (x^2 - y^2) + 2xyi,$$

where $u(x, y) = x^2 - y^2$ and $v(x, y) = 2xy$. However, function $g(z) = \sqrt{z}$ is a multiple-valued function because two values $+\sqrt{z}$ and $-\sqrt{z}$ can correspond to the same $z$. For example, for $z = i$, we know from the discussion earlier in this chapter that

$$\sqrt{i} = \frac{\sqrt{2}}{2}(1 + i), \quad -\frac{\sqrt{2}}{2}(1 + i).$$

In this case, there are two sets of $u$ and $v$ for the same $z$.

A function $f(z)$ is called analytic at $z_0$ if $f'(z)$ exists for all $z$ in some $\epsilon$-neighbourhood of $z_0$ or differentiable in some open disk $|z - z_0| < \epsilon$, no matter how small $\epsilon > 0$ can be. If $f(z) = u + iv$ is analytic at every point in a domain $\Omega$, then $u(x,y)$ and $v(x,y)$ satisfies the Cauchy-Riemann equations

$$\frac{\partial u}{\partial x} = \frac{\partial v}{\partial y}, \qquad \frac{\partial u}{\partial y} = -\frac{\partial v}{\partial x}. \tag{6.59}$$

Conversely, if $u$ and $v$ in $f(z) = u + iv$ satisfy the Cauchy-Riemann equation at all points in a domain, then the complex function $f(z)$ is analytic in the same domain.

For example, the elementary power function $w = z^n, (n > 1)$ is analytic on the whole plane, $w = \rho e^{i\phi}$, $z = re^{i\theta}$, then

$$\rho = r^n, \quad \phi = n\theta.$$

The logarithm $\ln z$ is also an elementary function such that

$$\ln z = \ln |z| + i \arg(z) = \ln r + i(\theta + w\pi k),$$

which has an infinite number of values, due to the multiple values of $\theta$, with the difference of $2\pi i k, (k = 0, \pm 1, \pm 2, ...)$. If we use the principal argument $\text{Arg}(z)$, then we have the principal logarithm function

$$\text{Ln}(z) = \ln |z| + i\text{Arg}(z) = \ln r + i\theta.$$

For example, as $-1 = e^{i\pi} = \cos(\pi) - i \sin \pi$, we have

$$\ln(-1) = \ln 1 + i\text{Arg}(-1) = 0 + i\pi = i\pi.$$

This means that the logarithm of a negative number is a pure imaginary number.

If we differentiate the Cauchy-Riemann equations and use $\partial^2 u/\partial x \partial y = \partial^2 u/\partial y \partial x$, we have

$$\frac{\partial^2 u}{\partial x^2} = \frac{\partial^2 v}{\partial x \partial y}, \quad \frac{\partial^2 v}{\partial y^2} = -\frac{\partial^2 v}{\partial y \partial x} = -\frac{\partial^2 v}{\partial x \partial y}. \tag{6.60}$$

Adding these two, we have

$$\frac{\partial^2 u}{\partial x^2} + \frac{\partial^2 u}{\partial y^2} = 0. \tag{6.61}$$

A similar argument for $v$ leads to the following theorem: For a given analytic function $f(z) = u + iv$, both $u$ and $v$ satisfy the Laplace equations

$$\frac{\partial^2 u}{\partial x^2} + \frac{\partial^2 u}{\partial y^2} = 0, \qquad \frac{\partial^2 v}{\partial x^2} + \frac{\partial^2 v}{\partial y^2} = 0. \tag{6.62}$$

This is to say, both real and imaginary parts of an analytic function are harmonic.

For $f(z) = z^3$, we know that $z = x + iy$, and

$$z^3 = (x + iy)^3 = x^3 + 3x^2(iy) + 3x(iy)^2 + (iy)^3$$

$$= x^3 + 3x^2yi + 3x(-1y^2) + (-i)y^3 = (x^3 - 3xy^2) + (3x^2y - y^3)i,$$

thus we have

$$\omega = u + iv = (x^3 - 3xy^2) + (3x^2y - y^3)i, \quad \text{or} \quad u = x^3 - 3xy^2, \quad v = 3x^2y - y^3.$$

Let us check if they satisfy the Laplace equation.

It is easy to show that

$$\frac{\partial u}{\partial x} = 3x^2 - 3y^2, \quad \frac{\partial^2 u}{\partial x^2} = 6x, \quad \frac{\partial u}{\partial y} = -6xy, \quad \frac{\partial^2 u}{\partial y^2} = -6x,$$

and

$$\frac{\partial v}{\partial x} = 6xy, \quad \frac{\partial^2 v}{\partial x^2} = 6y, \quad \frac{\partial v}{\partial y} = 3x^2 - 3y^2, \quad \frac{\partial^2 v}{\partial y^2} = -6y.$$

Thus, we have

$$\frac{\partial^2 u}{\partial x^2} + \frac{\partial^2 u}{\partial y^2} = 6x + (-6x) = 0, \quad \frac{\partial^2 v}{\partial x^2} + \frac{\partial^2 v}{\partial y^2} = 6y + (-6y) = 0,$$

which imply that they indeed satisfy the Laplace equation. This proves that $f(z) = z^3$ is an analytical function.

## 5. Complex Integrals

For the integral of a real univariate function, the integration is between two limits $x = a$ and $x = b$. That is

$$I = \int_a^b f(x)dx, \tag{6.63}$$

whose interval of integration can be considered as a straight line (or a path) linking $x = a$ and $x = b$. For example, we know that $\int_1^2 xdx = \frac{x^2}{2}\Big|_1^2 = \frac{2^2}{2} - \frac{1^2}{2} = 1.5$. However, if we try to extend this simple integral to the complex domain, we can write

$$I = \int_{z_1}^{z_2} zdz = \int_1^{2+i} zdz, \tag{6.64}$$

where two integration limits $z = z_1$ and $z = z_2$ corresponding to two points in the plane. Now the question is how to carry out the integral and what path we should take.

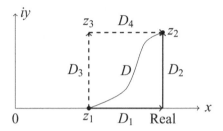

Figure 6.7: Complex integral and two integration paths from $z_1$ to $z_2$.

Suppose we naively use the same rule as we did for a real function, we may have

$$I = \int_1^{2+i} z\,dz = \frac{z^2}{2}\Big|_1^{2+i} = \frac{(2+i)^2}{2} - \frac{1^2}{2} = \frac{1}{2}[2+4i] = 1+2i. \tag{6.65}$$

Is this the right answer? Is this the right way to do it? Before we can answer these two questions, let us look at the integral again from a different perspective.

We know that $z = x + iy$, thus $dz = dx + i\,dy$. Since there are many ways to reach point $z_2 = 2 + i$ from point $z_1 = 1$, we can denote any of such paths as $D$ (see Fig. 6.7). Thus we have

$$I = \int_D z\,dz = \int_D (x+iy)(dx+i\,dy) = \int_D (x\,dx + ix\,dy + iy\,dx + i^2 y\,dy)$$

$$= \int_D (x\,dx - y\,dy) + i \int_D (y\,dx + x\,dy). \tag{6.66}$$

Now what path should we use? One simple way is to go first along a path $D_1$ from $z_1 = 1$ to $z_* = 2$ (horizontal with $dy = 0$ and $y = 0$) and then along a path $D_2$ from $z_* = 2$ to $z_2 = 2 + i$ (vertical with $dx = 0$ and $x = 2$, thus $y$ from 0 to 1). With this path, we have

$$I = \underbrace{\int_1^2 x\,dx}_{D_1} + \underbrace{\int_0^1 (-y)\,dy + i\int_0^1 2\,dy}_{D_2}$$

$$= \frac{x^2}{2}\Big|_1^2 - \frac{y^2}{2}\Big|_0^1 + 2iy\Big|_0^1 = \frac{3}{2} - \frac{1}{2}(1^2 - 0^2) + 2i(1-0) = \frac{3}{2} - \frac{1}{2} + 2i = 1+2i, \tag{6.67}$$

which is the answer $(1 + 2i)$ we obtained earlier. On the other hand, if we first go along a path $D_3$ vertically from $z_1 = 1$ to $z_3 = 1 + i$ (with $dx = 0$ and $x = 1$, $y$ from 0 to 1) and then along a path $D_4$ horizontally from $z_3 = 1 + i$ to $z_2 = 2 + i$ (with $dy = 0$ and

$y = 1$), we now have

$$I = \int_0^1 (-y)dy + i \int_0^1 1dy + \underbrace{\int_1^2 xdx + i \int_1^2 1dx}_{D_4} = -\frac{y^2}{2}\Big|_0^1 + iy\Big|_0^1 + \frac{x^2}{2}\Big|_1^2 + ix\Big|_1^2$$

$$\underbrace{\phantom{\int_0^1 (-y)dy + i \int_0^1 1dy}}_{D_3}$$

$$= -\frac{1}{2}(1^2 - 0^2) + i(1 - 0) + \frac{1}{2}(2^2 - 1^2) + i(2 - 1) = -\frac{1}{2} + i + \frac{3}{2} + i = 1 + 2i, \quad (6.68)$$

which is the same result as we obtained before. It seems that the integral value does not depend on its integral path or contour. In fact, for analytical functions such as $f(z) = z$, the integrals are path-independent, as dictated by Cauchy's integral theorem, to be introduced in the next section.

## 5.1. Cauchy's Integral Theorem

We say a path is simply closed if its end points and initial points coincide and the curve does not cross itself. For an analytic function $f(z) = u(x,y) + iv(x,y)$, the integral on a simply closed path

$$I = \int_\Gamma (u + iv)(dx + idy)] = \int_\Gamma (udx - vdy) + i \int_\Gamma (vdx + udy). \quad (6.69)$$

However, the formal proof and description of Cauchy's integral theorem requires Green's theorem, which states as follows. For a vector field $W(x,y) = P(x,y)dx + Q(x,y)dy$ that is continuously differentiable in a simply-connected domain $\Omega$ in the $z$-plane, a line integral of $W$ along a closed path $\Gamma$ satisfies

$$\oint_\Gamma \left[ P(x,y)dx + Q(x,y)dy \right] = \iint_\Omega \left( \frac{\partial Q}{\partial x} - \frac{\partial P}{\partial y} \right) dxdy, \quad (6.70)$$

which essentially links a surface integral with a line integral.

By using the Green theorem, this becomes

$$I = \iint_\Omega \left( -\frac{\partial u}{\partial y} - \frac{\partial v}{\partial x} \right) dxdy + i \iint_\Omega \left( \frac{\partial u}{\partial x} - \frac{\partial v}{\partial y} \right) dxdy. \quad (6.71)$$

From the Cauchy-Riemann equations (6.59), we know that both integrals are zero. Thus, we have Cauchy's Integral Theorem, which states that the integral of any analytic function $f(z)$ on a simply closed path $\Gamma$ in a simply connected domain $\Omega$ is zero:

$$\int_\Gamma f(z)dz = 0. \quad (6.72)$$

It is worth pointing out that the loop (or closed path) must be a positively oriented loop. Here, a positive oriented loop is a simple closed curve or path such that the curve interior is to the left when traveling along the curve. In most cases in practice,

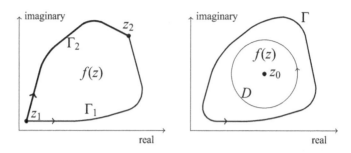

Figure 6.8: Integration paths (left) and Caucy's integral on a simple closed path (right).

we can achieve this by simply integrating anticlockwise along a chosen loop.

This theorem is very important, as it has interesting consequences. If the closed path is decomposed into two paths with reverse directions $\Gamma_1$ and $\Gamma_2$, then $\Gamma_1$ and $-\Gamma_2$ form a closed path $\Gamma = \Gamma_1 + (-\Gamma_2)$ such that $\int_\Gamma f(z)dz = 0$, which leads to

$$\int_{\Gamma_1} f(z)dz = \int_{\Gamma_2} f(z)dz. \tag{6.73}$$

That is to say that the integral over any curve between two points is independent of the path (see Fig. 6.8), as we have demonstrated in the previous example. This property becomes very useful for evaluation of integrals.

Let us introduce Cauchy's integral formula for complex integration without proof. The integral of $f(z)/(z - z_0)$ over any simply closed path $\Gamma$ enclosing a point $z_0$ in the domain $\Omega$ is

$$\frac{1}{2\pi i} \oint_\Gamma \frac{f(z)}{z - z_0} dz = f(z_0), \tag{6.74}$$

which is the basic form of the Cauchy integral formula. In addition, the same integral value can be obtained if the integration is carried out along a circle $D$ enclosed $z_0$. Therefore, we can choose any integration path as convenient and necessary.

As we can see that the integrals here are along a closed contour, such integrals are also called contour integrals.

## Example 6.6

Let us look at a simple example to calculate the contour integral of

$$I = \oint_\Gamma \frac{e^{\beta z}}{z} dz, \quad \beta > 0, \tag{6.75}$$

along a loop $\Gamma$ defined by a circle $x^2 + y^2 = 1$, we know that the integrand can be written as $e^{\beta z}/z = f(z)/(z - z_0)$ with $f(z) = e^{\beta z}$ and $z_0 = 0$. Thus, the above equation (6.74) becomes

$$\oint_{x^2+y^2=1} \frac{e^{\beta z}}{z} dz = \oint_{\Gamma} \frac{f(z)}{z - z_0} dz = 2\pi i f(z_0) = 2\pi i e^0 = 2\pi i. \qquad (6.76)$$

This integral along any other loops that enclose the origin $z = 0$ will also give the same integral value.

On the other hand, there is another way to calculate the same integral (6.75). Since $z = e^{i\theta} = \cos\theta + i\sin\theta$ on the unit circle $x^2 + y^2 = 1$ where $\theta$ varies from 0 to $2\pi$, we have $dz = d(e^{i\theta}) = ie^{i\theta}d\theta$. We have

$$I = 2\pi i = \oint_{x^2+y^2=1} \frac{e^{\beta z}}{z} dz = \int_0^{2\pi} \frac{e^{\beta(\cos\theta+i\sin\theta)}}{e^{i\theta}}(ie^{i\theta})d\theta$$

$$= i\int_0^{2\pi} e^{\beta\cos\theta} e^{i\beta\sin\theta} d\theta = i\int_0^{2\pi} e^{\beta\cos\theta}\Big[\cos(\beta\cos\theta) + i\sin(\beta\sin\theta)\Big]d\theta$$

$$= -\int_0^{2\pi} e^{\beta\cos\theta}\sin(\beta\sin\theta)d\theta + i\int_0^{2\pi} e^{\beta\cos\theta}\cos(\beta\sin\theta)d\theta, \qquad (6.77)$$

which leads to

$$\int_0^{2\pi} e^{\beta\cos\theta}\cos(\beta\sin\theta)d\theta = 2\pi. \qquad (6.78)$$

This integral may be very difficult to evaluate using other methods. This clearly demonstrates that complex integrals can sometimes provide an efficient way to calculate complicated integrals.

For higher derivatives, we have the generalized Cauchy integral formula:

$$\oint_{\Gamma} \frac{f(z)}{(z - z_0)^{n+1}} dz = \frac{2\pi i f^{(n)}(z_0)}{n!}, \qquad n > 0. \qquad (6.79)$$

## 5.2. Residue Theorem

A function $f(z)$ has a pole or singularity of order $n$ at $z = z_0$ if $f(z)$ is not analytic at $z = z_0$ but $(z - z_0)^n f(z)$ is analytic at $z = z_0$. This suggests that $f(z)$ can be expanded as a power series, called Laurant series

$$f(z) = \sum_{k=-n}^{\infty} \alpha_k(z - z_0)^k, \qquad (6.80)$$

where $\alpha_k$ are the coefficients. The most important coefficient is probably $\alpha_{-1}$ which is called the residue of $f(z)$ at the pole $z = z_0$. If $f(z)$ has a pole of order $N$ at $z_0$, the following formula gives a quick way to calculate the residue

$$\text{Res}\,f(z)|_{z_0} = \frac{1}{(N-1)!} \lim_{z \to z_0} \frac{d^{N-1}[(z-z_0)^N f(z)]}{dz^{N-1}}. \tag{6.81}$$

For any analytic $f(z)$ function in a domain $\Omega$ except for isolated singularities at finite points $z_1, z_2, ..., z_N$, the residue theorem states

$$\oint_\Gamma f(z)dz = 2\pi i \sum_{k=1}^N \text{Res}\,f(z)|_{z_k} = 2\pi i \Big[\text{Res}\,f(z_1) + \text{Res}\,f(z_2) + ... + \text{Res}\,f(z_N)\Big] \tag{6.82}$$

where $\Gamma$ is a simple closed path enclosing all these isolated points or poles.

Let us calculate the contour integral on a unit circle

$$I = \oint_{|z|=1} \frac{e^{3z}}{z^7} dz. \tag{6.83}$$

Since $n = 6$, $f(z) = e^{3z}$ and $z_0 = 0$, we have

$$I = \oint_{x^2+y^2=1} \frac{e^{3z}}{z^7} dz = \frac{2\pi i}{6!} f''''''(z_0) = \frac{2\pi i}{720} \cdot 3^6 e^{3z_0} = \frac{1458\pi i}{720} e^0 = \frac{81\pi i}{40}. \tag{6.84}$$

Complex numbers have a wide range of applications as many mathematical techniques become simpler and more powerful using complex numbers. For example, in the discrete Fourier series to be introduced later in this book, formulas are simpler when written in the form of complex numbers.

## Exercises

**6.1.** Simplify the following expressions:
- $i^{1001} + i^{100} + i^{10} + i + 1$
- $e^{\pi i} + e^{-i\pi} + e^{-2\pi i}$
- $\sinh(i\pi) - \cosh(i\pi) - \sinh(0)$
- $\cos(\pi) + i\sin(\pi) - i^{10}$
- $(1 + 2i)(3 + 4i)/(5 + 6i)$
- $|(3 + 4i)(1 - 2i)(2 + 4i)/(6 - 8i)|$

**6.2.** Compute $\ln(1 + i)$ and $\ln i$.

**6.3.** Show that the inverse of the tanh function is $\tanh^{-1}(x) = \frac{1}{2}\ln\left(\frac{1+x}{1-x}\right)$.

**6.4.** Calculate the following contour integral

$$I = \oint_D \frac{e^z}{(z-\pi)^3} dz,$$

where $D$ is a square path with four corners at $(\pm 10, \pm 10)$.

# CHAPTER 7

# Ordinary Differential Equations

Contents

## Key Points

- Introduce ordinary differential equations (ODEs) with an emphasis on linear ODEs.
- Discuss both first-order ODEs and second-order ODEs, together with their relevant solution techniques.
- Model harmonic motion using second-order ODEs.

Differential equations are a very useful tool in sciences and engineering. In fact, many processes in engineering can be modelled using differential equations. This chapter introduces all the fundamentals of ordinary differential equations.

## 1. Differential Equations

In the introduction of basic equations such as $x^3 - x = 0$, we know that the relationship is a function $f(x) = x^3 - x$ and the only unknown is $x$. The aim is to find values of $x$ which satisfy $f(x) = 0$. It is easy to verify that the equation has three solutions $x = 0, \pm 1$.

A differential equation, on the other hand, is a relationship that contains an unknown function and its derivatives. For example, the following equation

$$\frac{dy}{dx} = x^3 - x,$$ (7.1)

is a differential equation because it provides a relationship between the derivative $dy/dx$ and the function $f(x) = x^3 - x$. The unknown is a function $y(x)$ and the aim is to find a function $y(x)$ (not a simple value) which satisfies the above equation. Here $x$ is the independent variable.

From integration, we know that the gradient $dy/dx$ is $x^3 - x$, and the function $\frac{1}{4}x^4 - \frac{1}{2}x^2$ has the gradient $x^3 - x$. We can say that $y(x) = \frac{x^4}{4} - \frac{x^2}{2}$ satisfies the differential equation (7.1), and thus $y(x)$ is a solution to (7.1). Then the question is: Are there any other solutions to this equation? The answer is an infinite number. We know that there is a family of curves whose gradient is $x^3 - x$. The solution in general should contain an arbitrary constant $C$. That is to say, the general solution of (7.1) can be written as

$$y(x) = \frac{x^4}{4} - \frac{x^2}{2} + C.$$ (7.2)

Any solution that corresponds to a single specific curve is a particular solution. For example, both $x^4/4 - x^2/2$ and $x^4/4 - x^2/2 - 1$ are particular solutions.

Another important concept is the order of a differential equation. The order of a differential equation is the highest derivative of the unknown. For example, the order of (7.1) is 1 as the highest derivative is the gradient, so this equation is called a first-order differential equation. The following equation

$$\frac{d^2y(x)}{dx^2} - 2x\frac{dy(x)}{dx} + y(x) = x^2,$$ (7.3)

is the second-order differential equation as the highest derivative is the second derivative $d^2y/dx^2$.

All the above equations only contain first and/or second derivatives, and there is only a single independent variable $x$. Such differential equations are called ordinary differential equations (ODE).

## 2. First-Order Differential Equations

We know that $dy/dx = x^3 - x$ is a first-order ordinary differential equation, which can be generalized as

$$\frac{dy(x)}{dx} = f(x),$$ (7.4)

where $f(x)$ is a given function of $x$. Its solution can be obtained by simple integration

$$y(x) = \int f(x)dx + C. \tag{7.5}$$

However, this is a very special case of first-order ODEs.

The integration constant $C$ can be determined by an initial condition at $x = 0$. This initial condition often takes the following form:

$$y(x = 0) = y_0, \quad \text{or} \quad y(0) = y_0, \tag{7.6}$$

where $y_0$ is a known constant.

## Example 7.1

The discharge of a capacitor or a battery can be approximated by

$$\frac{dV(t)}{dt} = -rV(t),$$

where $r > 0$ is a constant. Suppose the initial voltage is $V_0 = V(t = 0) = 5$ volts with $r = 0.1$, what is the discharge curve $V(t)$?

We can re-arrange the above equation

$$\frac{1}{V(t)}\frac{dV(t)}{dt} = -r, \quad \text{or} \quad \frac{d\ln V(t)}{dt} = -r$$

Integrating with respect to $t$, we have

$$\ln V(t) = -rt + C,$$

where $C$ is an integration constant. Taking exponential of both sides, we have

$$e^{\ln V(t)} = V(t) = e^{-rt+C} = Ae^{-rt},$$

where $A = e^C$ is a constant.

From the initial condition $V(0) = V_0 = 5$ at $t = 0$, we have

$$V(t = 0) = V_0 = 5 = Ae^{-r\times 0},$$

which gives $A = 5$. Since $r = 0.1$, the final solution is

$$V(t) = 5e^{-rt} = 5e^{-0.1t}.$$

In general, the first-order ordinary differential equation can be written as

$$p(x)\frac{dy(x)}{dx} + q(x)y(x) = f(x), \tag{7.7}$$

or

$$p(x)y'(x) + q(x)y(x) = f(x). \tag{7.8}$$

It is obvious that $p(x) \neq 0$; otherwise, the above equation degenerates into an algebraic equation. So we divide both sides by $p(x)$, we have

$$y'(x) + a(x)y(x) = b(x), \quad a(x) = \frac{q(x)}{p(x)}, \quad b(x) = \frac{f(x)}{p(x)}. \tag{7.9}$$

Therefore, a first-order linear differential equation can generally be written as

$$y' + a(x)y = b(x), \tag{7.10}$$

where $a(x)$ and $b(x)$ are known functions of $x$.

Multiplying both sides of the equation by $\exp[\int a(x)dx]$, called the integrating factor, we have

$$y'e^{\int a(x)dx} + a(x)ye^{\int a(x)dx} = b(x)e^{\int a(x)dx}, \tag{7.11}$$

which can be written as

$$\left[ye^{\int a(x)dx}\right]' = b(x)e^{\int a(x)dx}. \tag{7.12}$$

By simple integration, we have

$$ye^{\int a(x)dx} = \int b(x)e^{\int a(x)dx}dx + C. \tag{7.13}$$

After dividing the integrating factor $\exp[\int a(x)dx]$, the final solution becomes

$$y(x) = e^{-\int a(x)dx} \int b(x)e^{\int a(x)dx}dx + Ce^{-\int a(x)dx}, \tag{7.14}$$

where $C$ is an integration constant to be determined by an initial condition at $x = 0$.

Let us use the above solution to study the cooling process of a hot object. Newton's law of cooling states that the rate of cooling is proportional to the difference between the temperature $T$ of the object and the ambient temperate $T_a$. That is

$$\frac{dT}{dt} = k(T_a = T) = -k(T - T_a), \tag{7.15}$$

where $t$ is time and $k > 0$ is a constant, depending on the material properties, geometry and size of the cooling object.

Writing (7.15) in a standard form as (7.10)

$$\frac{dT}{dt} + kT = kT_a, \tag{7.16}$$

we have

$$a = k, \quad b = kT_a. \tag{7.17}$$

Thus, from (7.14), we can write its solution as

$$T(t) = e^{-\int k dt} \int (kT_a)e^{\int k dt} dt + Ce^{-\int k dt} = e^{-kt} \int (kT_a)e^{kt} dt + Ce^{-kt}$$

$$= e^{-kt}\left[(kT_a)\frac{1}{k}e^{kt}\right] + Ce^{-kt} = T_a + Ce^{-kt}, \tag{7.18}$$

where $C$ is a constant to be determined by an initial condition. At $t = 0$, the initial temperature $T(t = 0) = T_0$, so we have

$$T_0 = T_a + Ce^{-k\times 0} = T_a + C, \tag{7.19}$$

which gives $C = T_0 - T_a$. Finally, the solution becomes

$$T(t) = T_a + (T_0 - T_a)e^{-kt}. \tag{7.20}$$

Obviously, when $t \to \infty$, $T(t) \to T_a$ as expected.

## Example 7.2

A cup of hot tea with an initial temperature $T = 80°C$ is left on a table with a room temperature $T_a = 20°C$. After 5 minutes, the temperature of the tea reduces to $65°C$, what is the temperature after 15 minutes? How long does it take to cool to $30°C$?
  Since $T_0 = 80$, $T_a = 20$, we have

$$T(t) = 20 + (80 - 20)e^{-kt} = 20 + 60e^{-kt}.$$

After $t = 5$ minutes, we have $T(5) = 65$, which leads to

$$65 = 20 + 60e^{-5k}, \quad \text{or} \quad \frac{45}{60} = e^{-5k}.$$

Taking the logarithm, we have

$$\ln(45/60) = -5k,$$

which gives

$$k = -\frac{\ln 0.75}{5} \approx 0.0575.$$

So after $t = 15$ minutes, the temperature becomes

$$T(15) = 20 + 60e^{-0.0575\times 15} \approx 45°C.$$

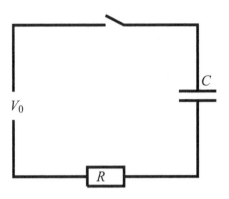

Figure 7.1: A simple RC circuit.

The time $t_*$ taken to cool to 30°C is

$$30 = 20 + 60e^{-0.0575t_*}, \quad \text{or} \quad \frac{10}{60} = e^{-0.0575t_*},$$

which means

$$t_* = \frac{\ln(1/6)}{-0.0575} \approx 31 \text{ minutes.}$$

Now let us use first-order differential equations to model a real system. The resistor-capacity (RC) circuit shown in Fig. 7.1 consists of a resistor (R) and a capacitor (C), connected to a voltage source $V_0$.

Let $V(t)$ be the voltages across the capacitor; we have

$$V(t) + V_R = V_0,$$

where $V_R$ is the voltage across the resistor. From the basic linear circuit theory, we know that the current through the circuit (when the switch is on) can be given by

$$i = C\frac{dV(t)}{dt}, \quad V_R = iR.$$

So we have

$$V + iR = V + RC\frac{dV}{dt} = V_0, \quad \text{or} \quad RC\frac{dV}{dt} + V = V_0.$$

This is essentially a first-order differential equation

$$\tau V'(t) + V(t) = V_0, \quad \tau = RC.$$

Here, $\tau$ is the time constant of the RC circuit.

## Example 7.3

By comparing with the general form of ODE (7.10), we have $a = 1/\tau$ and $b = V_0/\tau$ (constant). Since the independent variable is now $t$, the general solution becomes

$$y(t) = e^{-\int \frac{1}{\tau} dt} \int \frac{V_0}{\tau} e^{\int \frac{1}{\tau} dt} dt + C e^{-\int \frac{1}{\tau} dt}$$

$$= \frac{V_0}{\tau} e^{-t/\tau} \int e^{t/\tau} dt + C e^{-t/\tau} = \frac{V_0}{\tau} e^{-t/\tau} \left[\tau e^{t/\tau}\right] + C e^{-t/\tau} = V_0 + C e^{-t/\tau}.$$

Before the switch is on, the voltage across the capacitor is zero. That is, $V(t = 0) = 0$. Applying this initial condition to the above equation at $t = 0$, we have

$$y(0) = 0 = V_0 + C e^{-0/\tau} = V_0 + C,$$

which gives $C = -V_0$. So the final solution for the voltage variation is

$$V(t) = V_0 - V_0 e^{-t/\tau} = V_0(1 - e^{-t/\tau}).$$

We can see that $V(t) \to V_0$ as $t \to \infty$, which means that this capacitor is fully charged.

In Eq. (7.10), if $a(x)$ is a constant, the equation becomes a linear ordinary differential equation with constant coefficients.

## 3. Second-Order Equations

For second-order ordinary differential equations (ODEs), it is generally more tricky to find their general solutions. However, a special case with significantly practical importance and mathematical simplicity is the second-order linear differential equation with constant coefficients in the following form

$$\frac{d^2y}{dx^2} + b\frac{dy}{dx} + cy(x) = f(x), \tag{7.21}$$

where the coefficients $b$ and $c$ are constants, and $f(x)$ is a known function of $x$. Obviously, the more general form is

$$a\frac{d^2y}{dx^2} + b\frac{dy}{dx} + cy(x) = f(x). \tag{7.22}$$

However, if we divide both sides by $a$, we will reach our standard form. Here we assume $a \neq 0$. In a special case of $a = 0$, it reduces to a first-order linear differential equation, which has been discussed in the previous section. So we will start our discussion from (7.21).

A differential equation is said to be homogeneous if $f(x) = 0$. For a given generic second-order differential equation (7.21), a function that satisfies the homogeneous equation

$$\frac{d^2y}{dx^2} + b\frac{dy}{dx} + cy(x) = 0, \tag{7.23}$$

is called the complementary function, denoted by $y_c(x)$. Obviously, the complementary function $y_c$ alone cannot satisfy the original equation (7.21) because there is no way to produce the required $f(x)$ on the right-hand side. Therefore, we have to find a specific function, $y_*(x)$ called the particular integral, so that it indeed satisfies the original equation (7.21). The combined general solution

$$y(x) = y_c(x) + y_*(x), \tag{7.24}$$

will automatically satisfy the original equation (7.21). The general solution of (7.21) consists of two parts: the complementary function $y_c(x)$ and the particular integral $y_*(x)$. We can obtain these two parts separately, and simply add them together because the original equation is linear, so their solutions are linear combinations.

## 3.1. Solution Technique

First things first, how to obtain the complementary function? The general technique is to assume that it takes the form

$$y_c(x) = Ae^{\lambda x}, \tag{7.25}$$

where $A$ is a constant, and $\lambda$ is an exponent to be determined. Substituting this assumed form into the homogeneous equation and using both $y'_c = A\lambda e^{\lambda x}$ and $y''_c = A\lambda^2 e^{\lambda x}$, we have

$$A\lambda^2 e^{\lambda x} + bA\lambda e^{\lambda x} + cAe^{\lambda x} = 0. \tag{7.26}$$

Since $Ae^{\lambda x}$ should not be zero (otherwise, we have a trivial solution $y_c = 0$ everywhere), we can divide all the terms by $Ae^{\lambda x}$, and we have

$$\lambda^2 + b\lambda + c = 0, \tag{7.27}$$

which is the characteristic equation for the homogeneous equation. It is also called the auxiliary equation of the ODE. The solution of $\lambda$ in this case is simply

$$\lambda = \frac{-b \pm \sqrt{b^2 - 4c}}{2}. \tag{7.28}$$

For simplicity, we can take $A = 1$ as it does not affect the results.

From quadratic equations, we know that there are three possibilities for $\lambda$. They are: I) two real distinct roots, II) two identical roots, and III) two complex roots.

In the case of two different roots: $\lambda_1 \neq \lambda_2$. Then, both $e^{\lambda_1 x}$ and $e^{\lambda_2 x}$ satisfy the homogeneous equation, so their linear combination forms the complementary function

$$y_c(x) = Ae^{\lambda_1 x} + Be^{\lambda_2 x}, \tag{7.29}$$

where $A$ and $B$ are constants.

In the special case of identical roots $\lambda_1 = \lambda_2$, or

$$c = \lambda_1^2, \qquad b = -2\lambda_1. \tag{7.30}$$

we cannot simply write

$$y_c(x) = Ae^{\lambda_1 x} + Be^{\lambda_1 x} = (A + B)e^{\lambda_1 x}, \tag{7.31}$$

because it is still only one part of the complementary function $y_1 = Ce^{\lambda_1}$ where $C = A + B$ is just another constant. In this case, we should try a different combination, say, $y_2 = xe^{\lambda_1 x}$ to see if it satisfies the homogeneous equation or not. Since $y_2'(x) = e^{\lambda_1 x} + x\lambda_1 e^{\lambda_1 x}$, and $y_2''(x) = \lambda_1 e^{\lambda_1 x} + \lambda_1 e^{\lambda_1 x} + x\lambda_1^2 e^{\lambda_1 x}$, we have

$$y_2''(x) + by_2'(x) + cy_2(x)$$

$$= e^{\lambda_1 x}(2\lambda_1 + x\lambda_1^2)e^{\lambda_1 x} + be^{\lambda_1 x}(1 + x\lambda_1) + cxe^{\lambda_1 x}$$

$$= e^{\lambda_1 x}[(2\lambda_1 + b)] + xe^{\lambda_1 x}[\lambda_1^2 + b\lambda_1 + c] = 0, \tag{7.32}$$

where we have used $b + 2\lambda_1 = 0$ (identical roots) and $\lambda_1^2 + b\lambda_1 + c = 0$ (the auxiliary equation). This indeed implies that $xe^{\lambda_1 x}$ also satisfies the homogeneous equation. Therefore, the complementary function for the identical roots is

$$y_c(x) = Ae^{\lambda_1 x} + Bxe^{\lambda_1 x} = (A + Bx)e^{\lambda_1 x}. \tag{7.33}$$

Now let us use this technique to solve a second-order homogeneous ODE.

## Example 7.4

The second-order homogeneous equation

$$\frac{d^2 y}{dx^2} + \frac{dy}{dx} - 6y = 0,$$

has a corresponding auxiliary equation

$$\lambda^2 + \lambda - 6 = (\lambda - 2)(\lambda + 3) = 0.$$

It has two real roots $\lambda_1 = 2$, and $\lambda_2 = -3$. So the complementary function is

$$y_c(x) = Ae^{2x} + Be^{-3x}.$$

But for the differential equation

$$\frac{d^2y}{dx^2} + 6\frac{dy}{dx} + 9y = 0,$$

its auxiliary equation becomes

$$\lambda^2 + 6\lambda + 9 = 0,$$

which has two identical roots $\lambda_1 = \lambda_2 = -3$. The complementary function in this case can be written as

$$y_c(x) = (A + Bx)e^{-3x}.$$

As complex roots always come in pairs, the case of complex roots would give

$$\lambda_{1,2} = \alpha \pm i\beta, \tag{7.34}$$

where $\alpha$ and $\beta$ are real numbers. The complementary function becomes

$$y_c(x) = Ae^{(\alpha+i\beta)x} + Be^{(\alpha-i\beta)x} = Ae^{\alpha x}e^{i\beta x} + Be^{\alpha x}e^{-i\beta x} = e^{\alpha x}[Ae^{i\beta x} + Be^{-i\beta x}]$$

$$= e^{\alpha x}\{A[\cos(\beta x) + i\sin(\beta x)] + B[\cos(-\beta x) + i\sin(-\beta x)]\}$$

$$= e^{\alpha x}[(A + B)\cos(\beta x) + i(A - B)\sin(\beta x)] = e^{\alpha x}[C\cos\beta x + D\sin\beta x], \tag{7.35}$$

where we have used the Euler formula $e^{\theta i} = \cos\theta + i\sin\theta$ and also absorbed the constants $A$ and $B$ into $C = A + B$ and $D = (A - B)i$.

A special case is when $\alpha = 0$, so the roots are purely imaginary. We have $b = 0$, and $c = \beta^2$. Equation (7.21) in this case becomes

$$\frac{d^2y}{dx^2} + \beta^2 y = 0, \tag{7.36}$$

which is a differential equation for harmonic motions such as the oscillations of a pendulum or a small-amplitude seismic detector. Here $\beta$ is the angular frequency of the system.

For a simple pendulum of mass $m$ shown in Fig. 7.2, we now try to derive its equation of oscillations and its period.

Since the motion is circular, the tension or the centripetal force $T$ is thus given by

$$T = m\frac{v^2}{L} = m\dot{\theta}^2 L, \tag{7.37}$$

where $\dot{\theta} = d\theta/dt$ is the angular velocity and $v = \dot{\theta}L$ is the linear velocity.

Forces must be balanced both vertically and horizontally. The component of $T$ in

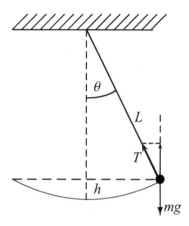

Figure 7.2: A simple pendulum and its harmonic motion.

the vertical direction is $T \cos \theta$ which must be equivalent to $mg$, though in the opposite direction. Here $g$ is the acceleration due to gravity. That is

$$T \cos \theta = mg.$$

Since $\theta$ is small or $\theta \ll 1$, we have $\cos \theta \approx 1$. This means that $T \approx mg$.

In the horizontal direction, Newton's second law $F = ma$ implies that the horizontal force $-T \sin \theta$ must be equal to the mass $m$ times the acceleration $L\frac{d^2\theta}{dt^2}$. Now we have

$$m(L\frac{d^2\theta}{dt^2}) = -T \sin \theta \approx -mg \sin \theta. \tag{7.38}$$

Dividing both sides by $mL$, we have

$$\frac{d^2\theta}{dt^2} + \frac{g}{L} \sin \theta = 0.$$

Since $\theta$ is small, we have $\sin \theta \approx \theta$. Therefore, we finally have

$$\frac{d^2\theta}{dt^2} + \frac{g}{L}\theta = 0. \tag{7.39}$$

This is the equation of motion for a simple pendulum. From equation (7.36), we know that the angular frequency is $\omega^2 = g/L$ or $\omega = \sqrt{g/L}$. Thus the period of the pendulum is

$$T = \frac{2\pi}{\omega} = 2\pi \sqrt{\frac{L}{g}}. \tag{7.40}$$

## Example 7.5

We can see that the period is independent of the bob mass. For $L = 1$ m and $g = 9.8$ m/s$^2$, the period is approximately [from Eq. (7.40)]

$$T = 2\pi \sqrt{\frac{1}{9.8}} \approx 2,$$

which is about 2 seconds. As we can see here, to double the period $T$ requires the quadruple of the length $L$.

Up to now, we have found only the complementary function. Now we will try to find the particular integral $y_*(x)$ for the original non-homogeneous equation (7.21). For particular integrals, we do not intend to find the general form; any specific function or integral that satisfies the original equation (7.21) will do. Before we can determine the particular integral, we have to use some trial functions, and such functions will have strong similarity to function $f(x)$.

For example, if $f(x)$ is a polynomial such as $\alpha x + \beta$; we will try a similar form $y_*(x) = ax + b$ and try to determine the coefficients. Let us demonstrate this by an example.

## Example 7.6

In order to solve the differential equation

$$\frac{d^2y}{dx^2} + \frac{dy}{dx} - 6y = x - 2,$$

we first find its complementary function. From the earlier example, we know that the complementary function can be written as

$$y_c(x) = Ae^{2x} + Be^{-3x}.$$

For the particular integral, we know that $f(x) = x - 2$, so we try the form

$$y_* = ax + b.$$

Thus $y_*' = a$ and $y_*'' = 0$. Substituting them into the original equation, we have

$$0 + a - 6(ax + b) = x - 2,$$

or

$$(-6a)x + (a - 6b) = x - 2.$$

As this equality must be true for any $x$, so the coefficients of the same power of $x$ on both sides of the equation should be equal. That is

$$-6a = 1, \qquad (a - 6b) = -2,$$

which gives $a = -\frac{1}{6}$, and $b = \frac{11}{36}$. So the general solution becomes

$$y(x) = Ae^{2x} + Be^{-3x} - \frac{x}{6} + \frac{11}{36}.$$

Similarly, if $f(x) = e^{\alpha x}$, we will try to $y_*(x) = ae^{\alpha x}$ so as to determine $a$. In addition, for $f(x) = \sin \alpha x$ or $\cos \alpha x$, we will attempt the general form $y_*(x) = a \cos \alpha x + b \sin \alpha x$. Again let us demonstrate this through an example.

## Example 7.7

Let us find the motion of a damped pendulum governed by

$$\frac{d^2y}{dt^2} + 4\frac{dy}{dt} + 5y = 40 \cos(3t),$$

with the initial conditions $y(0) = 0$ and $y'(0) = 0$. Its auxiliary equation becomes

$$\lambda^2 + 4\lambda + 5 = 0,$$

which has two complex solutions

$$\lambda = \frac{-4 \pm \sqrt{4^2 - 4 \times 5}}{2} = -2 \pm i.$$

Therefore, the complementary function becomes

$$y_c(x) = e^{-2t}(A \cos t + B \sin t),$$

where $A$ and $B$ are two undetermined constants.

Now we try to get the particular integral. Since $f(t) = 40 \cos(3t)$, so we try the similar form

$$y_* = C \cos(3t) + D \sin(3t).$$

Since

$$y'_* = -3C \sin(3t) + 3D \cos(3t), \quad y''_* = -9C \cos(3t) - 9D \sin(3t),$$

we have

$$-9[C \cos(3t) + D \sin(3t)] + 4[-3C \sin(3t) + 3D \cos(3t)] + 5[C \cos(3t) + D \sin(3t)] = 40 \cos(3t),$$

which leads to

$$(-9C + 12D + 5C)\cos(3t) + (-9D - 12C + 5D)\sin(3t) = 40\cos(3t).$$

By equating the coefficients for $\cos(3t)$ and $\sin(3t)$ on both sides, we have

$$-9C + 12D + 5C = 40, \quad -9D - 12C + 5D = 0,$$

which gives $C = -1$ and $D = 3$. The particular solution is

$$y_* = 3\sin(3t) - \cos(3t).$$

Therefore, the general solution becomes

$$y = e^{-2t}(A\cos t + B\sin t) + 3\sin(3t) - \cos(3t).$$

From the initial condition $y(0) = 0$, we have

$$y(0) = e^{-2 \times 0}(A\cos 0 + B\sin 0) + 3\sin(3 \times 0) - \cos(3 \times 0),$$

which gives $A = 1$. Using the first derivative

$$y'(t) = e^{-2t}[(B - 2A)\cos t - (A + 2B)\sin t] + 9\cos(3t) + 3\sin(3t),$$

and $y'(0) = 0$, we have

$$y'(0) = 0 = e^{-2 \times 0}[(B - 2)\cos 0 - (1 + 2B)\sin 0] + 9\cos(3 \times 0) + 3\sin(3 \times 0),$$

which gives $B = -7$. Therefore, the final solution becomes

$$y(t) = e^{-2t}(\cos t - 7\sin t) + 3\sin(3t) - \cos(3t). \tag{7.41}$$

This solution is shown in Fig. 7.3.

As time is sufficiently long (i.e., $t \to \infty$), we can see that the first two terms will approach zero (since $e^{-2t} \to 0$). Therefore, the solution will be dominated by the last two terms. That is, the long-term behaviour becomes

$$y(t \to \infty) = y_\infty = 3\sin(3t) - \cos(3t),$$

which is also shown in Fig. 7.3.

In an earlier chapter, we have shown that the following identity about the sum of two angles $A$ and $B$:

$$\cos(A + B) = \cos A \cos B - \sin A \sin B. \tag{7.42}$$

Multiplying both sides by a scale factor $R > 0$, we have

$$R\cos(A + B) = R\cos A \cos B - R\sin A \sin B. \tag{7.43}$$

If we set $A = \theta$ and $B = \phi$, we can write it

$$a\cos\theta + b\sin\theta = R\cos(\theta + \phi),$$

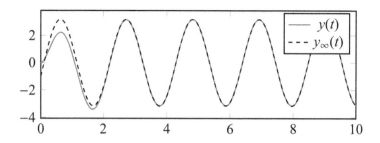

Figure 7.3: Solution with oscillations under a forcing term.

which gives

$$a = R\cos\phi, \quad b = -R\sin\phi. \tag{7.44}$$

Thus, we get

$$a^2 + b^2 = (R\cos\phi)^2 + (-R\sin\phi)^2 = R^2(\cos^2\phi + \sin^2\phi) = R^2, \tag{7.45}$$

where we have used $\cos^2\phi + \sin^2\phi = 1$. In addition, we have

$$\frac{b}{a} = \frac{-R\sin\phi}{R\cos\phi} = -\frac{\sin\phi}{\cos\phi} = -\tan\phi. \tag{7.46}$$

Therefore, we finally have

$$R = \sqrt{a^2 + b^2}, \quad \tan\phi = -\frac{b}{a}. \tag{7.47}$$

Using the above results, we have

$$y_* = 3\sin(3t) - \cos(3t) = -[\cos(3t) - 3\sin(3t)]$$

$$= -R\cos(t + \phi) = -\sqrt{10}\cos\left(3t + \tan^{-1}(3)\right), \tag{7.48}$$

where $R = \sqrt{(-1)^2 + 3^2} = \sqrt{10}$ and $\tan\phi = 3$.

## 3.2. Sturm-Liouville Eigenvalue Problem

One of the commonly used second-order ordinary differential equations is the Sturm-Liouville equation in the interval $x \in [a, b]$

$$\frac{d}{dx}[p(x)\frac{dy}{dx}] + q(x)y + \lambda r(x)y = 0, \tag{7.49}$$

with the boundary conditions

$$y(a) + \alpha y'(a) = 0, \qquad y(b) + \beta y'(b) = 0, \tag{7.50}$$

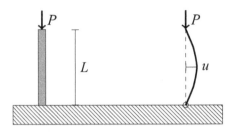

Figure 7.4: Euler buckling load $P$ of an elastic column.

where the known function $p(x)$ is differentiable, and the known functions $q(x), r(x)$ are continuous. The parameter $\lambda$ to be determined can only take certain values $\lambda_n$, called the eigenvalues, if the problem has solutions. For the obvious reason, this problem is called the Sturm-Liouville eigenvalue problem.

For each eigenvalue $\lambda_n$, there is a corresponding solution $\psi_{\lambda_n}$, called eigenfunctions. The Sturm-Liouville theory states that for two different eigenvalues $\lambda_m \neq \lambda_n$, their eigenfunctions are orthogonal. That is

$$\int_a^b \psi_{\lambda_m}(x)\psi_{\lambda_n}(x)r(x)dx = 0. \qquad (7.51)$$

or more generally

$$\int_a^b \psi_{\lambda_m}(x)\psi_{\lambda_n}(x)r(x)dx = \delta_{mn}. \qquad (7.52)$$

It is possible to arrange the eigenvalues in an increasing order

$$\lambda_1 < \lambda_2 < .... < \lambda_n < +\infty. \qquad (7.53)$$

As an example, let us look at the buckling of an Euler column, which is essentially an elastic rod with one pin-jointed end and the applied axial load $P$ at the other end (see Fig. 7.4). The column has a length of $L$. Its Young's modulus is $E$ and its second moment of area is $I = \int y^2 dA$, which is constant for a given geometry.

## Example 7.8

Let $u(x)$ be the transverse displacement; the Euler beam theory gives the following governing equation

$$M = -EI\frac{d^2u}{dx^2},$$

where $M$ is the bending moment. This moment must be balanced by the $Pu$ when the beam is in static balance. That is $M + Pu = 0$, or

$$EI\frac{d^2u}{dx^2} + Pu = 0,$$

which can be written as

$$\frac{d^2u}{dx^2} + \lambda^2 u = 0, \qquad \lambda^2 = \frac{P}{EI}.$$

This is essentially an eigenvalue problem with the eigenvalue $\alpha$ to be determined. Its general solution is

$$u = A \sin \lambda x + B \cos \lambda x. \tag{7.54}$$

Applying the boundary conditions, we have at the fixed end $u = 0$ at $x = 0$, which means that $B = 0$. At the other free end

$$u = 0, \qquad \text{(at } x = L), \qquad \text{so that} \quad A \sin(\lambda L) = 0. \tag{7.55}$$

Thus we have two kinds of solutions, either $A = 0$ or $\sin(\alpha L) = 0$. For $A = 0$, we have $u(x) = 0$ which is a trivial solution. So the non-trivial solution requires that

$$\sin(\lambda L) = 0,$$

or

$$\lambda L = 0 \text{ (trivial)}, \ \pi, \ 2\pi, \ ..., \ n\pi, \ ..., \quad \text{or} \quad \lambda = \frac{n\pi}{L}, \quad (n = 1, 2, 3, ...).$$

Here, $n = 1$ is the first buckling mode, while $n = 2, 3$ corresponds to the second and three buckling modes, respectively.

Therefore, the critical buckling load is

$$P = \lambda^2 EI = \frac{n^2 \pi^2 EI}{L^2}, \qquad (n = 1, 2, 3, ...). \tag{7.56}$$

The solutions have fixed mode shapes (sine functions), and the lowest eigenvalue when $n = 1$ is

$$P_* = \frac{\pi^2 EI}{L}, \tag{7.57}$$

which is the Euler buckling load for an elastic rod.

## 4. Higher-Order ODEs

Higher order ODEs are more complicated to solve even for linear cases. For the special case of higher-order ODEs where all the coefficients $a_n, ..., a_1, a_0$ are constants,

$$a_n y^{(n)} + ... + a_1 y' + a_0 y = f(x), \tag{7.58}$$

its general solution $y(x)$ consists of two parts: the complementary function $y_c(x)$ and the particular integral or particular solution $y_p^*(x)$. We have

$$y(x) = y_c(x) + y_p^*(x). \tag{7.59}$$

The complementary function which is the solution of the linear homogeneous equation with constant coefficients can be written in a generic form

$$a_n y_c^{(n)} + a_{n-1} y_c^{(n-1)} + \ldots + a_1 y_c' + a_0 = 0. \tag{7.60}$$

Assuming $y = A e^{\lambda x}$ where $A$ is a constant, we get the characteristic equation as a polynomial

$$a_n \lambda^n + a_{n-1} \lambda^{(n-1)} + \ldots + a_1 \lambda + a_0 = 0, \tag{7.61}$$

which has $n$ roots in the general case. Then, the solution can be expressed as the summation of various terms

$$y_c(x) = \sum_{k=1}^{n} c_k e^{\lambda_k x}$$

if the polynomial has $n$ distinct zeros $\lambda_1, \ldots \lambda_n$. Here, $c_k (k = 1, 2, \ldots, n)$ are undermined coefficients.

For complex roots, and complex roots always occur in pairs $\lambda = r \pm i\omega$, the corresponding linearly independent terms can then be replaced by $e^{rx}[A \cos(\omega x) + B \sin(\omega x)]$.

The particular solution $y_p^*(x)$ is any $y(x)$ that satisfies the original inhomogeneous equation (7.58). Depending on the form of the function $f(x)$, the particular solutions can take various forms. For most of the combinations of basic functions such as $\sin x, \cos x, e^{kx}$ and $x^n$, the method of the undetermined coefficients is widely used. For $f(x) = \sin(\alpha x)$ or $\cos(\alpha x)$, then we can try $y_p^* = A \sin \alpha x + B \sin \alpha x$. We then substitute it into the original equation (7.58) so that the coefficients $A$ and $B$ can be determined. For a polynomial $f(x) = x^p$ ($p = 0, 1, 2, \ldots, N$), we then try $y_p^* = A + Bx + \ldots + Qx^p$ (polynomial). For $f(x) = e^{kx} x^n$, $y_p^* = (A + Bx + \ldots Qx^n) e^{kx}$. Similarly, $f(x) = e^{kx} \sin \alpha x$ or $f(x) = e^{kx} \cos \alpha x$, we can use $y_p^* = e^{kx}(A \sin \alpha x + B \cos \alpha x)$. More general cases and their particular solutions can be found in various textbooks.

A very useful technique is to use the method of differential operator $D$. A differential operator $D$ is defined as

$$D \equiv \frac{d}{dx}. \tag{7.62}$$

Since we know that $De^{\lambda x} = \lambda e^{\lambda x}$ and $D^n e^{\lambda x} = \lambda^n e^{\lambda x}$, so they are equivalent to $D \mapsto \lambda$, and $D^n \mapsto \lambda^n$. Thus, any polynomial $P(D)$ will be mapped to $P(\lambda)$. On the other hand, the integral operator $D^{-1} = \int dx$ is just the inverse of differentiation.

The beauty of the differential operator form is that one can factorize it in the same way as for a polynomial, then solve each factor separately. The differential operator is very useful in finding both the complementary functions and particular integral.

## Example 7.9

To find the particular integral for the equation

$$y'''' + 3y = 7e^{3x}, \tag{7.63}$$

we get

$$(D^4 + 3)y_p^* = 7e^{3x}, \quad \text{or} \quad y_p^* = \frac{7}{D^4 + 3}e^{3x}. \tag{7.64}$$

Since $D^4 \mapsto \lambda^4 = 3^4$, we have

$$y_p^* = \frac{7e^{3x}}{3^4 + 3} = \frac{e^{3x}}{12}. \tag{7.65}$$

This method also works for $\sin x$, $\cos x$, $\sinh x$ and others, and this is because they are related to $e^{\lambda x}$ via $\sin \theta = \frac{1}{2i}(e^{i\theta} - e^{-i\theta})$ and $\cosh x = (e^x + e^{-x})/2$.

## 5. System of Linear ODEs

For a linear ODE of order $n$ (7.58), it can always be written as a linear system

$$\frac{dy}{dx} = y_1, \quad \frac{dy_1}{dx} = y_2, \quad ..., \quad \frac{dy_{n-1}}{dx} = y_n,$$

$$a_n(x)y'_{n-1} = -\left[a_{n-1}(x)y_{n-1} + ... + a_1(x)y_1 + a_0(x)y\right] + f(x), \tag{7.66}$$

which is a system for $u = [y \; y_1 \; y_2 \; ... \; y_{n-1}]^T$. If the independent variable $x$ does not appear explicitly in $y_i$, then the system is said to be autonomous with important properties. For simplicity and in keeping with the convention, we use $t = x$ and $\dot{u} = du/dt$ in our following discussion. A general linear system of $n$-th order can be written as

$$\begin{pmatrix} \dot{u}_1 \\ \dot{u}_2 \\ \vdots \\ \dot{u}_n \end{pmatrix} = \begin{pmatrix} a_{11} & a_{12} & \cdots & a_{1n} \\ a_{21} & a_{22} & \cdots & a_{2n} \\ \vdots & & & \vdots \\ a_{n1} & a_{n2} & \cdots & a_{nn} \end{pmatrix} \begin{pmatrix} u_1 \\ u_2 \\ \vdots \\ u_n \end{pmatrix}, \tag{7.67}$$

or

$$\dot{u} = Au. \tag{7.68}$$

If $\mathbf{u} = \mathbf{v}\exp(\lambda t)$, then this becomes an eigenvalue problem,

$$(\mathbf{A} - \lambda\mathbf{I})\mathbf{v} = \mathbf{0}, \tag{7.69}$$

which will have a non-null solution only if $\det(\mathbf{A} - \lambda\mathbf{I}) = 0$. The eigenvalues of the above system will control the dynamic behaviour of the system. The full mathematical analysis requires the theory of dynamic systems. Interested readers can refer to more advanced literature.

Higher-order differential equations (including nonlinear ODEs) can conveniently be written as a system of $n$ first-order differential equations. A system of ODEs is more suitable for mathematical analysis and numerical integration.

## 6. Harmonic Motions

## 6.1. Undamped Forced Oscillations

The simple system with a spring attached with a mass $m$ is a good example of harmonic motion (see Figure 7.5). If the spring stiffness constant is $k$, then the governing equation of the oscillations is a second-order ordinary differential equation for undamped forced harmonic motion. Using Newton's second law, we have

$$F(t) - ky = m\frac{d^2y(t)}{dt^2} = my'', \tag{7.70}$$

which can be written as

$$y'' + \omega_0^2 y = f(t), \quad \omega_0^2 = \frac{k}{m}, \quad f(t) = \frac{F(t)}{m}, \tag{7.71}$$

where $\omega_0$ is the angular frequency of the system, and $f(t)$ is the a known function of $t$. In the case of $f(t) = F_0\cos\omega t$, we have

$$y'' + \omega_0^2 y = F_0\cos\omega t, \tag{7.72}$$

where $\omega_0$ is the natural frequency of the system, and $F_0 m$ is the amplitude of external forcing.

The general solution $y(t) = y_c + y_p$ consists of a complementary function $y_c$ and a particular integral $y_p$. The complementary function $y_c$ satisfies the homogeneous equation

$$y'' + \omega_0^2 y = 0. \tag{7.73}$$

Its general solution is

$$y_c(t) = A\sin\omega_0 t + B\cos\omega_0 t, \tag{7.74}$$

where $A$ and $B$ are undetermined constants.

For the particular integral $y_p$, we have to consider two different cases $\omega \neq \omega_0$ and

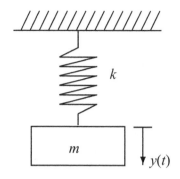

Figure 7.5: Undamped forced oscillations with a springer constant $k$ and mass $m$.

$\omega = \omega_0$ because for $\omega = \omega_0$ the standard particular $a \sin \omega t + b \cos \omega t$ does not work as it needs some modifications.

For $\omega \neq \omega_0$, we assume that $y_p = a \sin \omega t + b \cos \omega t$, and we have

$$y' = a\omega \cos \omega t - b\omega \sin \omega t, \quad y_p'' = -\omega^2(a \sin \omega t + b \cos \omega t). \qquad (7.75)$$

Substituting into Eq.(7.72), we have

$$-\omega^2(a \sin \omega t + b \cos \omega t) + \omega_0^2(a \sin \omega t + b \cos \omega t) = F_0 \cos(\omega t), \qquad (7.76)$$

which gives $a = 0$ and $b = F_0/(\omega_0^2 - \omega^2)$ by equating the coefficients of $\cos \omega t$ and $\sin \omega t$ on both sides. We thus obtain

$$y_p = \frac{F_0}{\omega_0^2 - \omega^2} \cos \omega t. \qquad (7.77)$$

Therefore, the general solution

$$y(t) = A \sin \omega_0 t + B \cos w_0 t + \frac{F_0}{\omega_0^2 - \omega^2} \cos \omega t. \qquad (7.78)$$

If we further assume that the system is initially at rest when the force starts to act, we have the initial conditions $y(0) = 0$ and $y'(0) = 0$. With these conditions, we have $A = 0$ and $B = -F_0/(\omega_0^2 - \omega^2)$ in the general solution. We now have

$$y(t) = \frac{F_0}{\omega_0^2 - \omega^2}(\cos \omega t - \cos \omega_0 t). \qquad (7.79)$$

Using $\cos \theta_1 - \cos \theta_2 = -2 \sin \frac{(\theta_1 + \theta_2)}{2} \sin \frac{(\theta_1 - \theta_2)}{2}$, we have

$$y(t) = \frac{2F_0}{\omega_0^2 - \omega^2} \sin \frac{(\omega - \omega_0)t}{2} \sin \frac{(\omega + \omega_0)t}{2} = A(t) \sin \frac{(\omega + \omega_0)t}{2} = A(t) \sin \tilde{\omega}t,$$

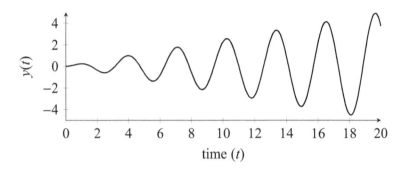

Figure 7.6: Oscillations become unbounded for $\omega \approx \omega_0 = 2$.

where

$$A(t) = \frac{2F_0}{\omega_0^2 - \omega^2} \sin \frac{(\omega - \omega_0)t}{2}, \qquad \tilde{\omega} = \frac{\omega + \omega_0}{2}.$$

As $|\omega - \omega_0| < |\omega + \omega_0|$, we can see that the oscillator oscillates with a major and fast frequency $\tilde{\omega} = (\omega + \omega_0)/2$, while its amplitude or envelope oscillates slowly with a frequency $\delta\omega = (\omega - \omega_0)/2$. This phenomenon is called 'beats'.

For the special case of $\omega = \omega_0$, the complementary function is the same as before, but the particular solution should take the following form

$$y_p = t(a \sin \omega t + b \cos \omega t), \tag{7.80}$$

which gives

$$y_p(t) = \frac{F_0}{2\omega_0} t \sin \omega_0 t. \tag{7.81}$$

The general solution is therefore

$$y(t) = A \sin \omega_0 t + B \cos \omega_0 t + \frac{F_0}{2\omega_0} t \sin \omega_0 t. \tag{7.82}$$

Similarly, the initial solution $y(0) = y'(0) = 0$ implies that $A = B = 0$. We now have

$$y(t) = \frac{F_0}{2\omega_0} t \sin \omega_0 t = A(t) \sin \omega_0 t, \tag{7.83}$$

where $A(t) = F_0 t/(2\omega_0)$. As the amplitude $A(t)$ increases with time as shown in Fig. 7.6, this phenomenon is called resonance, and the external forcing causes the oscillations to grow out of control when the forcing is acted at the natural frequency $\omega_0$ of the system.

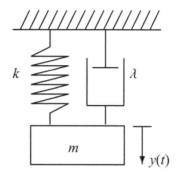

Figure 7.7: Damped harmonic motion with a spring constant $k$ and a dashpot rate $\lambda$.

## 6.2. Damped Forced Oscillations

As all the real systems have a certain degree of friction, thus damping should be included. An example of damping is shown in Fig. 7.7. With damping and external forcing $F(t)$, the force through the damper is $-\lambda y'(t)$, Newton's second law gives

$$F(t) - \lambda y'(t) - ky(t) = m\frac{d^2y(t)}{dt^2} = my''(t), \tag{7.84}$$

which can be written as

$$my'' + \lambda y'(t) + ky(t) = F(t). \tag{7.85}$$

Defining

$$\omega_0^2 = k/m \ \ \text{(natural frequency)}, \quad \xi = \frac{\lambda}{2m\omega_0} = \frac{\lambda}{2\sqrt{km}}, \tag{7.86}$$

we can rewrite the above equation as

$$y''(t) + 2\xi\omega_0 y'(t) + \omega_0^2 y(t) = \frac{F(t)}{m}. \tag{7.87}$$

In addition, if we write $f(t) = F(t)/m = F_0 \cos \omega t$ as the normalized forcing, we finally obtain

$$y''(t) + 2\xi\omega_0 y'(t) + \omega_0^2 y(t) = F_0 \cos \omega t, \tag{7.88}$$

where $\xi$ is the damping coefficient. In principle, one can try to solve this equation using the standard method, but it may become a little awkward as it involves complex numbers. In fact, there is quite an elegant method using the complex variables. In order to do this, we write the companion equation for equation (7.88) with a different forcing term

$$\eta''(t) + 2\xi\omega_0\eta'(t) + \omega_0^2\eta(t) = F_0 \sin \omega t. \tag{7.89}$$

Since $e^{i\omega t} = \cos \omega t + i \sin \omega t$, we can multiply (7.89) by $i$, and add it to (7.88), and we have

$$z''(t) + 2\xi\omega_0 z + \omega_0^2 z = F_0 e^{i\omega t}, \tag{7.90}$$

where $z(t) = y(t) + i\eta(t)$.

By solving this equation, we essentially solve both equations (7.88) and (7.89) at the same time if we can separate the real and imaginary parts. The complementary function corresponds to the transient part while the particular function corresponds to the steady state. For the transient part, the characteristic equation gives

$$\mu^2 + 2\xi\omega_0\mu + \omega_0^2 = 0, \tag{7.91}$$

or

$$\mu = \omega_0\left(-\xi \pm \sqrt{\xi^2 - 1}\right). \tag{7.92}$$

If $\xi \geq 1$, then $\mu < 0$. If $\xi < 1$, then

$$\mu = \omega_0\left(-\xi + i\sqrt{1 - \xi^2}\right), \tag{7.93}$$

and the real part is negative (i.e., $\Re(\mu) = -\omega_0\xi < 0$). In both cases $\Re(\mu) < 0$, thus the solution $z_c \propto e^{-\omega_0\xi t} \to 0$. In engineering, it is conventional to define a case of critical damping when $\xi = 1$.

For $\xi = 0$, we have $\mu = i\omega_0$, which corresponds to the harmonic oscillations without damping. For $\xi = 1$, $\mu = -\omega_0$, it is critical damping as the imaginary term is zero. The amplitude decreases exponentially at just the slowest possible manner without any oscillations. For $\xi < 1$, we get

$$\mu = -\omega_0\xi + i\omega_0\sqrt{1 - \xi^2}. \tag{7.94}$$

The real part corresponds to the exponential decrease of the amplitude and the imaginary part corresponds to oscillations. For this reason, it is called under-damped.

Finally, $\xi > 1$ leads to $\mu = \omega_0(-\xi \pm \sqrt{\xi^2 - 1}) < 0$. The imaginary part is zero (no oscillation). As the amplitude decreases much faster than that at the critical damping, this case is thus called over-damped. Figure 7.8 shows the characteristics of these three cases.

If we try the particular solution in the form $z = z_0 e^{i\omega t}$, we have

$$z'' + 2\xi\omega_0 z' + \omega_0^2 z = F_0 e^{i\omega t} \implies P(i\omega)z = F_0 e^{i\omega t}, \tag{7.95}$$

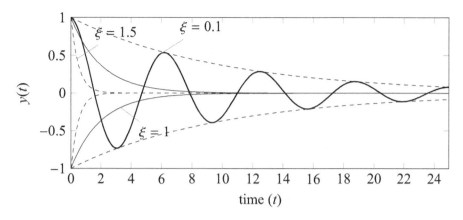

Figure 7.8: Critical damping ($\xi = 1$), under-damping ($\xi = 0.1$) and over-damping ($\xi = 1.5$).

and

$$P(i\omega) = (i\omega)^2 + 2\xi\omega_0(i\omega) + \omega_0^2 = (\omega_0^2 - \omega^2) + 2\xi\omega_0\omega i, \qquad (7.96)$$

which is essentially the characteristic polynomial. The general solution becomes

$$z(t) = \frac{F_0}{P(i\omega)}e^{i\omega t} = \frac{F_0}{[(\omega_0^2 - \omega^2) + 2i\xi\omega_0\omega]}e^{i\omega t}. \qquad (7.97)$$

It is conventional to call $H(i\omega) = 1/P(i\omega)$ the transfer function. We can always write the general solution $z = Ae^{i(\omega t + \phi)}$, where $A = |z|$ is the modulus and $\phi$ is the phase shift. Therefore, we have

$$z = Ae^{i(\omega t + \phi)}, \qquad (7.98)$$

where

$$A = \frac{F_0}{\sqrt{(\omega_0^2 - \omega^2)^2 + 4\xi^2\omega_0^2\omega^2}} = \frac{F_0/\omega_0^2}{\left\{\left[1 - (\omega/\omega_0)^2\right]^2 + (2\xi\omega/\omega_0)^2\right\}^{1/2}}, \qquad (7.99)$$

and

$$\phi = \tan^{-1}\left(\frac{-2\xi\omega_0\omega}{\omega_0^2 - \omega^2}\right) = \tan^{-1}\left\{\frac{-2\xi\omega/\omega_0}{1 - (\omega/\omega_0)^2}\right\}. \qquad (7.100)$$

As the amplitude of the forcing is $F_0$, the gain $G(\omega)$ of the oscillation is

$$G(\omega) = \frac{A}{F_0/\omega_0^2} = \frac{1}{\sqrt{[1 - (\omega/\omega_0)^2]^2 + 4\xi^2(\omega/\omega_0)^2}}, \qquad (7.101)$$

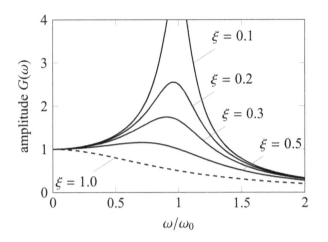

Figure 7.9: Variations of amplitude with frequency $\omega$ and damping coefficient $\xi$ for $\omega_0 = 2$.

which is shown in Figure 7.9.

Finally, the solution of Eq. (7.88) is the real part of (7.98). That is

$$y(t) = A \cos(\omega t + \phi). \tag{7.102}$$

Some special cases where $\omega \to 0$ and $\omega \to \infty$ are very interesting. For $\omega \ll \omega_0$, the driving force is at very low frequency, we have [from Eq.(7.99)]

$$A \to \frac{F_0}{\omega_0^2}, \qquad \phi \to 0, \qquad y(t) \approx \frac{F_0}{\omega_0^2} \cos(\omega t). \tag{7.103}$$

The system is in the same phase with the forcing.

If $\omega \gg \omega_0$, the forcing is at very high frequency. We have $A \to 0$, and $\phi \to -\pi$. The oscillator is completely out of phase with the forcing.

If $\omega \approx \omega_0$, we have

$$A \to \frac{F_0}{2\xi\omega_0^2}, \qquad \phi \to -\frac{\pi}{2}, \tag{7.104}$$

and the gain becomes

$$G(\omega) = \frac{A}{F_0/\omega_0^2} = \frac{1}{(2\xi)} = Q, \tag{7.105}$$

which is the quality factor $Q = 1/(2\xi)$. In addition, we have

$$y(t) = \frac{F_0}{2\xi\omega_0^2} \cos(\omega t - \frac{\pi}{2}) = \frac{F_0}{2\xi\omega_0^2} \sin \omega t. \tag{7.106}$$

At resonance, the response gain is maximum, which means that

$$\frac{dG(\omega)}{d\omega} = -\frac{4\xi^2 W - 2W(1 - W^2)}{\sqrt{(1 - W^2)^2 + 4\xi^2 W^2}} = 0, \quad (\text{here } W = \frac{\omega}{\omega_0}), \tag{7.107}$$

which gives

$$2W(2\xi^2 - 1 + W^2) = 0. \tag{7.108}$$

This means either $W = 0$ (trivial solution) or

$$2\xi^2 - 1 + W^2 = 0, \quad \text{or} \quad W = \frac{\omega}{\omega_0} = \sqrt{1 - 2\xi^2}. \tag{7.109}$$

This gives a resonance frequency $\omega_r = \omega_0 \sqrt{1 - 2\xi^2}$. At resonance, we have $W = \omega/\omega_0 = \sqrt{1 - 2\xi^2}$ and the gain/response becomes maximum

$$G(\omega) = \frac{1}{\sqrt{(1 - W^2)^2 + (2\xi W)^2}} = \frac{1}{\sqrt{4\xi^2(1 - \xi^2)}} = \frac{1}{2\xi\sqrt{1 - \xi^2}}. \tag{7.110}$$

Let us now look at a practical example.

## Example 7.10

A small machine of mass $m = 200$ kg is mounted to a wall. The connection to the wall can be modelled as a spring with a stiffness $k = 800$ N/m and a dashpot of rate $\lambda = 160$ Ns/m. The machine is vibrating under a harmonic forcing of $F(t) = 20 \cos \omega t$ with a peak force of 20 N and an angular frequency of $\omega = 1.86$ rad/s. What is the maximum of the vibrations and what is the phase angle $\phi$?

Since we know that $k = 800$ N/n, $m = 200$ kg and $\lambda = 160$ Ns/m, we have the natural frequency $\omega_0$

$$\omega_0 = \sqrt{\frac{k}{m}} = \sqrt{\frac{800}{200}} = 2 \text{ rad/s}, \quad \xi = \frac{\lambda}{2\sqrt{km}} = \frac{160}{2\sqrt{800 \times 200}} = 0.2.$$

In addition, we know that $\omega = 1.86$ m and $F_0 = 20/m = 20/200 = 0.1$. Now we have $\omega/\omega_0 = 1.86/2 = 0.93$. From Eq.(7.101), we have

$$G(\omega) = \frac{A}{F_0/\omega_0^2} = \frac{1}{\sqrt{[1 - (\omega/\omega_0)^2]^2 + (2\xi\omega/\omega_0)^2}} = \frac{1}{\sqrt{(1 - 0.93^2)^2 + (2 \times 0.2 \times 0.93)^2}} \approx 2.53.$$

Thus, the maximum amplitude of the vibrations is

$$A = G(\omega)F_0/\omega_0^2 = 2.53 \times \frac{0.1}{2^2} \approx 0.063 \text{ m},$$

which is about 6.3 cm. In addition, the resonance frequency $\omega_r$ is

$$\omega_r = \omega_0 \sqrt{1 - 2\xi^2} = 2 \times \sqrt{1 - 2 \times 0.2^2} \approx 1.92 \text{ rad/s},$$

which is different from both $\omega_0$ (natural frequency) and external forcing frequency $\omega$. Furthermore, the phase shift $\phi$ is

$$\phi = \tan^{-1}\left\{\frac{-2\xi\omega/\omega_0}{1 - (\omega/\omega_0)^2}\right\} = \tan^{-1}\left(\frac{-2 \times 0.2 \times 0.93}{1 - 0.93^2}\right) \approx \tan^{-1}(-2.75) \approx -70.0°,$$

which means that it is about 70° lag.

Obviously, differential equations are widely used and there are other solution techniques such as Laplace transform methods and numerical methods. We will introduce these topics in the coming chapters.

## Exercises

**7.1.** Solve $y'(x) = \cos(x) - \exp(-x)$ with the initial condition $y(0) = 1$.

**7.2.** Solve the following differential equations:
- $y''(x) + 5y'(x) + 4y(x) = 0$.
- $y'(x) + x^2 y(x) = 0$.
- $y''(x) + 4y(x) = 0$.

**7.3.** Solve $y'''(x) + y''(x) + 4y'(x) + 4y(x) = 0$.

**7.4.** Solve $y'(x) + y(x) = 1 - e^{-2x}$ with the initial condition $y(0) = 0$. What is the new solution if $y(0) = 1$?

**7.5.** A uniform cantilever beam of length $L$ is fixed at one end and the free end carries a point load $P$. Show that the deflection $u$ is governed by

$$EI\frac{d^2 u}{dx^2} = P(L - x),$$

where $x$ is the distance from the fixed end. $E$ is the Young modulus and $I$ the bending moment of inertia. Solve this equation with appropriate boundary conditions to show that the maximum deflection is $D_{\max} = PL^3/(3EI)$.

# CHAPTER 8

# Fourier Transform and Laplace Transform

Contents

## Key Points

- Functions, including non-smooth functions, can be represented by Fourier series. Examples of triangular waves and sawtooth waves are analyzed.
- Fourier transforms are introduced, together with their discrete forms and the fast Fourier transforms.
- Laplace transforms are introduced, together with their scaling properties. Examples of some commonly used Laplace transform pairs are also given,
- Other transforms such as Z-transforms are briefly introduced and their relationships to Laplace transforms are also discussed briefly.

Mathematical transformation is a method of changing one kind of function and equation into another, often simpler and easier to solve. The Fourier transform maps a function in the time domain such as a signal into another function in the frequency domain, which is commonly used in signal and image processing as well as structural analysis in civil engineering.

## 1. Fourier Series

From earlier discussions, we know that function $e^x$ can be expanded into a series in terms of a polynomial with terms of $x$, $x^2$, ..., $x^n$. In this case, we are in fact trying to expand it in terms of the basis functions $1$, $x$, $x^2$, ..., and $x^n$. There are many other basis functions. For example, the basis functions $\sin(n\pi t)$ and $\cos(n\pi t)$ are more widely used in signal processing. In general, this is essentially about the Fourier series.

### 1.1. Fourier Series

French mathematician Joseph Fourier at the turn of the 1800s was tackling a heat transfer problem and he approximated his functions as a series of trigonometrical functions.

For a function $f(t)$ on an interval $t \in [-T, T]$ where $T > 0$ is a finite constant or half period, the Fourier series is defined as

$$f(t) = \frac{a_0}{2} + a_1 \cos(\frac{\pi t}{T}) + b_1 \sin(\frac{\pi t}{T}) + a_2 \cos(\frac{2\pi t}{T}) + b_2 \sin(\frac{2\pi t}{T}) + \dots$$

$$= \frac{a_0}{2} + \sum_{n=1}^{\infty} \left[ a_n \cos(\frac{n\pi t}{T}) + b_n \sin(\frac{n\pi t}{T}) \right], \tag{8.1}$$

where

$$a_0 = \frac{1}{T} \int_{-T}^{T} f(t)dt, \qquad a_n = \frac{1}{T} \int_{-T}^{T} f(t) \cos(\frac{n\pi t}{T})dt, \tag{8.2}$$

and

$$b_n = \frac{1}{T} \int_{-T}^{T} f(t) \sin(\frac{n\pi t}{T})dt, \quad (n = 1, 2, \dots). \tag{8.3}$$

Here $a_n$ and $b_n$ are the Fourier coefficients of $f(t)$ on $[-T, T]$. It is worth pointing out that an implicit assumption is that $f(t)$ is periodic

$$f(t + 2T) = f(t), \tag{8.4}$$

which can also be seen from the fact that $\cos(n\pi t/T)$ and $\sin(n\pi t/T)$ are also periodic with a period of $2T$.

The function $f(t)$ can be continuous or piecewise continuous with a finite number of jump discontinuities. For a jump discontinuity at $t = t_0$, if both the left value $f'(t_0-)$ and the right value $f'(t_0+)$ exist with $f(t_0-) \neq f(t_0+)$, then the Fourier series converges to their average value. That is

$$f(t_0) = \frac{1}{2}\left[ f(t_0-) + f(t_0+) \right]. \tag{8.5}$$

In some textbooks, a special case of $T = \pi$ is often used. In this case, the formula becomes simpler and we have

$$f(t) = \frac{a_0}{2} + \sum_{n=1}^{\infty}(a_n \cos nt + b_n \sin nt), \quad (-\pi \le t \le \pi), \tag{8.6}$$

with the Fourier coefficients

$$a_0 = \frac{1}{\pi}\int_{-\pi}^{\pi} f(t)dt, \quad a_n = \frac{1}{\pi}\int_{-\pi}^{\pi} f(t)\cos nt dt, \quad b_n = \frac{1}{\pi}\int_{-\pi}^{\pi} f(t)\sin nt dt, \tag{8.7}$$

for all positive integers $n$ ($n = 1, 2, ...$).

## Example 8.1

The well-known square wave is defined by

$$f(t) = \begin{cases} 1 & \text{if } -\pi < t \le 0, \\ 0 & \text{if } 0 < t \le \pi. \end{cases}$$

It has a period of $2\pi$ and $f(t + 2\pi) = f(t)$.

It is easy to check that

$$a_0 = \frac{1}{\pi}\int_{-\pi}^{\pi} f(t)dt = \frac{1}{\pi}\int_{-\pi}^{0} 1 dt + \frac{1}{\pi}\int_{0}^{\pi} 0 dt = \frac{1}{\pi}[0 - (-\pi)] = 1.$$

For $a_n(n \ge 1)$, the coefficient can be obtained by direct integration. We have

$$a_n = \frac{1}{2}\int_{-\pi}^{\pi} f(t)\cos(nt)dt = \frac{1}{\pi}\int_{-\pi}^{0} 1\cos(nt)dt + \frac{1}{\pi}\int_{0}^{\pi} 0\cos(nt)dt$$

$$= \frac{1}{\pi}\left(\frac{\sin nt}{n}\right)\Big|_{-\pi}^{0} = \frac{1}{n\pi}[\sin 0 - \sin(-n\pi)] = 0,$$

where we have used $\sin(n\pi) = 0$ when $n$ is an integer.

Similarly, we have

$$b_n = \frac{1}{\pi}f(t)\sin(nt)dt = \frac{1}{\pi}\int_{-\pi}^{0} 1\sin(nt)dt + \frac{1}{\pi}\int_{0}^{\pi} 0\sin(nt)dt = \frac{1}{\pi}\int_{-\pi}^{0} \sin(nt)dt$$

$$= \frac{1}{\pi}\left(-\frac{\cos nt}{n}\right)\Big|_{-\pi}^{0} = -\frac{1}{n\pi}\left[\cos 0 - \cos(-n\pi)\right] = \frac{-1}{n\pi}[1 - (-1)^n] = \begin{cases} -\frac{2}{n\pi} & \text{for odd } n, \\ 0 & \text{for even } n, \end{cases}$$

where we have used $\cos(n\pi) = (-1)^n$. To avoid stating that $n$ is odd or even, we can set $n = 2k - 1$ and thus we have

$$b_n \to b_k = -\frac{(-1)^k 2}{(2k-1)\pi}, \quad (k = 1, 2, 3, ...).$$

From the above results, we can see that apart from $a_0 = 1$, all $a_n(n = 1, 2, ...)$ are zero.

Therefore, the Fourier series of the square wave becomes

$$f(t) = \frac{1}{2} - \sum_{k=1}^{\infty} \frac{(-1)^k 2}{(2k-1)\pi} \sin((2k-1)t)$$

$$= \frac{1}{2} - \frac{2}{\pi} \left[ \sin t + \frac{\sin 3t}{3} + \frac{\sin 5t}{5} + \frac{\sin 7t}{7} + \frac{\sin 9t}{9} + ... \right].$$

It seems that the calculations of the Fourier coefficients are quite tedious, but the good news is that for a given function we only need to do it once and the integrals involved are usually easy to obtain.

The Fourier series theory usually guarantees that such expansions are valid. However, it is also sometimes useful to check when appropriate. For example, we know that the above square wave has a discontinuity at $t = 0$ because $f(0-) = 1$ and $f(0+) = 0$. According to (8.5), the series should converge to $(1+0)/2 = 1/2$. Let us see if this is the case.

## Example 8.2

From the Fourier series obtained in the previous example, we can set $t = 0$ and we have

$$\frac{1}{2} - \frac{2}{\pi} \left[ \sin t + \frac{\sin 3t}{3} + \frac{\sin 5t}{5} + ... \right] = \frac{1}{2} - \frac{2}{\pi} \left[ \sin 0 + \frac{\sin 0}{3} + \frac{\sin 0}{5} + ... \right]$$

$$= \frac{1}{2} - \frac{2}{\pi} [0 + 0 + 0 + ...] = \frac{1}{2},$$

where we have used $\sin 0 = 0$. Thus, the series indeed converges to $1/2$ at $t = 0$.

## 1.2. Orthogonality and Fourier Coefficients

You may wonder how to derive the formulas for the coefficient $a_n$ and $b_n$? Before we proceed, let us prove the orthogonality relation

$$J = \int_{-T}^{T} \sin(\frac{n\pi t}{T}) \sin(\frac{m\pi t}{T}) dt = \begin{cases} 0 & (n \neq m) \\ T & (n = m) \end{cases}, \tag{8.8}$$

where $n$ and $m$ are integers.

From the trigonometrical functions, we know that

$$\cos(A + B) = \cos A \cos B - \sin A \sin B, \quad \cos(A - B) = \cos A \cos B + \sin A \sin B.$$

By subtracting, we have

$$\cos(A - B) - \cos(A + B) = 2 \sin A \sin B. \tag{8.9}$$

Now the orthogonality integral becomes

$$J = \int_{-T}^{T} \sin(\frac{n\pi t}{T}) \sin(\frac{m\pi t}{T}) dt = \frac{1}{2} \int_{-T}^{T} \{ \cos[\frac{(n - m)\pi t}{T}] - \cos[\frac{(n + m)\pi t}{T}] \} dt. \tag{8.10}$$

If $n \neq m$, we have

$$J = \frac{1}{2} \{ \frac{T}{(n - m)\pi} \sin[\frac{(n - m)\pi t}{T}] \Big|_{-T}^{T} - \frac{T}{(n + m)\pi} \sin[\frac{(n + m)\pi t}{T}] \Big|_{-T}^{T} \}$$

$$= \frac{1}{2} [ \frac{T}{(n - m)\pi} \times (0 - 0) - \frac{T}{(n + m)\pi} \times (0 - 0) ] = 0. \tag{8.11}$$

If $n = m$, we have

$$J = \frac{1}{2} \int_{-T}^{T} \{ 1 - \cos[\frac{2n\pi t}{T}] \} dt$$

$$= \frac{1}{2} \{ t \Big|_{-T}^{T} - \frac{T}{2n\pi} \sin[\frac{2n\pi t}{T}] \Big|_{-T}^{T} \} = \frac{1}{2} [2T - \frac{T}{2n\pi} \times 0] = T, \tag{8.12}$$

which proves the relationship (8.8).

Using similar calculations, we can easily prove the following orthogonality relations

$$\int_{-T}^{T} \cos(\frac{n\pi t}{T}) \cos(\frac{m\pi t}{T}) dt = \begin{cases} 0 & (n \neq m) \\ T & (n = m) \end{cases}, \tag{8.13}$$

and

$$\int_{-T}^{T} \sin(\frac{n\pi t}{T}) \cos(\frac{m\pi t}{T}) dt = 0, \qquad \text{for all } n \text{ and } m. \tag{8.14}$$

Now we can try to derive the expressions for coefficients $a_n$. Multiplying both sides of the Fourier series (8.1) by $\cos(m\pi t/T)$ and taking the integration from $-T$ to $T$, we have

$$\int_{-T}^{T} f(t) \cos(\frac{m\pi t}{T}) dt = \frac{a_0}{2} \int_{-T}^{T} \cos(\frac{m\pi t}{T}) dt$$

$$+ \sum_{n=1}^{\infty} \left\{ a_n \int_{-T}^{T} \cos(\frac{n\pi t}{T}) \cos(\frac{m\pi t}{T}) dt + b_n \int_{-T}^{T} \sin(\frac{n\pi t}{T}) \cos(\frac{m\pi t}{T}) dt \right\}.$$

Using the relations (8.13) and (8.14) as well as $\int_{-T}^{T} \cos(m\pi t/T) dt = 0$, we know that the only non-zero integral on the right-hand side is when $n = m$. Therefore, we get

$$\int_{-T}^{T} f(t) \cos(\frac{n\pi t}{T}) dt = 0 + [a_n T + b_n \times 0], \qquad (8.15)$$

which gives

$$a_n = \frac{1}{T} \int_{-T}^{T} f(t) \cos(\frac{n\pi t}{T}) dt, \qquad (8.16)$$

where $n = 1, 2, 3, \cdots$. Interestingly, when $n = 0$, it is still valid and becomes $a_0$ as $\cos 0 = 1$. That is

$$a_0 = \frac{1}{T} \int_{-T}^{T} f(t) dt. \qquad (8.17)$$

In fact, $a_0/2$ is the average of $f(t)$ over the period $2T$.

The coefficients $b_n$ can be obtained by multiplying $\sin(m\pi t/T)$ and following similar calculations. We have

$$b_n = \frac{1}{T} \int_{-T}^{T} f(t) \sin(\frac{n\pi t}{T}) dt, \quad (n = 1, 2, 3, ...). \qquad (8.18)$$

Fourier series in general tends to converge slowly. In order for a function $f(x)$ to be expanded properly, it must satisfy the following Dirichlet conditions:
- A piecewise function $f(x)$ must be periodic with at most a finite number of discontinuities,
- and/or a finite number of minima or maxima within one period.

In addition, the integral of $|f(x)|$ must converge. For example, these conditions suggest that $\ln(x)$ cannot be expanded into a Fourier series in the interval $[0, 1]$ as $\int_0^1 |\ln x| dx$ diverges.

The $n$th term of the Fourier series,

$$a_n \cos(n\pi t/T) + b_n \sin(n\pi t/T),$$

is called the $n$th harmonic. The energy of the $n$th harmonic is defined by $A_n^2 = a_n^2 + b_n^2$, and the sequence of $A_n^2$ forms the energy or power spectrum of the Fourier series.

From the coefficient $a_n$ and $b_n$, we can easily see that $b_n = 0$ for an even function $f(-t) = f(t)$ because $g(t) = f(t) \sin(n\pi t/T)$ is now an odd function $g(-t) = -g(t)$ due

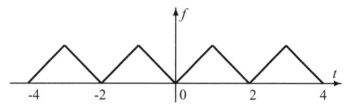

Figure 8.1: Triangular wave with a period of 2.

to the fact $\sin(2\pi t/T)$ is an odd function. We have

$$b_n = \frac{1}{T}\int_{-T}^{T} f(t)\sin(\frac{n\pi t}{T})dt = \frac{1}{T}\left[\int_{-T}^{0} g(t)dt + \int_{0}^{T} g(t)dt\right]$$

$$= \frac{1}{T}\left[\int_{0}^{T} g(-t)dt + \int_{0}^{T} g(t)dt\right] = \frac{1}{T}\int_{0}^{T}(-g(t)+g(t))dt = 0. \qquad (8.19)$$

Similarly, we have $a_0 = a_n = 0$ for an odd function $f(-t) = -f(t)$. In both cases, only one side $[0,T]$ of the integration is used due to symmetry.

Thus, for even function $f(t)$, we have the Fourier cosine series on [0,T]

$$f(t) = \frac{a_0}{2} + \sum_{n=1}^{\infty} a_n \cos(\frac{n\pi t}{T}), \quad \text{if } f(-t) = f(t). \qquad (8.20)$$

For odd function $f(t)$ (thus $f(-t) = -f(t)$), we have the sine series

$$f(t) = \sum_{n=1}^{\infty} \sin(\frac{n\pi t}{T}). \qquad (8.21)$$

## Example 8.3

The triangular wave is defined by $f(t) = |t|$ for $t \in [-1,1]$ with a period of 2 or $f(t+2) = f(t)$ shown in Fig. 8.1. Using the coefficients of the Fourier series, we have

$$a_0 = \int_{-1}^{1} |t|dt = \int_{-1}^{0}(-t)dt + \int_{0}^{1} tdt = 1.$$

Since both $|t|$ and $\cos(n\pi t)$ are even functions, we have for any $n \geq 1$,

$$a_n = \int_{-1}^{1} |t|\cos(n\pi t)dt = 2\int_{0}^{1} t\cos(n\pi t)dt$$

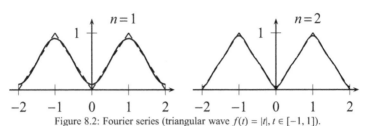

Figure 8.2: Fourier series (triangular wave $f(t) = |t|, t \in [-1, 1]$).

(a) first two terms ($n=1$); (b) first three terms ($n = 2$).

$$= 2\frac{t}{n\pi} \sin(n\pi t)\Big|_0^1 - \frac{2}{n\pi} \int_0^1 \sin(n\pi t)dt = \frac{2}{n^2\pi^2}[\cos(n\pi) - 1].$$

Because $|t| \sin(n\pi t)$ is an odd function, we have

$$b_n = \int_{-1}^1 |t| \sin(n\pi t)dt = 0.$$

Hence, the Fourier series for the triangular wave can be written as

$$f(t) = \frac{1}{2} + 2\sum_{n=1}^{\infty} \frac{\cos(n\pi) - 1}{n^2\pi^2} \cos(n\pi t) = \frac{1}{2} + \frac{4}{\pi^2} \sum_{n=1,3,5,...}^{\infty} \frac{(-1)^n}{n^2} \cos(n\pi t),$$

where we have used $\cos(n\pi) = (-1)^n$.

The first two terms, $f_n(t) = 1/2 - (4/\pi^2)\cos(\pi t)$, are shown in Fig. 8.2 where we can see that only a few terms are needed to produce a very good approximation.

Here we can see that the triangular wave with derivative discontinuity can be approximated well by two or three terms. This makes it easy for any mathematical analysis. Fourier series are widely applied in signal and image processing as well as telecommunications.

Sometimes, interesting results can be obtained from the infinite Fourier series when certain appropriate values are used. Let us look at an example.

## Example 8.4

From the triangular wave, we know that

$$f(t) = \frac{1}{2} + \frac{4}{\pi^2} \sum_{n=1,3,5,...}^{\infty} \cos(n\pi t) = \frac{1}{2} - \frac{4}{\pi^2}\Big[\cos(\pi t) + \frac{\cos(3\pi t)}{3^2} + \frac{\cos(5\pi t)}{5^2} + ...\Big].$$

For $t = 0$, we know that $f(0) = 0$ and $\cos(n\pi \times 0) = 1$, and we have

$$0 = f(0) = \frac{1}{2} - \frac{4}{\pi^2}\left[\cos 0 + \frac{\cos 0}{3^2} + \frac{\cos 0}{5^2} + \frac{\cos 0}{7^2} + ...\right] = \frac{1}{2} - \frac{4}{\pi^2}[1 + \frac{1}{3^2} + \frac{1}{5^2} + \frac{1}{7^2} + ...],$$

which gives

$$1 + \frac{1}{3^2} + \frac{1}{5^2} + \frac{1}{7^2} + ... + \frac{1}{(2n-1)^2} + ... = \frac{\pi^2}{8}.$$

This is the well-known Basel problem, first solved by Leonhard Euler in 1734.

Usually, most Fourier series can converge reasonably quickly and a few terms can often give good approximations. However, care must be taken when computing Fourier series, especially in the case when the Gibbs phenomenon may occur. Loosely speaking, the Gibbs phenomenon refers to the unusual oscillations of the partial sum near the discontinuity or jump of a piecewise periodic function. Before we proceed, let us look at an example of the sawtooth waves.

## Example 8.5

A sawtooth signal is a periodic function with a period of $2a$ where $a > 0$, and it has the following form:

$$f(t) = \begin{cases} t/a & -a \le t < a, \\ 0 & \text{otherwise.} \end{cases}$$

So we have

$$a_n = \frac{1}{a}\int_{-a}^{a} f(t)\cos(\frac{n\pi t}{a})dt = 0, \quad n \ge 0,$$

because the integrand $t\cos(n\pi t/a)$ is an odd function and thus its integral in the symmetrical domain is zero. For $b_n$ with $n \ge 1$, we have

$$b_n = \frac{1}{a}\int_{-a}^{a} \frac{t}{a}\sin(\frac{n\pi t}{a})dt = \frac{2}{a^2}\int_0^a t\sin(\frac{n\pi t}{a})dt = \frac{-2}{\pi n}\cos(n\pi) + \frac{2}{\pi^2 n^2}\sin(n\pi) = \frac{2(-1)^{n+1}}{\pi n}.$$

Here, we have used $\sin(n\pi) = 0$ and $\cos(n\pi) = (-1)^n$ as well as the fact that the integrand is an even function. So the Fourier series of the sawtooth wave is

$$f(t) = \frac{2}{\pi}\sum_{n=1}^{\infty} \frac{(-1)^{n+1}}{n}\sin(\frac{n\pi t}{a}).$$

In the case of $a = \pi$, we have

$$f(t) = \frac{2}{\pi} \sum_{n=1}^{\infty} \frac{(-1)^{n+1}}{n} \sin(nt) = \frac{2}{\pi}\left[ \sin(t) - \frac{1}{2}\sin(2t) + \frac{1}{3}\sin(3t) - \frac{1}{4}\sin(4t) + \ldots \right].$$

It seems that the larger the number of terms, the better approximation to the saw-tooth signal. However, unlike the previous example on the triangular wave, the approximation always overshoots by a few percent at some points. In fact, no matter how many terms we may use (10, 100 or 1000), there is always an overshoot at the discontinuity at $t = \pi$. This overshoot is known as the Gibbs phenomenon.

Such overshoot does not just occur for the sawtooth wave, and it also occurs in many other cases. For example, we can perform an exercise to show that a square wave with the function form

$$s(t) = \begin{cases} +1 & 0 \le t \le \pi, \\ -1 & -\pi \le t < 0, \end{cases}$$

has a period of $2\pi$ and the following Fourier series

$$s(t) = \frac{4}{\pi} \sum_{n=1,3,5,\ldots}^{\infty} \frac{\sin(nt)}{n} = \frac{4}{\pi} \sum_{k=1}^{\infty} \frac{\sin[(2k-1)t]}{(2k-1)}$$

$$= \frac{4}{\pi}\left[ \sin(t) + \frac{1}{3}\sin(3t) + \frac{1}{5}\sin(5t) + \frac{1}{7}\sin(7t) + \ldots \right].$$

The Gibbs phenomenon was discovered by J. W. Gibbs in 1899. For the square wave, rigorous mathematical analysis (which is beyond the scope of this book) shows that the overshoot is

$$E_o = \frac{1}{\pi} \int_0^\pi \frac{\sin(u)}{u} du - \frac{1}{2} = \frac{1}{\pi} \int_0^\pi \text{sinc}(u) du - \frac{1}{2} \approx 0.089490,$$

which is about 9% as shown in Fig. 8.3.

## 2. Fourier Transforms

In general, when the period $T$ becomes infinite, the Fourier coefficients of a function defined on the whole real axis $(-\infty, \infty)$ can be written as

$$a(\omega_n) = \int_{-T}^{T} f(t)\cos(\omega_n t)dt, \qquad b(\omega_n) = \int_{-T}^{T} f(t)\sin(\omega_n t)dt, \qquad (8.22)$$

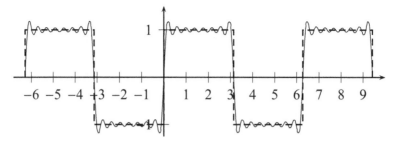

Figure 8.3: The square wave and its Fourier series. The overshoot is called the Gibbs phenomenon.

where $\omega_n = \frac{n\pi}{T}$ under the limits of $T \to \infty$ and $\omega_n \to 0$. If we further pose the constraint $\int_{-\infty}^{\infty} |f(t)| < \infty$, we get $a_0 \to 0$. In this case, the Fourier series becomes the Fourier integral

$$f(t) = \int_0^{\infty} [a(\omega)\cos(\omega t) + b(\omega)\sin(\omega t)]d\omega, \qquad (8.23)$$

where

$$a(\omega) = \frac{1}{\pi} \int_{-\infty}^{\infty} f(t)\cos(\omega t)dt, \quad b(\omega) = \frac{1}{\pi} \int_{-\infty}^{\infty} f(t)\sin(\omega t)dt. \qquad (8.24)$$

Following similar discussions above, even functions lead to Fourier cosine integrals and odd functions lead to Fourier sine integrals.

The Fourier transform $\mathcal{F}[f(t)]$ of $f(t)$ is defined as

$$F(\omega) = \mathcal{F}[f(t)] = \int_{-\infty}^{\infty} f(t)e^{-i\omega t}dt, \qquad (8.25)$$

and the inverse Fourier transform can be written as

$$f(t) = \mathcal{F}^{-1}[F(\omega)] = \frac{1}{2\pi} \int_{-\infty}^{\infty} F(\omega)e^{i\omega t}d\omega, \qquad (8.26)$$

where we have used $\exp[i\omega t] = \cos(\omega t) + i\sin(\omega t)$. The essence of such Fourier transform is to map the signal $f(t)$ in the time domain to a so-called spectrum in the frequency domain $F(\omega)$, which thus allows us to analyze the same signal from a different perspective.

It is worth pointing out that in some literature, the definition of Fourier transform may have a slightly different form with a factor of $1/\sqrt{2\pi}$.

Before we proceed, let us define the Dirac delta function $\delta$, which is the derivative of the Heaviside step function $H(x)$. That is

$$\delta(x) = \frac{dH(x)}{dx}, \quad H(x) = \begin{cases} 1, & \text{if } x > 0, \\ 0, & \text{if } x < 0. \end{cases} \tag{8.27}$$

This means that

$$\delta(x) = \begin{cases} +\infty & \text{if } x = 0, \\ 0 & \text{if } x \neq 0, \end{cases} \tag{8.28}$$

and

$$\delta(-x) = \delta(x), \quad \delta(x - a) = \delta(a - x), \tag{8.29}$$

for all $x$ and $a$. An interesting property is that

$$\int_{-\infty}^{+\infty} \delta(x)dx = 1. \tag{8.30}$$

In addition, $\delta(x)$ can be considered as the limit of the Gaussian distribution

$$\delta(x) = \lim_{\sigma \to 0} \frac{1}{\sqrt{2\pi\sigma^2}} e^{-x^2/(2\sigma^2)}, \tag{8.31}$$

which can also be defined as an integral

$$\delta(x) = \int_{-\infty}^{+\infty} e^{ixt} dt. \tag{8.32}$$

In practice, the peak amplitude of unit is often used for most calculations.
   Let us now try to derive the Fourier transform of a cosine function.

## Example 8.6

For $f(t) = \cos(\omega_0 t)$, using $\cos(\omega_0 t) = (e^{i\omega_0 t} + e^{-i\omega_0 t})/2$, we have

$$F(\omega) = \int_{-\infty}^{+\infty} f(t)e^{-i\omega t} dt = \int_{-\infty}^{+\infty} \frac{e^{i\omega_0 t} + e^{-i\omega_0 t}}{2} e^{-i\omega t} dt$$

$$= \frac{1}{2} \int_{-\infty}^{+\infty} e^{i(\omega_0 - \omega)t} dt + \frac{1}{2} \int_{-\infty}^{+\infty} e^{-i(\omega + \omega_0)t} dt.$$

Using the basic properties of delta function and (8.32), we get

$$F(\omega) = \mathcal{F}[\cos(\omega_0 t)] = \frac{1}{2}\delta(\omega - \omega_0) + \frac{1}{2}\delta(-\omega - \omega_0) = \frac{1}{2}\delta(\omega - \omega_0) + \frac{1}{2}\delta(\omega + \omega_0).$$

This means that the cosine signal $f(t) = \cos(\omega_0 t)$ becomes two vertical pulses, as shown in Fig. 8.4. In most textbooks, an unit impulse is represented by a delta function, implicitly assuming $\delta(0) = 1$.

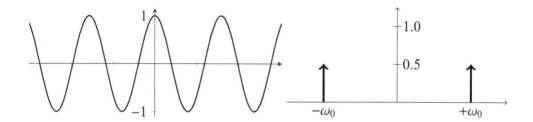

Figure 8.4: A cosine signal $\cos(\omega_0 t)$ is transformed into a spectrum in the frequency domain.

Up to now, we have used $t$ as the independent variable and $f(t)$ is a function of time $t$. However, Fourier transforms can also be applied to functions of spatial variable $x$ and other variables. Though there are differences between $x$ and $t$ in terms of physical variables, mathematically they are equivalent as long as their function forms and properties are the same. In addition, it is not required that the function $f(t)$ must be periodic for Fourier transforms.

Fourier transforms can have important properties such as linearity, shift and scaling. We will state the properties without any formal proof. Interested readers can refer to more advanced textbooks.

- Linearity: $\mathcal{F}[\alpha f(t) + \beta g(t)] = \alpha \mathcal{F}[f(t)] + \beta \mathcal{F}[g(t)]$.
- Scaling: $\mathcal{F}[f(\alpha t)] = \frac{1}{\alpha} F(\omega/\alpha)$.
- Shift: $\mathcal{F}[f(t - a)] = e^{-i\omega a} \mathcal{F}[f(t)]$.
- Derivative: $\mathcal{F}[(-it)^n f(t)] = \frac{d^n F(\omega)}{d\omega^n}$ and $\mathcal{F}[f'(t)] = i\omega \mathcal{F}[f(t)]$.
- Convolution: The convolution of two functions $f(t)$ and $g(t)$ is defined by

$$f(t) * h(t) = \int_{-\infty}^{\infty} f(\tau)g(t - \tau)d\tau. \tag{8.33}$$

Their Fourier transforms obey

$$\mathcal{F}[f(t) * g(t)] = \mathcal{F}[f(t)]\mathcal{F}[g(t)]. \tag{8.34}$$

Some of the properties can be easy to prove by direct calculations.

## Example 8.7

For example, by using $u = t - a$, we can show the shift property:

$$\mathcal{F}[f(t - a)] = \int_{-\infty}^{\infty} f(t - a)e^{-i\omega t}dt = \int_{-\infty}^{\infty} f(u)e^{-i\omega(u+a)}du$$

$$= e^{-i\omega a} \int_{-\infty}^{\infty} f(u)e^{-i\omega u}du = e^{-i\omega a}\mathcal{F}[f(t)]. \tag{8.35}$$

The convolution property can also be proved by direct integration. From

$$\mathcal{F}[f * g] = \int_{-\infty}^{\infty} \left[ \int_{-\infty}^{\infty} f(\tau)g(t-\tau)d\tau \right] e^{-i\omega t}dt, \tag{8.36}$$

we have (with $u = t - \tau$ or $t = u + \tau$)

$$\mathcal{F}[f * g] = \int_{-\infty}^{\infty} e^{-i\omega t}dt \int_{-\infty}^{\infty} [f(\tau)g(t-\tau)]d\tau = \int_{-\infty}^{\infty} f(\tau)d\tau \int_{-\infty}^{\infty} g(u)e^{-i\omega(u+\tau)}du$$

$$= \int_{-\infty}^{\infty} f(\tau)e^{-i\omega\tau}d\tau \int_{-\infty}^{\infty} g(u)e^{-i\omega u}du = \mathcal{F}[f(t)]\mathcal{F}[g(t)]. \tag{8.37}$$

There are some variations of the transforms such as the Fourier sine transform and the Fourier cosine transform. The Fourier transforms of some common functions, also called Fourier transform pairs, are listed in Table 8.1.

Table 8.1: Fourier Transforms.

| $f(t)$ | $F(\omega) = \mathcal{F}[f(t)]$ |
|---|---|
| $f(t - t_0)$ | $F(\omega)e^{-i\omega t_0}$ |
| $f(t)e^{-i\omega_0 t}$ | $F(\omega - \omega_0)$ |
| $\delta(t)$ | $1$ |
| $\cos(\omega_0 t)$ | $\pi[\delta(\omega - \omega_0) + \delta(\omega + \omega_0)]$ |
| $\sin(\omega_0 t)$ | $\frac{\pi}{i}[\delta(\omega - \omega_0) - \delta(\omega + \omega_0)]$ |
| $e^{i\omega_0 t}$ | $2\pi\delta(\omega - \omega_0)$ |
| $e^{-a|t|}$ | $\frac{2a}{a^2 + \omega^2}$ $(a > 0)$ |
| $u(t)e^{-at}$ | $\frac{1}{a + i\omega}$ $(a > 0)$ |
| $f'(t)$ | $i\omega F(\omega)$ |
| $f''(t)$ | $(i\omega)^2 F(\omega)$ |
| $tf(t)$ | $iF'(\omega)$ |
| $f(at)$ | $\frac{1}{|a|}F(\frac{\omega}{a})$ |

Discrete Fourier transforms, especially the Fast Fourier Transform (FFT), can be evaluated very efficiently in practice. In the rest of this chapter, we will introduce such transforms.

## 3. Discrete and Fast Fourier Transforms

Now we try to write the Fourier series (8.1) in a complex form using

$$\cos\theta = (e^{i\theta} + e^{-i\theta})/2$$

and

$$\sin\theta = (e^{i\theta} - e^{-i\theta})/2i,$$

we have the $n$th term

$$f_n(t) = a_n \cos(\frac{n\pi t}{T}) + b_n \sin(\frac{n\pi t}{T}) = \frac{a_n[e^{\frac{in\pi t}{T}} + e^{\frac{-in\pi t}{T}}]}{2} + \frac{b_n[e^{\frac{in\pi t}{T}} - e^{\frac{-in\pi t}{T}}]}{2i}$$

$$= \frac{(a_n - ib_n)}{2}e^{in\pi t/T} + \frac{(a_n + ib_n)}{2}e^{-in\pi t/T}. \tag{8.38}$$

If we define $\beta_n = \frac{(a_n - ib_n)}{2}$, and $\beta_{-n} = \frac{(a_n + ib_n)}{2}$ where $(n = 0, 1, 2, ...)$, and set $\beta_0 = a_0/2$, we get

$$f_n(t) = \beta_n e^{-in\pi t/T} + \beta_{-n}e^{-in\pi t/T}.$$

Therefore, the Fourier series can be written in the complex form

$$f(t) = \sum_{n=-\infty}^{\infty} \beta_n e^{i\pi nt/T}. \tag{8.39}$$

In signal processing, we are often interested in $f(t)$ in $[0, 2T]$ (rather than $[-T, T]$). Without loss of generality, we can set $T = \pi$. In this case, the Fourier coefficients become

$$\beta_n = \frac{1}{2\pi}\int_0^{2\pi} f(t)e^{-int}dt. \tag{8.40}$$

As it is not easy to compute these coefficients accurately, we often use the numerical integration to approximate the above integral with a step size $h = 2\pi/N$. This is equivalent to sampling $t$ with $N$ sample points $t_k = 2\pi k/N$ where $k = 0, 1, ..., N - 1$. Therefore, the coefficient can be estimated by

$$\beta_n = \frac{1}{2\pi}\int_0^{2\pi} f(t)e^{-int}dt \approx \frac{1}{N}\sum_{k=0}^{N-1} f(\frac{2\pi k}{N})e^{-2\pi ink/N}.$$

Once we know $\beta_n$, we know the whole spectrum. Let $f_k$ denote $f(2\pi k/N)$, we can define the discrete Fourier transform (DFT) as

$$F_n = \sum_{k=0}^{N-1} f_k e^{-\frac{2\pi ink}{N}}, \tag{8.41}$$

which is for periodic discrete signals $f(k)$ with a period of $N$. A periodic signal $f(k + N) = f(k)$ has a periodic spectrum $F(n + N) = F(n)$. The discrete Fourier transform consists of $N$ multiplications and $N - 1$ additions for each $F_n$, thus the computational complexity is of $O(N^2)$.

In fact, by rearranging the formulas, we can get a class of the Fast Fourier Transform (FFT) whose computational complexity is about $O(N \log(N))$. Using the notation $\omega = e^{-2\pi i/N}$ and $\omega^N = e^{2\pi i} = 1$, we can rewrite (8.41) as

$$F_n = \sum_{k=0}^{N-1} f_k \omega^{kn}, \quad n \in \mathbb{Z}. \tag{8.42}$$

It is worth pointing out that

$$\omega = e^{-2\pi i/N}$$

is the $N$th root of unity in the complex plane, thus the powers of $\omega$ always lie on a unit circle in the complex plane. Here, the computations only involve the summation and the power of $\omega$.

The inverse transform can be obtained by

$$f_k = \frac{1}{N} \sum_{n=0}^{N-1} F_n e^{2\pi i kn/N}, \quad k \in \mathbb{Z}. \tag{8.43}$$

If we are more interested in the range of $n = 0, 1, ..., N - 1$ because the sequence is $N$-periodic, then the above equation can be rewritten as a matrix form

$$\begin{pmatrix} F_0 \\ F_0 \\ \vdots \\ F_{N-1} \end{pmatrix} = \begin{pmatrix} 1 & 1 & 1 & \cdots & 1 \\ 1 & \omega & \omega^2 & \cdots & w^{N-1} \\ 1 & \omega^2 & \omega^4 & \cdots & \omega^{2(N-1)} \\ 1 & \omega^{N-1} & \omega^{2(N-1)} & \cdots & \omega^{(N-1)^2} \end{pmatrix} \begin{pmatrix} f_0 \\ f_1 \\ \vdots \\ f_{N-1} \end{pmatrix}.$$

As we know from matrix algebra, the original signal $f_k$ can conveniently be reconstructed by the inverse of the matrix.

This usually requires a lot of computations; however, in the case when $N$ can be factorized, some of the calculations can be decomposed into different steps and many of the calculations become unnecessary. In this case, we often use $N = 2^m$ where $m$ is a positive integer; it becomes the so-called FFT, and the computational complexity is now reduced to $2N \log_2(N)$. For example, when $N = 2^{20}$, FFT will reduce the computational time from three weeks for full Fourier transforms to less than a minute for FFT on modern desktop computers.

There is a huge amount of literature about FFT, filter design, signal reconstruction and their applications in signal and image processing.

## 4. Laplace Transform

The Laplace transform $\mathcal{L}[f(t)]$ of a function $f(t)$ is defined as

$$F(s) = \mathcal{L}[f(t)] = \int_0^\infty f(t)e^{-st}dt, \tag{8.44}$$

where $s > 0$ (or strictly the real part $\mathcal{R}(s) > 0$). The inverse Laplace transform of $F(s)$ is $f(t)$, and is denoted as $\mathcal{L}^{-1}[F(s)]$, or $f(t) = \mathcal{L}^{-1}[F(s)]$.

In theory, the inverse Laplace transform should be written as

$$f(t) = \frac{1}{2\pi i} \int_{\sigma - i\infty}^{\sigma + i\infty} F(s)e^{st}ds,$$

where $\sigma$ should be large enough so that $F(s)$ is defined for $\mathcal{R}(s) \geq \sigma$. In fact, $s$ should be treated as complex $s = \sigma + i\omega$ as we will see in the final section of this chapter. However, the inverse Laplace transform is not the focus here.

In comparison with Fourier transforms, we can consider that the Laplace transform is an extension of Fourier transforms by setting $s = i\omega$, or more generally $s = \sigma + i\omega$, for a signal $f(t)$ with $t \geq 0$.

The Laplace transforms of most simple functions can be obtained by direct integration.

### Example 8.8

For simple functions $t$ and $e^{\alpha t}$, we have

$$\mathcal{L}[t] = \int_0^\infty te^{-st}dt = \int_0^\infty \frac{1}{s}e^{-st}dt + \left[-\frac{t}{s}e^{-st}\right]_0^\infty = \frac{1}{s^2},$$

and

$$\mathcal{L}[e^{\alpha t}] = \int_0^\infty e^{\alpha t}e^{-st}dt = \int_0^\infty e^{-(s-\alpha)t}dt = \left[-\frac{1}{s-\alpha}e^{-(s-\alpha)t}\right]_0^\infty = \frac{1}{s-\alpha}.$$

Conversely, we have

$$\mathcal{L}^{-1}[\frac{1}{s^2}] = t, \quad \mathcal{L}^{-1}[\frac{1}{s-\alpha}] = e^{\alpha t}.$$

For the Dirac $\delta$-function, we have its Laplace transform

$$\mathcal{L}[\delta(t)] = \int_0^\infty \delta(t)e^{-st}dt = e^{-st}\Big|_{t=0}^\infty = 1. \tag{8.45}$$

Similarly, for a (unit) step function $f(t) = 1$ for $t \geq 0$, we have

$$\mathcal{L}(s) = \int_0^\infty e^{-st} dt = -\frac{1}{s} e^{-st} \Big|_0^\infty = \frac{1}{s}. \tag{8.46}$$

Here, we have implicitly assumed that $\exp(-st) \to 0$ as $t \to \infty$ and also implicitly assumed that $\mathcal{R}(s) > 0$. That is,

$$\mathcal{L}(1) = \int_0^\infty 1 e^{-st} dt = \frac{1}{s}. \tag{8.47}$$

## Example 8.9

For $f(t) = t^n$ where $n \geq 1$, we can use integration by parts by setting $u = t^n$ and $v' = e^{-st}$, and we have

$$\mathcal{L}(f(t)) = \int_0^\infty t^n e^{-st} dt$$

$$= -\frac{t^n e^{-st}}{s} \Big|_0^\infty + \frac{n}{s} \int_0^\infty t^{n-1} e^{-st} dt = \frac{n}{s} \mathcal{L}(t^{n-1}) = \ldots = \frac{n!}{s^{n+1}},$$

where we have assumed that $\mathcal{R}(s) > 0$.

Sometimes it is necessary to use other formulas and their combinations to obtain the Laplace transform of some functions.

## Example 8.10

In order to obtain the Laplace transform of $f(t) = \cos \omega t$, we shall first write

$$f(t) = \cos \omega t = \frac{1}{2}(e^{i\omega t} + e^{-i\omega t}).$$

Then, we have

$$\mathcal{L}[f(t)] = F(s) = \int_0^\infty [\frac{1}{2}(e^{i\omega t} + e^{-i\omega t})] e^{-st} dt = \frac{1}{2}[\int_0^\infty e^{-(s-i\omega)t} dt + \int_0^\infty e^{-(s+i\omega)t} dt]$$

$$= \frac{1}{2} \int_0^\infty e^{-u} \frac{du}{s - i\omega} + \frac{1}{2} \int_0^\infty e^{-u} \frac{du}{s + i\omega} = \frac{1}{2(s - i\omega)} \int_0^\infty e^{-u} du + \frac{1}{2(s + i\omega)} \int_0^\infty e^{-u} du$$

$$= \frac{1}{2}[\frac{1}{s - i\omega} + \frac{1}{s + i\omega}] = \frac{s}{s^2 + \omega^2}.$$

Here we have used $u = (s - i\omega)t$ assuming $s$ and $\omega$ are constants or parameters. We have also used the fact that $\int_0^\infty e^{-u}du = 1$.

Similarly, for the sine function, we have

$$\mathcal{L}[\sin(\omega t)] = \frac{\omega}{s^2 + \omega^2}. \tag{8.48}$$

## 4.1. Laplace Transform Pairs

The Laplace transform of a function can often be obtained by direct integration. However, the inverse Laplace transform is usually more complicated. It often involves the partial fractions of polynomials and usage of different rules of Laplace transforms.

The original function $f(t)$ and its Laplace transform $F(s)$ form a Laplace pair. The Laplace transform pairs of common functions are listed below in Table 8.2.

Table 8.2: Laplace Transform pairs.

| Function $f(t)$ | Laplace Transform $F(s)$ |
|---|---|
| 1 | $\frac{1}{s}$ |
| $\delta(t)$ | 1 |
| $t^n, n > 0$ | $\frac{n!}{s^{n+1}}$ |
| $\cos(\alpha t)$ | $\frac{s}{s^2+\alpha^2}$ |
| $\sin(\alpha t)$ | $\frac{\alpha}{s^2+\alpha^2}$ |
| $e^{\alpha t}$ | $\frac{1}{s-\alpha}$ |
| $t^{1/2}$ | $\frac{1}{2}(\frac{\pi}{s^3})^{1/2}$ |
| $t^{-1/2}$ | $\sqrt{\frac{\pi}{s}}$ |
| $t^n f(t)$ | $(-1)^n \frac{d^n F(s)}{ds^n}$ |
| $\cos(\alpha t + \beta)$ | $\frac{s\cos(\beta)-\alpha\sin(\beta)}{s^2+\alpha^2}$ |
| $\sinh(\alpha t)$ | $\frac{\alpha}{s^2-\alpha^2}$ |
| $\cosh(\alpha t)$ | $\frac{s}{s^2-\alpha^2}$ |
| $\mathrm{erfc}(\frac{\alpha}{2\sqrt{t}})$ | $\frac{1}{s}e^{-\alpha\sqrt{s}}$ |
| $\frac{1}{\sqrt{\pi t}}e^{-\frac{\alpha^2}{4t}}$ | $\frac{1}{\sqrt{s}}e^{-\alpha\sqrt{s}}$ |
| $\sin \alpha \sqrt{t}$ | $\frac{\alpha}{2}\sqrt{\frac{\pi}{s^3}}e^{-\frac{\alpha^2}{4s}}$ |
| $\frac{1-e^{-\alpha t}}{t}$ $(\alpha > 0)$ | $\ln(1 + \frac{\alpha}{s})$ |
| $\frac{1}{\alpha-\beta}(e^{\alpha t} - e^{\beta t})$ $(\alpha \neq \beta)$ | $\frac{1}{(s-\alpha)(s-\beta)}$ |

Both Fourier and Laplace transforms follow the convolution theorem. For two functions $f$ and $g$, their convolution $f * g$ is defined by

$$f * g = \int_0^t f(t - \alpha)g(\alpha)d\alpha. \tag{8.49}$$

and their Laplace transforms follow

$$\mathcal{L}[f(t) * g(t)] = F(s)G(s), \tag{8.50}$$

$$\mathcal{L}^{-1}[F(s)G(s)] = \int_0^t f(t - \alpha)g(\alpha)d\alpha. \tag{8.51}$$

## 4.2. Scalings and Properties

From the basic definition, it is straightforward to prove that the Laplace transform has the following properties:

$$\mathcal{L}[\alpha f(t) + \beta g(t)] = \alpha\mathcal{L}[f(t)] + \beta\mathcal{L}[g(t)], \tag{8.52}$$

$$\mathcal{L}[e^{\alpha t} f(t)] = F(s - \alpha), \quad s > \alpha, \quad F(s) = \mathcal{L}[f(t)], \tag{8.53}$$

$$\mathcal{L}[f(t - \alpha)] = e^{-\alpha s}\mathcal{L}[f(t)], \tag{8.54}$$

$$\mathcal{L}[f'(t)] = s\mathcal{L}[f(t)] - f(0), \quad \mathcal{L}[\int_0^t f(\tau)d\tau] = \frac{1}{s}\mathcal{L}[f], \tag{8.55}$$

These properties are easy to verify, and we will show how to derive some of these properties below. For example, it is straightforward to verify that

$$\mathcal{L}[e^{\alpha t} f(t)] = \int_0^\infty f(t)e^{\alpha t}e^{-st}dt = \int_0^\infty f(t)e^{-(s-a)t}dt = F(s - a). \tag{8.56}$$

For a function $g(t) = f(at)$ where $a > 0$, we can set $\tau = at$, and thus we have $d\tau = adt$ (or $dt = d\tau/a$). We finally get

$$G(s) = \int_0^\infty f(at)e^{-st}dt = \frac{1}{a}\int_0^\infty f(\tau)e^{s\tau/a}d\tau = \frac{1}{a}F(\frac{s}{a}). \tag{8.57}$$

The properties of Laplace transforms can be used to simplify the calculations of Laplace transforms for certain complex expressions. Let us see an example.

## Example 8.11

From $\mathcal{L}(\cos t) = \frac{s}{s^2+1}$ and using the above equation, we have

$$\mathcal{L}[e^{-t}\cos(t)] = \frac{s + 1}{(s + 1)^2 + 1} = \frac{s + 1}{s^2 + 2s + 2}. \tag{8.58}$$

For a more complicated expression such as

$$f(t) = e^{-t} + e^{-t}\cos t + \sinh(at) + 2\sqrt{t},$$

we can use the above linear properties and we have

$$\mathcal{L}[f(t)] = \mathcal{L}[e^{-t}] + \mathcal{L}[e^{-t}\cos t] + \mathcal{L}[\sinh(at)] + 2\mathcal{L}[\sqrt{t}]$$

$$= \frac{1}{s+1} + \frac{s+1}{(s+1)^2+1} + \frac{a}{s^2-a^2} + 2\frac{\sqrt{\pi}}{2s^{3/2}} = \frac{1}{s+1} + \frac{s+1}{s^2+2s+2} + \frac{a}{s^2-a^2} + \frac{\sqrt{\pi}}{s^{3/2}},$$

where we have used the result of $\mathcal{L}[e^{-t}\cos t]$.

For a given time delay, $w(t) = f(t - T)$ where $t \geq T$, we have

$$W(s) = \int_0^\infty w(t)e^{-st}dt = \int_T^\infty f(t-T)e^{-st}dt$$

$$= \int_0^\infty f(\tau)e^{-s(\tau+T)}d\tau = e^{-sT}F(s). \tag{8.59}$$

## Example 8.12

Using the above properties, it is easy to obtain the following:

$$\mathcal{L}[(e^{3t}(t^2+1)] = \frac{2}{(s-3)^3} + \frac{1}{(s-3)}. \tag{8.60}$$

because

$$\mathcal{L}[(t^2+1)] = \frac{2}{s^3} + \frac{1}{s}, \quad \mathcal{L}[e^{3t}(t^2+1)] = \frac{2}{(s-3)^3} + \frac{1}{s-3}. \tag{8.61}$$

Similarly, we have

$$\mathcal{L}[e^{at}\cos \omega t] = \frac{(s-a)}{(s-a)^2+\omega^2}, \tag{8.62}$$

and

$$\mathcal{L}[t^3 + 2t + 7] = \mathcal{L}(t^3) + 5\mathcal{L}(t) + 7\mathcal{L}(1) = \frac{3!}{s^4} + \frac{2}{s^2} + \frac{7}{s}. \tag{8.63}$$

## 4.3. Derivatives and Integrals

It is straightforward to check that

$$\mathcal{L}(f') = \int_0^\infty f'(t)e^{-st}dt = e^{-st}f(t)\Big|_0^\infty - \int_0^\infty f(t)(-se^{-st})dt$$

$$= \lim_{t\to\infty}[f(t)e^{-st}] - f(0) + s\int_0^\infty f(t)e^{-st}dt = sF(s) - f(0), \qquad (8.64)$$

where we have used two facts: $s$ is independent of $t$ and $e^{-st}f \to 0$ as $t \to \infty$.

Following the similar procedure, we can prove that

$$\mathcal{L}(f'') = s\mathcal{L}(f') - f'(0) = s[sF(s) - f(0)] - f'(0)$$

$$= s^2F(s) - sf(0) - f'(0). \qquad (8.65)$$

Since $\sin(\omega t) = -(\cos\omega t)'/\omega$, we have

$$\mathcal{L}(\sin\omega t) = -\frac{1}{\omega}\mathcal{L}(\cos\omega t) = -\frac{1}{\omega}\Big[s(\frac{s}{s^2+\omega^2}) - 1\Big] = \frac{\omega}{s^2+\omega^2}. \qquad (8.66)$$

Furthermore, for $g(t) = tf(t)$, we have

$$\mathcal{L}[g(t)] = -F'(s). \qquad (8.67)$$

We can prove this formula by differentiating (with respect to $s$)

$$F'(s) = \Big[\int_0^\infty f(t)e^{-st}\Big]' = \int_0^\infty \Big(\frac{d[f(t)e^{-st}]}{ds}\Big)dt = \int_0^\infty (-t)e^{-st}f(t)dt = -\mathcal{L}[tf(t)].$$
$$(8.68)$$

That is

$$\mathcal{L}[tf(t)] = -F'(s). \qquad (8.69)$$

On the other hand, for integration $h(t) = \int_0^t f(\tau)d\tau$, we have

$$H(s) = \int_0^\infty \Big(\int_0^t f(\tau)d\tau\Big)e^{-st}dt = \int_0^\infty \Big[\int_0^t f(\tau)e^{-st}dt\Big]d\tau. \qquad (8.70)$$

Because we are integrating horizontally first over a triangular area over $0 \leq \tau < t$, it is possible to switch the order to integrating vertically first. This means that we have changed the integral limits and we have

$$\int_{\tau=0}^\infty \int_{t=\tau}^\infty f(\tau)e^{-st}dtd\tau = \int_{\tau=0}^\infty f(\tau)\Big(\int_{t=\tau}^\infty e^{-st}dt\Big)d\tau$$

$$= \int_0^\infty f(\tau)(\frac{1}{s})e^{-s\tau}d\tau = \frac{F(s)}{s}. \qquad (8.71)$$

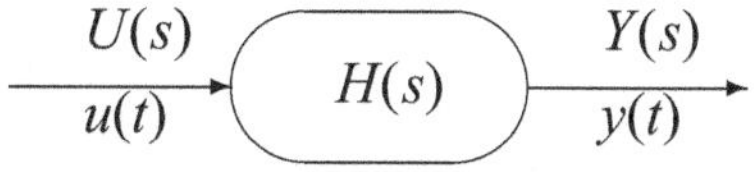

Figure 8.5: System response $Y(s) = H(s)U(s)$ for a given input $u(t)$.

That is

$$\mathcal{L}\left(\int_0^t f(\tau)d\tau\right) = \frac{F(s)}{s}. \tag{8.72}$$

So with Laplace transforms, differentiation becomes multiplication, while integration becomes division, which makes the system analysis in control theory much easier.

Transfer functions are used widely in control systems when describing the characteristics and response behaviour of linear systems. Laplace transforms typically transform the differential equation that describes the system into a polynomial transfer function in the state space or in the complex frequency domain.

A transfer function $H(s)$ is a representation of a system in the $s$-domain, often in terms of Laplace transforms. From any input $u(t)$ to a system (see Fig. 8.5), the output or response of the system $y(t)$ can be obtained by

$$Y(s) = H(s)U(s), \tag{8.73}$$

where $U(s) = \mathcal{L}(u(t))$ and $Y(s) = \mathcal{L}(y(t))$ are the Laplace transforms of $u(t)$ and $y(t)$, respectively. Therefore, $H(s)$ is a system function. The behaviour and characteristics of a system can be fully described by its transfer function $H(s)$.

At the same time, a system (such as a damped spring system or a robotic arm) can be modelled as a second-order ordinary differential equation

$$my''(t) + by'(t) + ky(t) = u(t), \tag{8.74}$$

where $u(t)$ is the driving force of a unit-step type. This can be converted into a transfer function

$$G(s) = \frac{Y(s)}{U(s)} = \frac{1}{ms^2 + bs + k}. \tag{8.75}$$

## Example 8.13

An RLC circuit can be modelled as

$$L\frac{\partial i}{\partial t} + Ri + \frac{1}{C}\int i dt = V, \tag{8.76}$$

or

$$Lq''(t) + Rq'(t) + \frac{q(t)}{C} = V, \quad i = dq/dt, \tag{8.77}$$

where $R$, $L$ and $C$ are constants. This differential equation is equivalent to the following transfer function

$$G(s) = \frac{Q(s)}{V(s)} = \frac{1}{Ls^2 + Rs + 1/C}. \tag{8.78}$$

It is worth pointing out that transfer functions are usually about zero initial states and thus all initial conditions are zero.

## 5. Solving ODE via Laplace Transform

The beauty of the Laplace transform method is to transform an ordinary differential equation (ODE) into an algebraic equation. The solution in the $s$-domain can be obtained by solving the algebraic equation, and the final solution in the original domain can be obtained by the inverse Laplace transform after some algebraic manipulations. Let us look at two examples.

### Example 8.14

For a first-order ODE

$$y'(t) + y(t) = u(t), \quad y(0) = 0,$$

where $u(t)$ is a unit step function ($u = 1$ for $t \geq 0$ and $u = 0$ when $t < 0$). Taking Laplace transform, we have

$$sY(s) - y(0) + Y(s) = \frac{1}{s}, \tag{8.79}$$

where we have used $\mathcal{L}(u(t)) = 1/s$. The solution $Y(s)$ is

$$Y(s) = \frac{1/s}{s+1} = \frac{1}{s(s+1)} = \frac{1}{s} - \frac{1}{s+1}. \tag{8.80}$$

The true solution $y(t)$ can be obtained by transforming $Y(s)$ back

$$y(t) = \mathcal{L}^{-1}[\frac{1}{s} - \frac{1}{s+1}] = \mathcal{L}^{-1}(\frac{1}{s}) - \mathcal{L}^{-1}(\frac{1}{s+1}) = 1 - e^{-t}. \tag{8.81}$$

This ODE is relatively simple. To solve the higher-order differential equations, the procedure is the same, though the calculations can be more complicated.

## 6. Z-Transform

The Laplace transform $X(s)$ works for a continuous-time signal $x(t)$. That is,

$$X(s) = \int_0^\infty x(t)e^{-st}dt. \tag{8.82}$$

When signal $x(t)$ is sampled, at $t = nT_s$ where $T_s$ is a fixed constant and $n = 0, 1, 2, ...,$ we can denote the signal as $x(n)$. If we let $T_s = 1$, the above equation (8.82) becomes

$$x(e^s) = \sum_{n=0}^\infty x(n)e^{-sn}. \tag{8.83}$$

Setting $z = e^s$, we have

$$X(z) = \sum_{n=0}^\infty x(n)z^{-n}, \tag{8.84}$$

which is a one-sided z-transform. It is straightforward to extend this to a two-sided z-transform

$$X(z) = \sum_{n=-\infty}^\infty x(n)z^{-n}. \tag{8.85}$$

Obviously, if we let the signal $x(n) = 0$ for $n < 0$, then we have the original one-sided z-transform, as usual. Therefore, the z-transform can be considered as a discrete form of the Laplace transform.

The z-transform has many properties similar to those of the Laplace transform, including linearity.

For two signals $x_1(n)$ and $x_2(n)$, their transforms $X_1(z) = \mathcal{Z}(x_1)$ and $X_2(z) = \mathcal{Z}(x_2)$ obey

$$\mathcal{Z}(\alpha x_1 + \beta x_2) = \alpha X_1(z) + \beta X_2(z). \tag{8.86}$$

For a time shift $k$, we can use $m = n - k$ and we have

$$\mathcal{Z}[x(n-k)] = \sum_{n=-\infty}^\infty x(n-k)z^{-n} = \sum_{m=-\infty}^\infty x(m)z^{-(m+k)}$$

$$= z^{-k} \sum_{m=-\infty}^\infty x(m)z^{-m} = z^{-k}X(z). \tag{8.87}$$

Other properties can be derived, in a similar way, from their corresponding Laplace transform counterparts.

## 7. Relationships between Fourier, Laplace and Z-transforms

It is worth pointing out that there are intrinsic links between Fourier transforms, Laplace transforms and z-transforms. Using $z = e^s$ and $s = \sigma + i\omega$, we have

$$z = e^s = e^\sigma e^{i\omega} = re^{i2\pi f}, \quad \text{so} \quad r = e^\sigma, \quad \omega = 2\pi f. \tag{8.88}$$

When $r = e^\sigma = 1$ (or $\sigma = 0$, corresponding to a unit circle in the z-plane or complex plane), the z-transform becomes the (discrete) Fourier transform.

$$X(z = e^{-i2\pi f}) = \sum_{n=-\infty}^{\infty} x(n)e^{-i2\pi f n}. \tag{8.89}$$

As a result, we can consider the z-transform as a generalized form of the Fourier transform (of a sampled signal). In addition, the z-transform is the Laplace transform of a discrete signal.

We also already know that the Laplace transform can be considered as a one-sided transform of the Fourier transform when $\sigma = 0$ so that $s = \sigma + i\omega = i\omega$.

The relationships between the Fourier transform, the Laplace transform and the z-transform can be summarized as follows:

- The Fourier transform is the Laplace transform when evaluating along the $s = i\omega$ axis.
- The z-transform is the Laplace transform for a sampled signal, and becomes the latter when $T_s \to 0$.
- The discrete-time Fourier transform (DTFT) can be obtained by substituting $z = e^{i\omega}$ in the z-transform.

## Exercises

**8.1.** Show that the Fourier series of $f(t) = |\sin t|$ is

$$f(t) = \frac{2}{\pi}\Big[1 - 2\sum_{n=1}^{\infty} \frac{\cos(2nt)}{4n^2 - 1}\Big].$$

Discuss the Fourier series for $g(t) = Af(t)$ where $A$ is a constant.

**8.2.** Show that $\mathcal{L}[\sinh(at)] = a/(s^2 - a^2)$.

**8.3.** Show that

$$\mathcal{L}[\sin(at + b)] = \frac{s\sin b + a\cos b}{(s^2 + a^2)},$$

where $a$ and $b$ are constants.

**8.4.** Solve $y'(t) - y(t) = e^{-t}$ with $y(0) = 0$ using the Laplace transform technique.

# CHAPTER 9

# Statistics and Curve Fitting

Contents

## Key Points

- Random variables and their probability distributions are introduced, including binomial, Poisson and Gaussian distributions and others.
- Key concepts such as mean, expectation and variance are also briefly presented with examples.
- Linear regression and least-squares method as a curve-fitting technique are explained in detail.

## 1. Random Variables, Means and Variance

Randomness such as roulette-rolling and noise arises from the lack of information, or incomplete knowledge of reality. It can also come from the intrinsic complexity, diversity and perturbations of the system. The theory of probability is mainly the study of random phenomena so as to find non-random regularity. Probability $P$ is a number or an expected frequency assigned to an event $A$ that indicates how likely it is that the event will occur when a random experiment is performed. This probability is often written as $P(A)$ to show that the probability $P$ is associated with event $A$.

For a discrete random variable $X$ with distinct values such as the number of cars passing through a junction, each value $x_i$ may occur with a certain probability $p(x_i)$.

In other words, the probability varies and is associated with its corresponding random variable. Traditionally, an uppercase letter such as $X$ is used to denote the random variable, while a lowercase letter such as $x_i$ to represent its values. For example, if $X$ means a coin-flipping event, then $x_i = 0, 1$. A probability function $p(x_i)$ is a function that assigns probabilities to all the discrete values $x_i$ of the random variable $X$. As an event must occur inside a sample space, the requirement that all the probabilities must be summed to one leads to

$$\sum_{i=1}^{n} p(x_i) = 1. \tag{9.1}$$

For example, the outcomes of tossing a fair coin form a sample space. The outcome of a head (H) is an event with a probability of $P(H) = 1/2$, and the outcome of a tail (T) is also an event with a probability of $P(T) = 1/2$. The sum of both probabilities should be one; that is

$$P(H) + P(T) = \frac{1}{2} + \frac{1}{2} = 1. \tag{9.2}$$

The cumulative probability function of $X$ is defined by

$$P(X \le x) = \sum_{x_i < x} p(x_i). \tag{9.3}$$

For a continuous random variable $X$ that takes a continuous range of values (such as the level of noise), its distribution is continuous and the probability density function $p(x)$ is defined for a range of values $x \in [a, b]$ for given limits $a$ and $b$ [or even over the whole real axis $x \in (-\infty, \infty)$]. In this case, we always use the interval $(x, x + dx]$ so that $p(x)$ is the probability that the random variable $X$ takes the value $x < X \le x + dx$ is

$$\Phi(x) = P(x < X \le x + dx) = p(x)dx. \tag{9.4}$$

As all the probabilities of the distribution shall be added to unity, we have

$$\int_{a}^{b} p(x)dx = 1. \tag{9.5}$$

The cumulative probability function becomes

$$\Phi(x) = P(X \le x) = \int_{a}^{x} p(x)dx, \tag{9.6}$$

which is the definite integral of the probability density function between the lower limit $a$ up to the present value $X = x$.

Two main measures for a random variable $X$ with a given probability distribution

$p(x)$ are its mean and variance. The mean $\mu$ or expectation of $E[X]$ is defined by

$$\mu \equiv E[X] \equiv <X> = \int xp(x)dx, \tag{9.7}$$

for a continuous distribution and the integration is within the integration limits. If the random variable is discrete, then the integration becomes the weighted sum

$$E[X] = \sum_i x_i p(x_i). \tag{9.8}$$

The variance $\text{var}[X] = \sigma^2$ is the expectation value of the deviation squared $(X - \mu)^2$. That is

$$\sigma^2 \equiv \text{var}[X] = E[(X - \mu)^2] = \int (x - \mu)^2 p(x)dx. \tag{9.9}$$

The square root of the variance $\sigma = \sqrt{\text{var}[X]}$ is called the standard deviation, which is simply $\sigma$.

## Example 9.1

A simple distribution is the so-called uniform distribution, which has a probability density

$$p(x) = \frac{1}{b - a}, \quad a \le x \le b, \; b > a > 0,$$

which means that $p(x)$ is simply a constant.

The mean can be calculated by

$$\mu = \int_a^b xp(x)dx = \int_a^b \frac{x}{(b - a)}dx = \frac{1}{(b - a)} \int_a^b xdx$$

$$= \frac{1}{(b - a)}\left[\frac{x^2}{2}\right]_a^b = \frac{1}{(b - a)} \cdot \left[\frac{b^2}{2} - \frac{a^2}{2}\right] = \frac{a + b}{2}.$$

In addition, the variance is

$$\sigma^2 = E[(X - \mu)^2] = \int_a^b (x - \mu)^2 p(x)dx = \int_a^b \left[x - \frac{(a + b)}{2}\right]^2 \cdot \frac{1}{(b - a)}dx$$

$$= \frac{1}{(b - a)} \int_a^b \left[x^2 - (a + b)x + \frac{(a + b)^2}{4}\right]dx = \frac{1}{(b - a)}\left[\frac{x^3}{3}\bigg|_a^b - \frac{(a + b)}{2}x^2\bigg|_a^b + \frac{(a + b)^2}{4}x\bigg|_a^b\right]$$

$$= \frac{1}{12(b - a)}(b^3 + 3a^2 b - 3ab^3 - a^3) = \frac{1}{12(b - a)}(b - a)^3 = \frac{(b - a)^2}{12}.$$

For a discrete distribution, the variance simply becomes the following sum

$$\sigma^2 = \sum_i (x - \mu)^2 p(x_i). \tag{9.10}$$

In addition, any other formulas for a continuous distribution can be converted to their counterpart for a discrete distribution if the integration is replaced by the sum. Therefore, we will mainly focus on the continuous distribution in the rest of the section.

From the above definitions, it is straightforward to prove

$$E[\alpha x + \beta] = \alpha E[X] + \beta, \qquad E[X^2] = \mu^2 + \sigma^2, \tag{9.11}$$

and

$$\text{var}[\alpha x + \beta] = \alpha^2 \text{var}[X], \tag{9.12}$$

where $\alpha$ and $\beta$ are constants.

Other frequently used measures are the mode and median. The mode of a distribution is defined by the value at which the probability density function $p(x)$ is the maximum. For an even number of data sets, the mode may have two values. The median $m$ of a distribution corresponds to the value at which the cumulative probability function $\Phi(m) = 1/2$. The upper and lower quartiles $Q_U$ and $Q_L$ are defined by $\Phi(Q_U) = 3/4$ and $\Phi(Q_L) = 1/4$.

## 2. Binomial and Poisson Distributions

A discrete random variable is said to follow the binomial distribution $B(n, p)$ if its probability distribution is given by

$$B(n, p) = {}^nC_x p^x (1 - p)^{n-x}, \qquad {}^nC_x = \binom{n}{x} = \frac{n!}{x!(n - x)!}, \tag{9.13}$$

where $x = 0, 1, 2, ..., n$ are the values that the random variable $X$ may take, and $n$ is the number of trials. There are only two possible outcomes: success or failure. $p$ is the probability of a so-called 'success' of the outcome. Subsequently, the probability of the failure of a trial is $q = 1 - p$. Therefore, $B(n, p)$ represents the probability of $x$ successes and $n - x$ failures in $n$ trials. The coefficients come from the coefficients of the binomial expansions

$$(p + q)^n = \sum_{x=0}^{n} {}^nC_x p^x q^{n-x} = 1, \tag{9.14}$$

which is exactly the requirement that all the probabilities should be summed to unity.

## Example 9.2

Tossing a fair coin 40 times, the probability of getting 30 heads is $B(n, 1/2)$. Since $p = 1/2$, $n = 40$ and $x = 30$, then we have

$$\binom{40}{30}\left(\frac{1}{2}\right)^{30}\left(\frac{1}{2}\right)^{10} = 847660528 \times \frac{1}{2^{40}} \approx 0.00077.$$

It is straightforward to prove that $\mu = E[X] = np$ and $\sigma^2 = npq = np(1 - p)$ for a binomial distribution, and thus their proofs are left as an exercise.

Another related distribution is the geometric distribution whose probability function is defined by

$$P(X = n) = pq^{n-1} = p(1 - p)^{n-1}, \tag{9.15}$$

where $n \geq 1$. This distribution is used to calculate the first success, thus the first $n - 1$ trials must be a failure if $n$ trials are needed to observe the first success. The mean and variance of this distribution are $\mu = 1/p$ and $\sigma^2 = (1 - p)/p^2$.

The Poisson distribution can be thought of as the limit of the binomial distribution when the number of trial is very large $n \to \infty$ and the probability $p \to 0$ (small probability) with the constraint that $\lambda = np$ is finite. For this reason, it is often called the distribution for small-probability events. Typically, it is concerned with the number of events that occur in a certain time interval (e.g., number of telephone calls in an hour) or spatial area. The Poisson distribution is

$$P(X = x) = \frac{\lambda^x e^{-\lambda}}{x!}, \qquad \lambda > 0, \tag{9.16}$$

where $x = 0, 1, 2, ..., n$ and $\lambda$ is the mean of the distribution.

Obviously, the sum of all the probabilities must be equal to one. That is

$$\sum_{x=0}^{\infty} \frac{\lambda^x e^{-\lambda}}{x!} = \frac{\lambda^0 e^{-\lambda}}{0!} + \frac{\lambda^1 e^{-\lambda}}{1!} + \frac{\lambda^2 e^{-\lambda}}{2!} + \frac{\lambda^3 e^{-\lambda}}{3!} + ...$$

$$= e^{-\lambda}[1 + \lambda + \frac{\lambda^2}{2!} + \frac{\lambda^3}{3!} + ...] = e^{-\lambda}e^{\lambda} = e^{-\lambda+\lambda} = e^0 = 1. \tag{9.17}$$

Many stochastic processes such as the number of phone calls in a call centre and the number of cars passing through a junction obey the Poisson distribution. If we are concerned with Poisson distribution with a time interval $t$, $\lambda$ will be the arrival rate per unit time. However, in general, we should use $x = \lambda t$ to replace $x$ when dealing with the arrivals in a fixed period $t$. Thus, the Poisson distribution becomes

$$P(X = n) = \frac{(\lambda t)^n e^{-\lambda t}}{n!}. \tag{9.18}$$

## Example 9.3

For example, if you receive 3 emails per hour on average, if you attend a 2-hour lesson, what is the probability of getting exactly 5 emails after the 2-hour session? Since $\lambda = 3$, $t = 2$ and $n = 5$, we have

$$P(n = 5) = \frac{(\lambda t)^n e^{-\lambda t}}{n!} = \frac{(3 \times 2)^5 e^{-3 \times 2}}{5!} \approx 0.16.$$

The probability of no email after 2 hours is

$$P(n = 0) = \frac{(3 \times 2)^0 e^{-3 \times 2}}{0!} \approx 0.0025.$$

As the Poisson distribution is so widely used for describing discrete events, let us look at another example in a different context.

## Example 9.4

Suppose your experiment requires some resistors and the resistors (by a manufacturer) have a 1.5% probability of defects. If you buy 200 resistors, what is the probability of this batch containing exactly 2 defective resistors?

The parameter $\lambda$ can be calculated as

$$\lambda = 200 \times 1.5\% = 200 \times 0.015 = 3.$$

So the probability of getting exactly 2 defective resistors is

$$P(X = 2) = \frac{3^2 e^{-3}}{2!} \approx 0.224.$$

The probability of at least one defective resistor is

$$P(X \geq 1) = 1 - P(X = 0) = 1 - \frac{3^0 e^{-3}}{0!} \approx 0.95.$$

Using the definitions of mean and variance, it is straightforward to prove that $\mu = \lambda$ and $\sigma^2 = \lambda$ for the Poisson distribution.

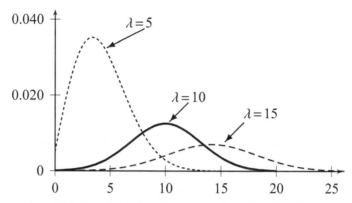

Figure 9.1: Poisson distributions for different values of $\lambda = 5, 10, 15$.

## Example 9.5

The mean or expectation $E(X)$ can be calculated by

$$E(X) = \sum_{x=0}^{\infty} xP(X = x) = \sum_{x=0}^{\infty} x\frac{\lambda^x e^{-\lambda}}{x!}$$

$$= 0 \times e^{-\lambda} + 1 \times (\lambda e^{-\lambda}) + 2 \times (\frac{\lambda^2 e^{-\lambda}}{2!}) + 3 \times (\frac{\lambda^3 e^{-\lambda}}{3!} + ...)$$

$$= \lambda e^{-\lambda}[1 + \lambda + \frac{\lambda^2}{2!} + \frac{\lambda^3}{3!} + ...] = \lambda e^{-\lambda}e^{+\lambda} = \lambda.$$

The parameter $\lambda$ controls the location of the peak as shown in Fig. 9.1, and $\lambda$ essentially describes a Poisson distribution uniquely. Therefore, some textbooks use Poisson($\lambda$) to denote a Poisson distribution with parameter $\lambda$.

## 3. Gaussian Distribution

The Gaussian distribution or normal distribution is the most important continuous distribution in probability and it has a wide range of applications. For a continuous random variable $X$, the probability density function (PDF) of a Gaussian distribution is given by

$$p(x) = \frac{1}{\sigma\sqrt{2\pi}}e^{-\frac{(x-\mu)^2}{2\sigma^2}}, \tag{9.19}$$

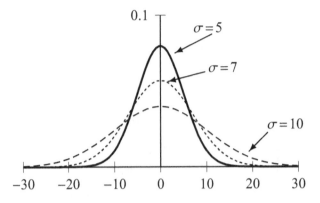

Figure 9.2: Gaussian distributions for $\sigma = 5, 7, 10$.

where $\sigma^2 = \text{var}[X]$ is the variance and $\mu = E[X]$ is the mean of the Gaussian distribution. From the Gaussian integral discussed in Chapter 5, it is straightforward to verify that

$$\int_{-\infty}^{\infty} p(x)dx = 1, \tag{9.20}$$

and this is exactly the reason why the factor $1/\sqrt{2\pi}$ is required in the normalization of all the probabilities.

The probability function reaches a peak at $x = \mu$ and the variance $\sigma^2$ controls the width of the peak (see Fig. 9.2) where $\mu = 0$ is used.

The cumulative probability function (CPF) for a normal distribution is the integral of $p(x)$, which is defined by

$$\Phi(x) = P(X < x) = \frac{1}{\sqrt{2\pi\sigma^2}} \int_{-\infty}^{x} e^{-\frac{(\zeta-\mu)^2}{2\sigma^2}} d\zeta. \tag{9.21}$$

Using the error function defined in early chapters, we can write the above equation as

$$\Phi(x) = \frac{1}{\sqrt{2}}\left[1 + \text{erf}\left(\frac{x-\mu}{\sqrt{2}\sigma}\right)\right], \tag{9.22}$$

where

$$\text{erf}(x) = \frac{2}{\sqrt{\pi}} \int_{0}^{x} e^{-\zeta^2} d\zeta. \tag{9.23}$$

The Gaussian distribution can be considered as the limit of the Poisson distribution when $\lambda \gg 1$. Using the Sterling's approximation $x! \sim \sqrt{2\pi x}(x/e)^x$ for $x \gg 1$, and setting $\mu = \lambda$ and $\sigma^2 = \lambda$, it can be verified that the Poisson distribution can be written as a Gaussian distribution

$$P(x) \approx \frac{1}{\sqrt{2\pi\lambda}} e^{-\frac{(x-\mu)^2}{2\lambda}},$$

(9.24)

where $\mu = \lambda$. In statistical applications, the normal distribution is often written as $N(\mu, \sigma^2)$ or $N(\mu, \sigma)$ to emphasize that the probability density function depends on two parameters $\mu$ and $\sigma$.

The standard normal distribution is a normal distribution $N(\mu, \sigma^2)$ with a mean of $\mu = 0$ and standard deviation $\sigma = 1$, that is $N(0, 1)$. This is useful to normalize or standardize data for statistical analysis. If we define a normalized variable

$$\xi = \frac{x - \mu}{\sigma},$$

(9.25)

it is equivalent to giving a score so as to place the data above or below the mean in the unit of standard deviation. In terms of the area under the probability density function, $\xi$ sorts where the data falls. It is worth pointing out that some books define $z = \xi = (x - \mu)/\sigma$ in this case, and call the standard normal distribution the Z distribution.

Table 9.1 Function $\phi$ defined by (9.27).

| $\xi$ | $\phi(\xi)$ | $\xi$ | $\phi(\xi)$ |
|---|---|---|---|
| 0.0 | 0.500 | 1.0 | 0.841 |
| 0.1 | 0.540 | 1.1 | 0.864 |
| 0.2 | 0.579 | 1.2 | 0.885 |
| 0.3 | 0.618 | 1.3 | 0.903 |
| 0.4 | 0.655 | 1.4 | 0.919 |
| 0.5 | 0.692 | 1.5 | 0.933 |
| 0.6 | 0.726 | 1.6 | 0.945 |
| 0.7 | 0.758 | 1.7 | 0.955 |
| 0.8 | 0.788 | 1.8 | 0.964 |
| 0.9 | 0.816 | 1.9 | 0.971 |

Now the probability density function of the standard normal distribution becomes

$$p(x) = \frac{1}{\sqrt{2\pi}} e^{-\frac{\xi^2}{2}}.$$

(9.26)

Its cumulative probability function is

$$\phi(\xi) = \frac{1}{\sqrt{2\pi}} \int_{-\infty}^{\xi} e^{-\frac{\xi^2}{2}} d\xi = \frac{1}{2}\left[1 + \text{erf}(\frac{\xi}{\sqrt{2}})\right].$$

(9.27)

As the calculations of $\phi$ and the error function involve numerical integration, it is a traditional practice to tabulate $\phi$ in a table (see Table 9.1) so that we do not have to calculate their values each time we use this distribution.

## 4. Other Distributions

There are a number of other important distributions such as the exponential distribution, log-normal distribution, uniform distribution and the $\chi^2$-distribution. As we mentioned earlier, the uniform distribution has a probability density function

$$p = \frac{1}{\beta - \alpha}, \qquad x \in [\alpha, \beta], \tag{9.28}$$

whose mean is $E[X] = (\alpha + \beta)/2$ and variance is $\sigma^2 = (\beta - \alpha)^2/12$.

The exponential distribution has the following probability density function

$$f(x) = \lambda e^{-\lambda x}, \quad \lambda > 0, \quad (x > 0), \tag{9.29}$$

and $f(x) = 0$ for $x \le 0$. Its mean and variance are

$$\mu = 1/\lambda, \qquad \sigma^2 = 1/\lambda^2. \tag{9.30}$$

## Example 9.6

The expectation $E(X)$ of an exponential distribution is

$$\mu = E(X) = \int_{-\infty}^{\infty} x\lambda e^{-\lambda x} dx = \int_{0}^{\infty} x\lambda e^{-\lambda x} dx$$

$$= \left[ -xe^{-\lambda x} - \frac{1}{\lambda} e^{-\lambda x} \right]_{0}^{\infty} = \frac{1}{\lambda}.$$

For $E(X^2)$, we have

$$E(X^2) = \int_{0}^{\infty} x^2 \lambda e^{-\lambda x} dx = \left[ -x^2 e^{-\lambda x} \right]_{0}^{\infty} + 2 \int_{0}^{\infty} x e^{-\lambda x} dx$$

$$= \left[ -x^2 e^{-\lambda x} \right]_{0}^{\infty} + \left[ -\frac{2x}{\lambda} e^{-\lambda x} - \frac{2}{\lambda^2} e^{-\lambda x} \right]_{0}^{\infty} = \frac{2}{\lambda^2}.$$

Here, we have used the fact that $x$ and $x^2$ grow slower than $\exp(-\lambda x)$ decreases. That is, $x \exp(-\lambda x) \to 0$ and $x^2 \exp(-\lambda x) \to 0$ when $x \to \infty$.

Since $E(X^2) = \mu^2 + \sigma^2 = \mu^2 + \text{Var}(X)$ from (9.11), we have

$$\text{Var}(X) = \frac{2}{\lambda^2} - (\frac{1}{\lambda})^2 = \frac{1}{\lambda^2}.$$

Exponential distributions are widely used in queuing theory and when simulating discrete events. For example, the arrival process of customers in a bank is a Poisson process and the time interval between arrivals (or inter-arrival time) obeys an exponential distribution.

Before we proceed, let us define the convolution of two functions $f(t)$ and $g(t)$ as

$$(f * g)(t) \equiv f(t) * g(t) \equiv \int_{-\infty}^{+\infty} f(\tau)g(t - \tau)d\tau = \int_{-\infty}^{+\infty} f(t - \tau)g(\tau)d\tau, \qquad (9.31)$$

which gives two ways to calculate the same convolution integral.

Sometimes, we may have to calculate the probability distribution $f_S(s)$ of the sum of two independent random variables $X$ and $Y$ (i.e., $S = X + Y$), given the probability distribution $f_X(x)$ of $X$ and probability distribution $f_Y(y)$ of $Y$. The probability $f_S$ is the convolution of $f_X$ and $f_Y$. That is

$$f_S(s) = (f_X * f_Y)(s) = \int_{-\infty}^{+\infty} f_X(s - x)f_Y(x)dx. \qquad (9.32)$$

## Example 9.7

If both $X$ and $Y$ are uniform distributions with

$$f_X(x) = f_Y(x) = \begin{cases} 1 & \text{if } x \in [0, 1] \\ 0 & \text{otherwise,} \end{cases}$$

the sum $S = X + Y$ obeys the probability

$$f_S(s) = \int_{-\infty}^{+\infty} f_X(s - x)f_Y(x)dx = \int_0^1 f_X(s - x)1 dx, \qquad \text{for} \quad 0 \le x \le 1,$$

where we have used the fact that $f_Y(x) = 1$ for $x \in [0, 1]$. Since the values of $x$ are between 0 and 1, the values of s can be between 0 and 2. That is $0 \le s \le 2$. Thus, to calculate the above integral, we have two different cases: $0 \le s \le 1$ and $1 < s \le 2$.

When $0 \le s \le 1$, $0 \le s - x \le 1$ which means that $f_X(s - x) = 1$. Thus, the convolution integral becomes

$$f_S(s) = \int_0^s dx = s.$$

However, when $1 \le s \le 2$, $s - x$ is between $s - 1$ and 1 because $0 \le x \le 1$ and thus $f_X(s - x) = 1$. So the convolution integral for this case becomes

$$f_S(s) = \int_{s-1}^1 dx = x \Big|_{s-1}^1 = 1 - (s - 1) = 2 - s.$$

Therefore, the probability is

$$f_S(s) = \begin{cases} s & (0 \le s \le 1), \\ 2 - s & (1 \le s \le 2), \\ 0 & (s \notin [0, 2]), \end{cases}$$

which is a triangular distribution with the peak at $s = 1$.

This example, though for simple uniform distributions, requires care when doing integration. Let us look at another example where $X$ and $Y$ both obey Gaussian normal distributions.

## Example 9.8

If both $X$ and $Y$ obeys the same normal distribution

$$f_X(x) = f_Y(y) = \frac{1}{\sqrt{2\pi}} e^{-x^2/2}, \quad -\infty < x < +\infty,$$

the sum $S = X + Y$ obeys

$$f_S(s) = \int_{-\infty}^{+\infty} f_X(s - x) f_x(x) dx = \int_{-\infty}^{+\infty} \left( \frac{1}{\sqrt{2\pi}} e^{-(s-x)^2/2} \right) \left( \frac{1}{\sqrt{2\pi}} e^{-x^2} \right) dx$$

$$= \frac{1}{2\pi} \int_{-\infty}^{+\infty} e^{-(s-x)^2/2} e^{-x^2/2} dx.$$

Since $-[(s - x)^2 - x^2]/2 = -[(x - s/2)^2 + \frac{s^2}{4}]$, we have

$$f_S(s) = \frac{1}{2\pi} e^{-s^2/4} \left[ \int_{-\infty}^{\infty} e^{-(x-s/2)^2} dx \right].$$

From the Gaussian integral in Chapter 5 (treating $s$ as a constant), we know that

$$\int_{-\infty}^{+\infty} e^{-(x-s/2)^2} dx = \sqrt{\pi}.$$

Thus, we have

$$f_S(s) = \frac{1}{2\pi} e^{-s^2/4} \sqrt{\pi} = \frac{1}{\sqrt{4\pi}} e^{-s^2/4}.$$

This is also a normal distribution with a zero mean, but its standard derivation is $\sqrt{2}$.

In fact, for two normally distributed random variables with

$$f_X(x) = \frac{1}{\sqrt{2\pi\sigma_1^2}} e^{-(x-\mu_1)^2/(2\sigma_1^2)}, \quad f_Y(y) = \frac{1}{\sqrt{2\pi\sigma_2^2}} e^{-(y-\mu_2)^2/(2\sigma_2^2)}, \tag{9.33}$$

their sum $S = X + Y$ obeys

$$f_S(s) = \frac{1}{\sqrt{2\pi(\sigma_1^2 + \sigma_2^2)}} \exp\left\{ -\frac{[s - (\mu_1 + \mu_2)]^2}{2(\sigma_1^2 + \sigma_2^2)} \right\}, \tag{9.34}$$

which is a Gaussian distribution with a mean $\mu = \mu_1 + \mu_2$ and a variance $\sigma^2 = \sigma_1^2 + \sigma_2^2$.

If the sum of two identical, independent random variables obeys the same distribution, such a distribution is called a stable distribution. Therefore, from the above example, we know that Gaussian distributions are stable distributions. Other stable distributions include Cauchy distribution and Lévy distribution.

## 5. The Central Limit Theorem

The most important theorem in probability is the central limit theorem, which concerns a large number of trials and explains why the normal distribution occurs so widely.

Let $X_i(i = 1, 2, ..., n)$ be $n$ independent random variables, each of which is defined by a probability density function $p_i(x)$ with a corresponding mean $\mu_i$ and a variance $\sigma_i^2$. The sum of all these random variables

$$\Theta = \sum_{i=1}^{n} X_i = X_1 + X_2 + ... + X_n, \tag{9.35}$$

is also a random variable whose distribution approaches the Gaussian distribution as $n \to \infty$. Its mean $E[\Theta]$ and variance $\text{var}[\Theta]$ are given by

$$E[\Theta] = \sum_{i=1}^{n} E[X_i] = \sum_{i=1}^{n} \mu_i, \tag{9.36}$$

and

$$\text{var}[\Theta] = \sum_{i=1}^{n} \text{var}[X_i] = \sum_{i=1}^{n} \sigma_i^2. \tag{9.37}$$

The proof of this theorem is beyond the scope of this book as it involves the moment generating functions, characteristics functions and other techniques. In engineering mathematics, we simply use these important results for statistical analysis.

In the special case when all the variables $X_i$ are described by the same probability density function with the same mean $\mu$ and variance $\sigma^2$, these results become

$$E[\Theta] = n\mu, \qquad \text{var}[\Theta] = n\sigma^2. \tag{9.38}$$

By defining a new variable

$$\xi_n = \frac{\Theta - n\mu}{\sigma \sqrt{n}}, \tag{9.39}$$

then the distribution of $\xi_n$ converges towards the standard normal distribution $N(0, 1)$ as $n \to \infty$.

## 6. Weibull Distribution

Although the distribution functions of the real-world random processes are dominated by the Gaussian or normal distribution, however, there are some cases where other distributions describe the related phenomena more accurately. Weibull's distribution is such a distribution with many applications in areas such as reliability analysis, engineering design and quality assessment. Therefore, it deserves a special introduction in detail. This distribution was originally developed by Swedish physicist, A. Weibull in 1939, to try to explain a fact that was well-known but unexplained at that time, that the relative strength of a specimen decreases with its increasing dimension. Since then, it has been applied to study many real-world stochastic processes, even including the distributions of wind speeds, rainfalls, energy resources and earthquakes.

Weibull's distribution is a three-parameter distribution given by

$$p(x, \lambda, \beta, n) = \begin{cases} \frac{n}{\lambda}(\frac{x-\beta}{\lambda})^{n-1} \exp[-(\frac{x-\beta}{\lambda})^n] & (x \geq \beta), \\ 0 & (x < \beta), \end{cases} \tag{9.40}$$

where $\lambda$ is a scaling parameter, and $n$ is the shape parameter, often referred to as the Weibull modulus. The parameter $\beta$ is the threshold of the distribution. For example, $n = 1$ gives an exponential distribution. When $n > 3.5$, the distribution can be used to approximate a Gaussian distribution.

By straightforward integration, we have the cumulative probability density distribution

$$\Phi(x, \lambda, \beta, n) = 1 - e^{-(\frac{x-\beta}{\lambda})^n}. \tag{9.41}$$

For the fixed values $\lambda = 1$ and $\beta = 0$, the variation of $n$ will give a broad range of shapes and can be used to approximate various distributions as shown in Fig. 9.3.

In reliability analysis, especially for a large infrastructure such as a dam or a tall

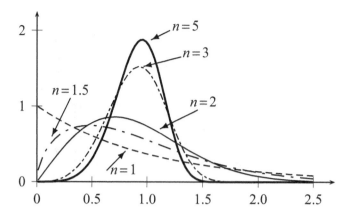

Figure 9.3: Weibull density function for different values of $\lambda$.

building under stress, the survival probability is more conveniently represented as

$$P_s(V) = \exp\left[\int_V -(\frac{\sigma}{\sigma_0})^n \frac{dV}{V_0}\right],$$  (9.42)

where $V$ is the volume of the system. $\sigma_0$ is the failure stress (either tensile or shear) for the reference volume $V_0$. The failure probability is

$$P_f(V) = 1 - P_s(V).$$  (9.43)

For constant stress $\sigma$ over the whole volume $V$, we simply have

$$P_s(V) = \exp\left[-(\frac{\sigma}{\sigma_0})^n \frac{V}{V_0}\right].$$  (9.44)

At the reference point $\sigma = \sigma_0$ and $V = V_0$ often obtained using laboratory tests, we have

$$P_s(V_0) = e^{-1} \approx 0.3679.$$  (9.45)

As the stress becomes extreme, $\sigma \to \infty$, then $P_s \to 0$ and $P_f \to 1$.

## Example 9.9

A batch of some critical components (such as turbine blades) were manufactured. Assuming that the distribution is a Weibull distribution with a characteristic stress $\sigma_0 = 240$ MPa, a shape parameter $n = 6$ and $V = V_0$, what is the service stress that can be applied so as to ensure that no less than 99.99% of all components will survive?

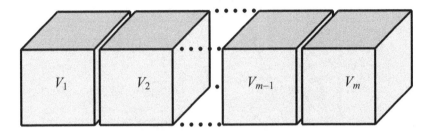

Figure 9.4: The weakest link analogy.

The survival probability is 99.99%, so the probability of failure is

$$1 - \exp[-(\frac{\sigma}{240})^6] < 0.01\% = 10^{-4}.$$

Thus, we have

$$1 - 10^{-4} < \exp[-(\frac{\sigma}{240})^6], \quad \text{or} \quad (\frac{\sigma}{240})^6 \leq 10^{-4},$$

where we have used $\ln(1 + x) \approx x$ when $x \to 0$. The above inequality requires that

$$\sigma \leq \sqrt[6]{10^{-4} \times 240^6} \approx 51.7 \text{ MPa.}$$

Suppose during the manufacturing process, the materials' properties can vary about 10%, and thus the minimum characteristic stress becomes $\sigma_0 = 215$ MPa instead of 240 MPa. How much will the above results change? The new allowable service stress becomes

$$\sigma \leq \sqrt[6]{10^{-4} \times 215^6} \approx 46.3 \text{ MPa.}$$

However, if the previous stress level 51.7 MPa is still applied, then the failure probability or reliability will become

$$P = 1 - \exp[-(51.7/215)^6] \approx 1.9 \times 10^{-4},$$

which is almost twice the previous probability.

The fundamental idea of this volume-related probability function is the weakest link theory. The larger the volume of a system, the more likely it is to have critical flaws that could cause potential failure. We can consider that the whole large volume $V$ consists of $m$ small volumes/blocks $V_1, V_2, ..., V_m$ and these small blocks are glued together (see Fig. 9.4), thus the probability of the survival of the whole system is equivalent to the survival of all the subsystem blocks. If any one of these blocks fails, the system is considered flawed and thus failed. In the simplest case, $V_1 = V_2 = ... = V_m = V_0$ and $m = V/V_0$, the survival probability of the whole system under constant stress $\sigma$ is

$$P_s(V) = P_s(mV_0) = \overbrace{P_s(V_0) \times P_s(V_0) \times ... \times P_s(V_0)}^{m}$$

$$= [P_s(V_0)]^m = [e^{-(\frac{\sigma}{\sigma_0})^m}]^{\frac{V}{V_0}} = \exp\left[-\frac{V}{V_0}(\frac{\sigma}{\sigma_0})^m\right]. \tag{9.46}$$

There are a whole range of statistical methods based on solid probability theory. However, we will not discuss any theory further, as our focus in this book is engineering mathematics and applicable numerical methods. Interested readers can refer to more advanced literature.

## 7. Sample Mean and Variance

If a sample consists of $n$ independent observations $x_1, x_2, ..., x_n$ on a random variable $x$ such as the noise level on a road or the price of a cup of coffee, two important and commonly used parameters are sample mean and sample variance, which can easily be estimated from the sample. The sample mean is calculated by

$$\bar{x} \equiv <x> = \frac{1}{n}(x_1 + x_2 + ... + x_n) = \frac{1}{n}\sum_{i=1}^{n} x_i, \tag{9.47}$$

which is essentially the arithmetic average of the values $x_i$.

The sample variance $S^2$ is defined by

$$S^2 = \frac{1}{n-1}\sum_{i=1}^{n}(x_i - \bar{x})^2. \tag{9.48}$$

Let us look at an example.

### Example 9.10

The compressive strengths of a batch of high-strength concrete mixtures have been measured. The readings are (in MPa):

$$66, \ 73, \ 73, \ 74, \ 83, \ 70, \ 69, \ 77, \ 72, \ 75.$$

From the data, we know that $n = 10$ and the mode is 73 as 73 appears twice (all the rest only appears once).

The sample mean of the marks is

$$\bar{x} = \frac{1}{10}(x_1 + x_2 + ... + x_{10})$$

$$= \frac{1}{10}(66 + 73 + 73 + 74 + 83 + 70 + 69 + 77 + 72 + 75) = \frac{732}{10} = 73.2.$$

The corresponding sample variance can be calculated by

$$S^2 = \frac{1}{n-1} \sum_{i=1}^{n} (x_i - \bar{x})^2$$

$$= \frac{1}{10-1} \sum_{i=1}^{10} (x_i - 73.2)^2 = \frac{1}{9}[(66 - 73.2)^2 + (73 - 73.2)^2 + \ldots + (75 - 73.2)^2]$$

$$= \frac{1}{9}[(-7.2)^2 + (-0.2)^2 + \ldots + (1.8)^2] = \frac{195.6}{9} \approx 21.73.$$

Thus, the standard derivation is

$$S = \sqrt{S^2} \approx \sqrt{21.73} \approx 4.662.$$

Generally speaking, if $u$ is a linear combination of $n$ independent random variables $y_1, y_2, \ldots, y_n$ and each random variable $y_i$ has an individual mean $\mu_i$ and a corresponding variance $\sigma_i^2$, we have the linear combination

$$u = \sum_{i=1}^{n} \alpha_i y_i = \alpha_1 y_1 + \alpha_2 y_2 + \ldots + \alpha_n y_n, \tag{9.49}$$

where the parameters $\alpha_i (i = 1, 2, \ldots, n)$ are the weighting coefficients. From the central limit theorem, we have the mean $\mu_u$ of the linear combination

$$\mu_u = E(u) = E(\sum_{i=1}^{n} \alpha_i y_i) = \sum_{i=1}^{n} \alpha E(y_i) = \sum \alpha_i \mu_i. \tag{9.50}$$

Then, the variance $\sigma_u^2$ of the combination is

$$\sigma_u^2 = E[(u - \mu_u)^2] = E\Big[ \sum_{i=1}^{n} \alpha_i (y_i - \mu_i)^2 \Big], \tag{9.51}$$

which can be expanded as

$$\sigma_u^2 = \sum_{i=1}^{n} \alpha_i^2 E[(y_i - \mu_i)^2] + \sum_{i,j=1; i \neq j}^{n} \alpha_i \alpha_j E[(y_i - \mu_i)(y_j - \mu_j)], \tag{9.52}$$

where $E[(y_i - \mu_i)^2] = \sigma_i^2$. Since $y_i$ and $y_j$ are independent, we have

$$E[(y_i - \mu_i)(y_j - \mu_j)] = E[(y_i - \mu_i)]E[(y_j - \mu_j)] = 0. \tag{9.53}$$

Therefore, we get

$$\sigma_u^2 = \sum_{i=1}^{n} \alpha_i^2 \sigma_i^2. \tag{9.54}$$

The sample mean defined in Eq. (9.47) can also be viewed as a linear combination of all the $x_i$ assuming that each of which has the same mean $\mu_i = \mu$ and variance $\sigma_i^2 = \sigma^2$, and the same weighting coefficient $\alpha_i = 1/n$. Hence, the sample mean is an unbiased estimate of the sample due to the fact $\mu_{\bar{x}} = \sum_{i=1}^{n} \mu/n = \mu$. In this case, however, we have the variance

$$\sigma_{\bar{x}}^2 = \sum_{i=1}^{n} \frac{1}{n^2} \sigma^2 = \frac{\sigma^2}{n}, \tag{9.55}$$

which means the variance becomes smaller as the size $n$ of the sample increases by a factor of $1/n$.

For the sample variance $S^2$ defined earlier by

$$S^2 = \frac{1}{n-1} \sum_{i=1}^{n} (x_i - \bar{x})^2, \tag{9.56}$$

we can see that the factor is $1/(n-1)$ not $1/n$ because only $1/(n-1)$ will give the correct and unbiased estimate of the variance. From the probability theory in the previous chapter, we know that $E[x^2] = \mu^2 + \sigma^2$. The mean of the sample variance is

$$\mu_{S^2} = E[\frac{1}{n-1} \sum_{i=1}^{n} (x_i - \bar{x})^2] = \frac{1}{n-1} \sum_{i=1}^{n} E[(x_i^2 - n\bar{x}^2)]. \tag{9.57}$$

Using $E[\bar{x}^2] = \mu^2 + \sigma^2/n$, we get

$$\mu_{S^2} = \frac{1}{n-1} \sum_{i=1}^{n} \{E[x_i^2] - nE[\bar{x}^2]\}$$

$$= \frac{1}{n-1} \left[ n(\mu^2 + \sigma^2) - n(\mu^2 + \frac{\sigma^2}{n}) \right] = \sigma^2. \tag{9.58}$$

Obviously, if we used the factor $1/n$ instead of $1/(n-1)$, we would get $\mu_{S^2} = \frac{n-1}{n}\sigma^2 < \sigma^2$, which would underestimate the sample variance. The other way to think about the factor $1/(n-1)$ is that we need at least one value to estimate the mean, and we need at least 2 values to estimate the variance. Thus, for $n$ observations, only $n-1$ different values of variance can be obtained to estimate the total sample variance.

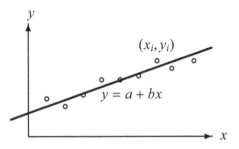

Figure 9.5: Least square and the best fit line.

## 8. Method of Least Squares

For a sample of $n$ values $x_1, x_2, ..., x_n$ of a random variable $X$ whose probability density function $p(x)$ depends on a set of $k$ parameters $\beta_1, ..., \beta_k$, the joint probability is the product of all the probabilities, that is

$$\Phi(\beta_1, ..., \beta_k) = \prod_{i=1}^{n} p(x_i, \beta_1, ..., \beta_k)$$

$$= p(x_1, \beta_1, ..., \beta_k) \cdot p(x_2, \beta_1, ..., \beta_k) \cdots p(x_n, \beta_1, ..., \beta_k), \qquad (9.59)$$

where $\Pi$ means the product of all its components. For example, $\Pi_{n=1}^{3} n = 1 \times 2 \times 3$. The essence of the maximum likelihood is to maximize $\Phi$ by choosing the parameters $\beta_j$. As the sample can be considered as given values, the maximum likelihood requires the following stationary conditions:

$$\frac{\partial \Phi}{\partial \beta_j} = 0, \qquad (j = 1, 2, ..., k), \qquad (9.60)$$

whose solutions for $\beta_j$ are the maximum likelihood estimates.

### 8.1. Linear Regression and Correlation Coefficient

For experiments and observations, we usually plot one variable such as pressure or price $y$ against another variable $x$ such as time or spatial coordinates. We try to present the data in such a way that we can see some trend in the data.

For a set of $n$ data points $(x_i, y_i)$, the usual practice is to try to draw a straight line $y = a + bx$ so that it represents the major trend. Such a line is often called the regression line or the best fit line as shown in Fig. 9.5.

The method of linear least squares is to try to determine the two parameters, $a$ (intercept) and $b$ (slope), for the regression line from $n$ data points, assuming that $x_i$

are known more precisely, and the values of $y_i$ obey a normal distribution around the potentially best fit line with a variance $\sigma^2$. So we have the probability

$$P = \prod_{i=1}^{n} p(y_i) = A \exp\left\{ -\frac{1}{2\sigma^2} \sum_{i=1}^{n} [y_i - f(x_i)]^2 \right\}, \tag{9.61}$$

where $A$ is a constant, and $f(x)$ is the function for the regression [$f(x) = a + bx$ for the linear regression]. It is worth pointing out that the exponent

$$\psi = \sum_{i=1}^{n} [y_i - f(x_i)]^2/\sigma^2$$

is similar to the quantity $\chi_n^2$ defined in the $\chi^2$-distribution.

The maximization of $\Phi$ is equivalent to the minimization of $\psi$. In order to minimize $\psi$ as a function of $a$ and $b$ (via $f(x) = a + bx$), its derivatives should be zero. That is

$$\frac{\partial \psi}{\partial a} = -2 \sum_{i=1}^{n} [y - (a + bx_i)] = 0, \tag{9.62}$$

and

$$\frac{\partial \psi}{\partial b} = -2 \sum_{i=1}^{n} x_i[y_i - (a + bx_i)] = 0, \tag{9.63}$$

where we have used $\sigma^2 \neq 0$ and thus omitted this factor.

By expanding these equations, we have

$$na + b \sum_{i=1}^{n} x_i = \sum_{i=1}^{n} y_i, \tag{9.64}$$

and

$$a \sum_{i=1}^{n} x_i + b \sum_{i=1}^{n} x_i^2 = \sum_{i=1}^{n} x_i y_i, \tag{9.65}$$

which is a system of linear equations for $a$ and $b$, and it is straightforward to obtain the solutions as

$$a = \frac{1}{n}\left[\sum_{i=1}^{n} y_i - b \sum_{i=1}^{n} x_i\right] = \bar{y} - b\bar{x}, \quad b = \frac{n \sum_{i=1}^{n} x_i y_i - (\sum_{i=1}^{n} x_i)(\sum_{i=1}^{n} y_i)}{n \sum_{i=1}^{n} x_i^2 - (\sum_{i=1}^{n} x_i)^2}, \tag{9.66}$$

where

$$\bar{x} = \frac{1}{n} \sum_{i=1}^{n} x_i, \qquad \bar{y} = \frac{1}{n} \sum_{i=1}^{n} y_i. \tag{9.67}$$

If we use the following notations

$$K_x = \sum_{i=1}^{n} x_i, \qquad K_y = \sum_{i=1}^{n} y_i, \tag{9.68}$$

and

$$K_{xx} = \sum_{i=1}^{n} x_i^2, \qquad K_{yy} = \sum_{i=1}^{n} y_i^2, \qquad K_{xy} = \sum_{i=1}^{n} x_i y_i, \tag{9.69}$$

then the above equations for $a$ and $b$ become

$$a = \frac{K_{xx}K_y - K_x K_{xy}}{nK_{xx} - (K_x)^2}, \tag{9.70}$$

and

$$b = \frac{nK_{xy} - K_x K_y}{nK_{xx} - (K_x)^2}. \tag{9.71}$$

The residual error is defined by

$$\epsilon_i = y_i - (a + bx_i), \tag{9.72}$$

whose sample mean is given by

$$\mu_\epsilon = \frac{1}{n}\sum_{i=1}^{n} \epsilon_i = \frac{1}{n}y_i - a - b\frac{1}{n}\sum_{i=1}^{n} x_i = \bar{y} - a - b\bar{x} = 0. \tag{9.73}$$

The sample variance $S^2$ is

$$S^2 = \frac{1}{n-2}\sum_{i=1}^{n}[y_i - (a + bx_i)]^2, \tag{9.74}$$

where the factor $1/(n-2)$ comes from the fact that two constraints are needed for the best fit, and therefore the residuals have $n-2$ degrees of freedom.

The correlation coefficient $r_{x,y}$ is a very useful parameter for finding any potential relationship between two sets of data $x_i$ and $y_i$ for two random variables $x$ and $y$, respectively. If $x$ has a mean $\mu_x$ and a sample variance $S_x^2$, and $y$ has a mean $\mu_y$ and a sample variance $S_y^2$, the correlation coefficient is defined by

$$r_{x,y} = \frac{\text{cov}(x,y)}{S_x S_y} = \frac{E[xy] - \mu_x\mu_y}{S_x S_y}, \tag{9.75}$$

where

$$\text{cov}(x,y) = E[(x - \mu_x)(y - \mu_y)], \tag{9.76}$$

is the covariance. It is worth pointing out that $S_x$ and $S_y$ must be sample means;

otherwise, the result is incorrect.

If the two variables are independent or $\text{cov}(x, y) = 0$, there is no correlation between them ($r_{x,y} = 0$). If $r_{x,y}^2 = 1$, then there is a linear relationship between these two variables. $r_{x,y} = 1$ is an increasing linear relationship where the increase of one variable will lead to the increase of another. On the other hand, $r_{x,y} = -1$ is a decreasing relationship when one increases while the other decreases.

For a set of $n$ data points $(x_i, y_i)$, the correlation coefficient can be calculated by

$$r_{x,y} = \frac{n \sum_{i=1}^{n} x_i y_i - \sum_{i=1}^{n} x_i \sum_{i=1}^{n} y_i}{\sqrt{\left[n \sum x_i^2 - (\sum_{i=1}^{n} x_i)^2\right]\left[n \sum_{i=1}^{n} y_i^2 - (\sum_{i=1}^{n} y_i)^2\right]}},$$

or

$$r_{x,y} = \frac{n K_{xy} - K_x K_y}{\sqrt{(n K_{xx} - K_x^2)(n K_{yy} - K_y^2)}}, \tag{9.77}$$

where $K_{yy} = \sum_{i=1}^{n} y_i^2$.

Now let us look at an example about the hardness and tensile stress of materials. Is there any relationship between the Vickers hardness $H_v$ and the tensile strength $Y$ in MPa? The data for a set of random samples are given in Table 9.1.

Table 9.1: Hardness and tensile strength of materials samples.

| Hardness | Tensile Strength | Hardness | Tensile Strength |
|----------|------------------|----------|------------------|
| 90 | 270 | 300 | 910 |
| 110 | 330 | 350 | 1080 |
| 140 | 410 | 400 | 1270 |
| 170 | 520 | 450 | 1450 |
| 190 | 560 | 490 | 1590 |
| 225 | 670 | 550 | 1810 |
| 250 | 750 | 650 | 2180 |

Now let us try to do a linear regression in the following:

$$Y = a + bH.$$

where $Y$ is the tensile strength and $H$ is the Vickers hardness.

## Example 9.11

From the data in Table 9.1 with $n = 14$, we can calculate

$$K_H = \sum_{i=1}^{14} H_i = 90 + 110 + \ldots + 650 = 4365,$$

$$K_Y = \sum_{i=1}^{14} Y_i = 270 + 330 + \ldots + 2180 = 13800,$$

$$K_{HY} = \sum_{i=1}^{14} H_i Y_i = 90 * 270 + \ldots + 650 * 2180 = 5654150,$$

$$K_{HH} = \sum_{i=1}^{14} H_i^2 = 90^2 + \ldots + 650^2 = 1758025,$$

and

$$K_{YY} = \sum_{i=1}^{14} Y_i^2 = 270^2 + \ldots + 2180^2 = 18211800.$$

Thus, we get

$$a = \frac{K_{HH}K_Y - K_H K_{HY}}{nK_{HH} - K_H^2}$$

$$= \frac{1758025 \times 13800 - 4365 \times 5654150}{14 \times 1758025 - 4365^2} \approx -75.48,$$

and

$$b = \frac{nK_{HY} - K_H K_Y}{nK_{HH} - K_H^2}$$

$$= \frac{14 \times 5654150 - 4365 \times 13800}{14 \times 1758025 - 4365^2} \approx 3.404.$$

So the regression line becomes

$$Y = -75.48 + 3.404H.$$

Therefore, their correlation coefficient $r$ is given by

$$r = \frac{nK_{HY} - K_H K_Y}{\sqrt{(nK_{HH} - K_H^2)(nK_{YY} - K_Y^2)}}$$

$$= \frac{14 \times 5654150 - 4365 \times 13800}{\sqrt{(14 \times 1758025 - 4365^2)(14 \times 18211800 - 13800^2)}} \approx 0.99903.$$

This is a relatively strong correlation indeed.

It is worth pointing out that this conclusion is based on a small set of samples. In fact, we know from materials science that there is a strong relationship and thus a conversion table between the Vickers hardness $H_v$ and tensile strength.

The above formulations are based on the fact that the curve-fitting function $y = f(x) = ax + b$ is linear in terms of the independent variable $x$ and the parameters ($a$ and $b$). Here, the key linearity is about parameters, but not about the basis function $x$. Thus, the above technique can still be applicable to both $f(x) = ax^2 + b$ and $g(x) = a + b\sin(x)$ functions with some minor adjustments to be discussed later in this chapter. However, if we have a function in the form

$$y = \ln(ax + b),$$

then the above technique cannot be applied directly, and some linearization approximations should be used.

## 8.2. Linearization

Sometimes, some obviously nonlinear functions can be transformed into linear forms so as to carry out linear regression, instead of more complicated nonlinear regression. However, there is no general formula for such linearization and thus it is often necessary to deal with each case individually. This can be illustrated by some examples.

### Example 9.12

For example, the following nonlinear function

$$f(x) = \alpha e^{-\beta x}, \tag{9.78}$$

can be transformed into a linear form by taking logarithms of both sides. We have

$$\ln f(x) = \ln(\alpha) - \beta x, \tag{9.79}$$

which is equivalent to $y = a + bx$ if we let $y = \ln f(x)$, $a = \ln(\alpha)$ and $b = -\beta$.

In addition, the following function

$$f(x) = \alpha e^{-\beta x + \gamma} = A e^{-\beta x},$$

where $A = \alpha e^{\gamma}$ is essentially the same as the above function.

Similarly, function

$$f(x) = \alpha x^{\beta}, \tag{9.80}$$

can also be transformed into

$$\ln[f(x)] = \ln(\alpha) + \beta \ln(x), \tag{9.81}$$

which is a linear regression $y = a + b\zeta$ between $y = \ln[f(x)]$ and $\zeta = \ln(x)$ where $a = \ln(\alpha)$ and $b = \beta$.

Furthermore, function

$$f(x) = \alpha \beta^x, \tag{9.82}$$

can also be converted into the standard linear form

$$\ln f(x) = \ln \alpha + x \ln \beta, \tag{9.83}$$

by letting $y = \ln[f(x)]$, $a = \ln \alpha$, and $b = \ln \beta$.

It is worth pointing out that the data points involving zeros should be taken out due to the potential singularity of the logarithm. Fortunately, these points rarely occur in the regression for the functions in the above form.

## Example 9.13

If a set of data can fit to a nonlinear function

$$y = ax \exp(-x/b),$$

in the range of $(0, \infty)$, it is then possible to convert it to a linear regression.

As $x = 0$ is just a single point, so we can leave this out. For $x \neq 0$, we can divide both sides by $x$, we have

$$\frac{y}{x} = a \exp(-x/b).$$

Taking the logarithm of both sides, we have

$$\ln \frac{y}{x} = \ln a - \frac{1}{b}x,$$

which is a linear regression of $y/x$ versus $x$.

## 9. Generalized Linear Regression

The most widely used linear regression is the so-called generalized least square as a linear combination of basis functions. Fitting to a polynomial of degree $p$

$$y(x) = \alpha_0 + \alpha_1 x + \alpha_2 x^2 + .. + \alpha_p x^p, \tag{9.84}$$

is probably the most widely used. This is equivalent to the regression to the linear combination of the basis functions $1, x, x, ...,$ and $x^p$. However, there is no particular reason why we have to use these basis functions. In fact, the basis functions can be any arbitrary known functions such as $\sin(x)$, $\cos(x)$ and even $\exp(x)$, and the main requirement is that they can be explicitly expressed as basis functions. In this sense, the generalized least square can be written as

$$y(x) = \sum_{j=0}^{p} \alpha_j f_j(x), \tag{9.85}$$

where the basis functions $f_j$ are known functions of $x$. Now the sum of least squares is defined as

$$\psi = \sum_{i=1}^{n} \frac{[y_i - \sum_{j=0}^{p} \alpha_j f_j(x_i)]^2}{\sigma_i^2}, \tag{9.86}$$

where $\sigma_i (i = 1, 2, ..., n)$ are the standard deviations of the $i$th data point at $(x_i, y_i)$. There are $n$ data points in total. In order to determine the coefficients uniquely, it requires that

$$n \geq p + 1. \tag{9.87}$$

In the case of unknown standard deviations $\sigma_i$, we can always set all the values $\sigma_i$ as the same constant $\sigma_i = \sigma = 1$.

Let $\mathbf{D} = [D_{ij}]$ be the design matrix which is given by

$$D_{ij} = \frac{f_j(x_i)}{\sigma_i}. \tag{9.88}$$

The minimum of $\psi$ is determined by

$$\frac{\partial \psi}{\partial \alpha_j} = 0, \qquad (j = 0, 1, ..., p). \tag{9.89}$$

That is

$$\sum_{i=1}^{n} \frac{f_k(x_i)}{\sigma_i^2} \left[ y_i - \sum_{j=0}^{p} \alpha_j f_j(x_i) \right] = 0, \qquad k = 0, ..., p. \tag{9.90}$$

Rearranging the terms and interchanging the order of summations, we have

$$\sum_{j=0}^{p}\sum_{i=1}^{n}\frac{\alpha_j f_j(x_i)f_k(x_i)}{\sigma_i^2} = \sum_{i=1}^{n}\frac{y_i f_k(x_i)}{\sigma_i^2}, \tag{9.91}$$

which can be written compactly as the following matrix equation

$$\sum_{j=0}^{p} A_{kj}\alpha_j = b_k, \tag{9.92}$$

or

$$\mathbf{A}\alpha = \mathbf{b}, \tag{9.93}$$

where

$$\mathbf{A} = \mathbf{D}^{\mathrm{T}} \cdot \mathbf{D},$$

is a $(p+1) \times (p+1)$ matrix. That is

$$A_{kj} = \sum_{i=1}^{n}\frac{f_k(x_i)f_j(x_i)}{\sigma_i^2}. \tag{9.94}$$

Here $b_k$ is a column vector given by

$$b_k = \sum_{i=1}^{n}\frac{y_i f_k(x_i)}{\sigma_i^2}, \tag{9.95}$$

where $(k = 0, ..., p)$. Equation (9.92) is a linear system of the so-called normal equations which can be solved using the standard methods for solving linear systems. The solution of the coefficients is $\alpha = \mathbf{A}^{-1}\mathbf{b}$ or

$$\alpha_k = \sum_{j=0}^{p}[A]_{kj}^{-1}b_j, \qquad (k = 0, ..., p), \tag{9.96}$$

where $\mathbf{A}^{-1} = [A]_{ij}^{-1}$.

A special case of the generalized linear least squares is the so-called polynomial least squares when the basis functions are simple power functions $f_i(x) = x^i$, $(i = 0, 1, ..., p)$. That is

$$f_i(x) = 1, \ x, \ x^2, ..., \ x^p. \tag{9.97}$$

For simplicity, we assume that $\sigma_i = \sigma = 1$. The matrix equation (9.92) simply

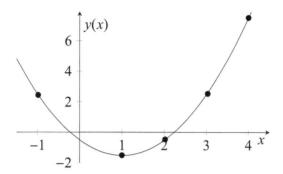

Figure 9.6: Best fit curve for $y(x) = x^2 - 2x - \frac{1}{2}$ with 2.5% noise.

becomes

$$
\begin{pmatrix}
\sum_{i=1}^{n} 1 & \sum_{i=1}^{n} x_i & \cdots & \sum_{i=1}^{n} x_i^p \\
\sum_{i=1}^{n} x_i & \sum_{i=1}^{n} x_i^2 & \cdots & \sum_{i=1}^{n} x_i^{p+1} \\
\vdots & & \ddots & \vdots \\
\sum_{i=1}^{n} x_i^p & \sum_{i=1}^{n} x_i^{p+1} & \cdots & \sum_{i=1}^{n} x_i^{2p}
\end{pmatrix}
\begin{pmatrix}
\alpha_0 \\
\alpha_1 \\
\vdots \\
\alpha_p
\end{pmatrix}
=
\begin{pmatrix}
\sum_{i=1}^{n} y_i \\
\sum_{i=1}^{n} x_i y_i \\
\vdots \\
\sum_{i=1}^{n} x_i^p y_i
\end{pmatrix}.
$$

In the simplest case when $p = 1$, it becomes the standard linear regression

$$y = \alpha_0 + \alpha_1 x = a + bx.$$

Now we have

$$
\begin{pmatrix}
n & \sum_{i=1}^{n} x_i \\
\sum_{i=1}^{n} x_i & \sum_{i=1}^{n} x_i^2
\end{pmatrix}
\begin{pmatrix}
\alpha_0 \\
\alpha_1
\end{pmatrix}
=
\begin{pmatrix}
\sum_{i=1}^{n} y_i \\
\sum_{i=1}^{n} x_i y_i
\end{pmatrix}.
\tag{9.98}
$$

Its solution is

$$
\begin{pmatrix}
\alpha_0 \\
\alpha_1
\end{pmatrix}
= \frac{1}{\Delta}
\begin{pmatrix}
\sum_{i=1}^{n} x_i^2 & -\sum_{i=1}^{n} x_i \\
-\sum_{i=1}^{n} x_i & n
\end{pmatrix}
\begin{pmatrix}
\sum_{i=1}^{n} y_i \\
\sum_{i=1}^{n} x_i y_i
\end{pmatrix}
$$

$$
= \frac{1}{\Delta}
\begin{pmatrix}
(\sum_{i=1}^{n} x_i^2)(\sum_{i=1}^{n} y_i) - (\sum_{i=1}^{n} x_i)(\sum_{i=1}^{n} x_i y_i) \\
n \sum_{i=1}^{n} x_i y_i - (\sum_{i=1}^{n} x_i)(\sum_{i=1}^{n} y_i)
\end{pmatrix}, \quad
\Delta = n \sum_{i=1}^{n} x_i^2 - (\sum_{i=1}^{n} x_i)^2. \tag{9.99}
$$

These are exactly the same coefficients as those in Eq. (9.71).

## Example 9.14

We now use a quadratic function to best fit the following data (as shown in Fig. 9.5):

$$x : -0.98, \quad 1.00, \quad 2.02, \quad 3.03, \quad 4.00$$

$$y : \quad 2.44, \quad -1.51, \quad -0.47, \quad 2.54, \quad 7.52$$

For the formula $y = \alpha_0 + \alpha_1 x + \alpha_2 x^2$, we have

$$
\begin{pmatrix}
n & \sum_{i=1}^{n} x_i & \sum_{i=1}^{n} x_i^2 \\
\sum_{i=1}^{n} x_i & \sum_{i=1}^{n} x_i^2 & \sum_{i=1}^{n} x_i^3 \\
\sum_{i=1}^{n} x_i^2 & \sum_{i=1}^{n} x_i^3 & \sum_{i=1}^{n} x_i^4
\end{pmatrix}
\begin{pmatrix}
\alpha_0 \\
\alpha_1 \\
\alpha_2
\end{pmatrix}
=
\begin{pmatrix}
\sum_{i=1}^{n} y_i \\
\sum_{i=1}^{n} x_i y_i \\
\sum_{i=1}^{n} x_i^2 y_i
\end{pmatrix}.
$$

Using the data set, we have $n = 5$, $\sum_{i=1}^{n} x_i = 9.07$ and $\sum_{i=1}^{n} y_i = 10.52$. Other quantities can be calculated in a similar way. Therefore, we have

$$
\begin{pmatrix}
5.0000 & 9.0700 & 31.2217 \\
9.0700 & 31.2217 & 100.119 \\
31.2217 & 100.119 & 358.861
\end{pmatrix}
\begin{pmatrix}
\alpha_0 \\
\alpha_0 \\
\alpha_2
\end{pmatrix}
=
\begin{pmatrix}
10.52 \\
32.9256 \\
142.5551
\end{pmatrix}.
$$

By direct inversion, we have

$$
\begin{pmatrix}
\alpha_0 \\
\alpha_1 \\
\alpha_2
\end{pmatrix}
=
\begin{pmatrix}
-0.5055 \\
-2.0262 \\
1.0065
\end{pmatrix}.
$$

Finally, the best fit equation is

$$y(x) = -0.5055 - 2.0262x + 1.0065x^2,$$

which is quite close to the formula $y = x^2 - 2x - 1/2$ used to generate the original data with a random component of about 2.5%.

## Exercises

**9.1.** For geometrical distribution $P(n) = p(1 - p)^n$ (where $0 < p < 1$ and $n = 0, 1, 2, ...$), find the mean and the variance of this distribution.

**9.2.** For two independent variables $U$ and $V$ that obey Poisson distribution $P_k(\lambda_1)$ and $P_k(\lambda_2)$, respectively, show that the sum $S = U + V$ will obey $P_k(\lambda_1 + \lambda_2)$.

**9.3.** Student $A$ receives on average one email per hour, while Student B receives 2 emails per hour. What is the probability of receiving a total of exactly 4 emails in one hour? What is the probability of receiving no emails after a two-hour lesson?

**9.4.** Use linearization to derive formulas for fitting $y = \exp[ax^2 + b \sin x]$.

# CHAPTER 10

# Partial Differential Equations

Contents

## Key Points

- The basic concepts of partial differential equations (PDEs) are explained, including the first-order PDEs and second-order PDEs with some examples.
- Classifications of second-order PDEs are introduced with examples about parabolic equations, hyperbolic equations and elliptical equations.
- Basic techniques for solving linear PDEs are explained with a few worked examples.
- Integral equations are also briefly introduced with some examples.

Most physical and chemical processes can be modelled by physical laws such as the conservation of mass, energy and momentum. Such laws are often expressed in terms of partial differential equations that describe the variations of some measurable physical quantities with space and time. Therefore, PDEs are widely used in almost all disciplines of sciences and engineering as well as economics and finance.

## 1. Introduction

A partial differential equation (PDE) is a relationship containing one or more partial derivatives. Similar to the ordinary differential equation, the highest $n$th partial derivative is referred to as the order $n$ of the partial differential equation. The general form of a partial differential equation can be written as

$$\psi(u, x, y, ..., \frac{\partial u}{\partial x}, \frac{\partial u}{\partial y}, \frac{\partial^2 u}{\partial x^2}, \frac{\partial^2 u}{\partial y^2}, \frac{\partial^2 u}{\partial x \partial y}, ...) = 0. \tag{10.1}$$

where $u$ is the dependent variable, and $x, y, ...$ are the independent variables.

## Example 10.1

For example, the following equation

$$\frac{\partial u}{\partial x} + \frac{\partial u}{\partial y} = 0,$$

is a simple first-order PDE that is also linear. Here, the unknown function $u(x, y)$ is essentially a function of two independent variables $x$ and $y$.

In addition, the well-known inviscid Burgers' equation

$$\frac{\partial u}{\partial t} + u\frac{\partial u}{\partial x} = 0,$$

is a simple nonlinear PDE, which can lead to a shock wave solution.

A relatively general example of partial differential equations is the linear first-order partial differential equation, which can be written as

$$a(x, y)\frac{\partial u}{\partial x} + b(x, y)\frac{\partial u}{\partial y} = f(x, y). \tag{10.2}$$

for two independent variables and one dependent variable $u$. If the right-hand side is zero or simply $f(x, y) = 0$, then the equation is said to be homogeneous. The equation is said to be linear if $a, b$ and $f$ are functions of $x, y$ only, not $u$ itself.

For simplicity in notation in the studies of PDEs, compact subscript forms are often used in the literature. They are

$$u_x \equiv \frac{\partial u}{\partial x}, \quad u_y \equiv \frac{\partial u}{\partial y}, \quad u_{xx} \equiv \frac{\partial^2 u}{\partial x^2}, \quad u_{xy} \equiv \frac{\partial^2 u}{\partial x \partial y}, \quad ... \tag{10.3}$$

and thus we can write (10.2) as

$$au_x + bu_y = f. \tag{10.4}$$

In the rest of the chapters in this book, we will use this notation whenever appropriate.

## 2. First-Order PDEs

A first-order linear partial differential equation can be written as

$$a(x,y)u_x + b(x,y)u_y = f(x,y), \qquad (10.5)$$

which can be solved using the method of characteristics in terms of a parameter $t$

$$\frac{dx}{dt} = a(x,y), \quad \frac{dy}{dt} = b(x,y), \qquad (10.6)$$

and

$$\frac{du}{dt} = f(x,y). \qquad (10.7)$$

They essentially form a system of first-order ordinary differential equations.

### Example 10.2

One of the simplest examples of first-order linear partial differential equations is the first-order hyperbolic equation

$$u_x + vu_y = 0, \qquad (10.8)$$

where $v > 0$ is a constant. Obviously, it is easy to check that a constant $u = K$ (constant) is a trial solution to this equation.

Its characteristics are

$$\frac{dx}{dt} = 1, \quad \frac{dy}{dt} = v.$$

The integration with respect to $t$ gives

$$x = t + A_1, \quad y = vt + A_2,$$

where $A_1$ and $A_2$ are two integration constants. If we multiply $x$ by $v$, we have

$$u = y - vx = (vt + A_2) - v(t + A_1) = A_2 - vA_1 = \text{constant},$$

which is also a solution of the original PDE.

If the initial shape is $u(x,0) = \psi(x)$, then $u(x,y) = \psi(y - vx)$ at $x$ (being a time). Therefore, the shape of the wave does not change with time though its position is constantly changing.

## 3. Classification of Second-Order PDEs

A linear second-order partial differential equation can be written in the generic form in terms of two independent variables $x$ and $y$,

$$au_{xx} + bu_{xy} + cu_{yy} + gu_x + hu_y + ku = f, \tag{10.9}$$

where $a, b, c, g, h, k$ and $f$ are functions of $x$ and $y$ only. If $f(x, y, u)$ is also a function of $u$, then we say that this equation is quasi-linear.

- If $\Delta = b^2 - 4ac < 0$, the equation is elliptic. One famous example is the Laplace equation

$$\frac{\partial^2 u}{\partial x^2} + \frac{\partial^2 u}{\partial y^2} = 0, \quad \text{or} \quad u_{xx} + u_{yy} = 0. \tag{10.10}$$

  In this case, we have $a = c = 1$ and $b = 0$, and it is obvious $\Delta = b^2 - 4ac = -4 < 0$.

- If $\Delta > 0$, it is hyperbolic. A good example is the wave equation

$$\frac{\partial^2 u}{\partial t^2} = v^2 \frac{\partial^2 u}{\partial x^2},$$

  where $v > 0$ is a speed constant. This PDE can be written as $u_{tt} = v^2 u_{xx}$. In this case, $a = 1$, $b = 0$ and $c = -v^2$ so that $\Delta = b^2 - 4ac = 4v^2 > 0$.

- If $\Delta = 0$, it is parabolic. A good example of this type is the diffusion and/or heat conduction

$$\frac{\partial u}{\partial t} = \kappa \frac{\partial^2 u}{\partial x^2},$$

  where $\kappa > 0$ is the diffusion constant. This equation can also be written as $u_t = \kappa u_{xx}$. In this case, we have $a = 0$, $b = 0$, $c = -\kappa$, which means that $\Delta = b^2 - 4ac = 0$.

However, the type of a PDE is not always clear and sometimes, a mixed type can occur, depending on the parameters and domain. Let us look at an example.

## Example 10.3

The following PDE

$$\frac{\partial^2 u}{\partial x^2} + x \frac{\partial^2 u}{\partial y^2} = 0,$$

is a mixed type. It is elliptic if $x > 0$ and hyperbolic if $x < 0$. However,

$$\frac{\partial^2 u}{\partial x^2} + x^2 \frac{\partial^2 u}{\partial y^2} = 0,$$

is always elliptic in the real domain.

## 4. Classic PDEs

Three types of classic partial differential equations are widely used and they occur in a vast range of applications. In fact, almost all books or studies on partial differential equations will have to deal with these three types of basic partial differential equations.

### Laplace's and Poisson's Equation

In heat transfer problems, the steady state of heat conduction with a source is governed by the Poisson equation

$$k\nabla^2 u = f(x, y, t), \qquad (x, y) \in \Omega, \tag{10.11}$$

or

$$u_{xx} + u_{yy} = q(x, y, t), \tag{10.12}$$

for two independent variables $x$ and $y$. Here $k$ is thermal diffusivity and $f(x, y, t)$ is the heat source. $\Omega$ is the domain of interest, usually a physical region. If there is no heat source ($q = f/\kappa = 0$), it becomes the Laplace equation.

### Example 10.4

Laplace's equation in a two-dimensional domain takes the form

$$\frac{\partial^2 u}{\partial x^2} + \frac{\partial^2 u}{\partial y^2} = 0,$$

while its three-dimensional case becomes

$$\frac{\partial^2 u}{\partial x^2} + \frac{\partial^2 u}{\partial y^2} + \frac{\partial^2 u}{\partial z^2} = 0.$$

The solution of an equation is said to be harmonic if it satisfies Laplace's equation. As a useful exercise, you can verify that

$$u(x, y, z) = \frac{1}{\sqrt{x^2 + y^2 + z^2}},$$

indeed satisfies the Laplace equation.

In order to determine the temperature $u$ completely, the appropriate boundary conditions are needed. A simple boundary condition is to specify the temperature $u = u_0$ on the boundary $\partial\Omega$. This type of problem is the Dirichlet problem.

On the other hand, if the temperature is not known, but the gradient $\partial u/\partial n$ is known on the boundary where $\mathbf{n}$ is the outward-pointing unit normal, this forms the Neumann problem. Furthermore, some problems may have a mixed type of boundary conditions in the combination of

$$\alpha u + \beta \frac{\partial u}{\partial \mathbf{n}} = \gamma,$$

which naturally occurs as a radiation or cooling boundary condition.

## Parabolic Equation

Time-dependent problems, such as diffusion and transient heat conduction, are governed by the parabolic equation

$$u_t = k u_{xx}. \tag{10.13}$$

Written in the $n$-dimensional case $x_1 = x$, $x_2 = y$, $x_3 = z$, ..., it can be extended to the reaction-diffusion equation

$$u_t = k\nabla^2 u + f(u, x_1, .., x_n, t). \tag{10.14}$$

For example, the diffusion equation in 2D can be written as

$$\frac{\partial u}{\partial t} = D\left(\frac{\partial^2 u}{\partial x^2} + \frac{\partial^2 u}{\partial y^2}\right),$$

where $D$ is the diffusion coefficient of the media. The 3D version of the diffusion equation becomes

$$\frac{\partial u}{\partial t} = D\nabla^2 u = D\left(\frac{\partial^2 u}{\partial x^2} + \frac{\partial^2 u}{\partial y^2} + \frac{\partial^2 u}{\partial z^2}\right).$$

## Hyperbolic Equation

The vibration of strings and travelling waves (such as sound in the air) are governed by the hyperbolic wave equation in a three-dimensional space in the following form

$$\frac{\partial^2 u}{\partial t^2} = c^2 \nabla^2 u = c^2\left(\frac{\partial^2 u}{\partial x^2} + \frac{\partial^2 u}{\partial y^2} + \frac{\partial^2 u}{\partial z^2}\right), \tag{10.15}$$

where $c > 0$ is the so-called wave speed.

## 5. Solution Techniques

Each type of equation usually requires a different solution technique. However, there are some methods that work for most of the linearly partial differential equations with appropriate boundary conditions on a regular domain. These methods include separation of variables, method of series expansion and transform methods such as the Laplace and Fourier transforms.

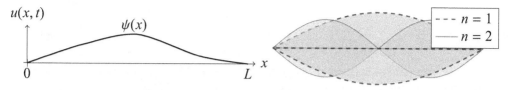

Figure 10.1: Vibrations of a string with $L$ and initial profile $\psi(x)$ (left) and different vibration modes [$n = 1$ (first mode) and $n = 2$ (second mode)].

## 5.1. Separation of Variables

The separation of variables attempts a solution of the form

$$u = X(x)Y(y)Z(z)T(t), \tag{10.16}$$

where $X(x), Y(y), Z(z), T(t)$ are functions of $x, y, z, t$, respectively. By determining these functions that satisfy the partial differential equation and the required boundary conditions in terms of eigenvalue problems, the solution of the original problem is then obtained.

The vibrations of a string are governed by the following one-dimensional (1D) wave equation

$$\frac{\partial^2 u}{\partial t^2} = v_0^2 \frac{\partial^2 u}{\partial x^2}, \quad \text{or} \quad u_{tt} = v_0^2 u_{xx}, \tag{10.17}$$

where $u(x, t)$ is the displacement of the string at $x$ and time $t$ (see Fig. 10.1), and $v_0$ is the wave speed. Let us try to solve the following problem

$$\frac{\partial^2 u}{\partial t^2} = v_0^2 \frac{\partial^2 u}{\partial x^2}, \quad 0 \leq x \leq L, \quad 0 \leq t < +\infty, \tag{10.18}$$

subject to the initial conditions

$$u(x, 0) = \psi(x), \quad u_t(x, 0) = \frac{\partial u(x, 0)}{\partial t} = 0, \tag{10.19}$$

and boundary conditions

$$u(0, t) = 0, \quad u(L, t) = 0. \tag{10.20}$$

To solve the above problem, let $u(x, t) = X(x)T(t)$ using separation of variables. We have $u_{tt} = XT''(t)$ and $u_{xx} = X''T$ and thus the wave equation becomes

$$XT'' = v_0^2 X''T, \tag{10.21}$$

or

$$\frac{1}{v_0^2}\frac{T''}{T} = \frac{X''}{X}.$$ (10.22)

Since the left-hand side depends only on $t$ and the right-hand side depends only on $x$, both sides must be equal to a constant (say, $\beta$). Thus, we have

$$\frac{1}{v_0^2}\frac{T''}{T} = \beta = \frac{X''}{X},$$

which is equivalent to two equations:

$$X'' = \beta X, \quad T'' = v_0^2 \beta T.$$ (10.23)

As will become clear later, this constant $\beta$ should be negative, and thus it is more convenient to write $\beta = -\lambda^2$. Now the above two equations become

$$X'' + \lambda^2 X = 0, \quad T'' + v_0^2 \lambda^2 T = 0.$$ (10.24)

Now we can rewrite the boundary conditions $u(0, t) = u(L, t) = 0$ as

$$X(0) = 0, \quad X(L) = 0.$$

The equation $X'' + \lambda^2 X = 0$ is a harmonic equation whose general solution is

$$X = A \sin(\lambda x) + B \cos(\lambda x),$$ (10.25)

where $A$ and $B$ are unknown constants to be determined by boundary conditions. From $X(0) = 0$, we have

$$X(0) = 0 = A \sin(0) + B \cos(0) = B, \quad \Longrightarrow \quad B = 0.$$

Similarly, from $X(L) = 0$, we have

$$X(L) = 0 = A \sin \lambda L,$$

which means either $A = 0$ (trivial solution because it implies $X = 0$ everywhere) or

$$\sin \lambda L = 0.$$

This is equivalent to the eigenvalues

$$\lambda L = n\pi, \quad (n = 1, 2, 3, ...), \quad \text{or} \quad \lambda = \frac{n\pi}{L}, \quad (n = 1, 2, 3, ...).$$ (10.26)

This is essentially the spatial frequency of the standing waves on the string. Since $A \neq 0$, we can take $A = 1$ without the loss of generality. Therefore, we have

$$X(x) = \sin\left(\frac{n\pi x}{L}\right).$$ (10.27)

Now the equation for $T$ becomes

$$T'' = -\lambda^2 v_0^2 T = -\left(\frac{n\pi v_0}{L}\right)^2 T,$$

which can be written as

$$T'' = -\omega_n^2 T, \quad \omega_n = \frac{n\pi v_0}{L}, \quad (n = 1, 2, 3, ...).$$

Its general solution is

$$T = C \sin \omega_n t + D \cos \omega_n t, \tag{10.28}$$

where $C$ and $D$ are undetermined constants. Here $\omega_n$ is the angular frequency of the vibrations.

Combining with the solution $X$, we have the fundamental solution

$$u(x, t) = X(x)T(t) = \sin\left(\frac{n\pi x}{L}\right)\left[C \sin \omega_n t + D \cos \omega_n t\right]. \tag{10.29}$$

However, as $n = 1, 2, 3, ...$ correspond to different vibration modes with different frequencies, the solution for $u$ should in general be the superposition of different modes, and thus the general solution is

$$u = \sum_{n=1}^{\infty} \sin \lambda x\left[C_n \sin \omega_n t + D_n \cos \omega_n t\right]$$

$$= \sum_{n=1}^{\infty} \sin\left(\frac{n\pi x}{L}\right)\left[C_n \sin\left(\frac{n\pi v_0 t}{L}\right) + D_n \cos\left(\frac{n\pi v_0 t}{L}\right)\right], \tag{10.30}$$

where $C_n$ and $D_n$ are the undetermined coefficients.

So far we have not used the initial conditions (10.19) yet. From $u(x, t = 0) = \psi(x)$, we get

$$u(x, 0) = \psi(x) = \sum_{n=1}^{\infty} \sin\left(\frac{n\pi x}{L}\right)\left[C_n \sin(0) + D_n \cos(0)\right], \tag{10.31}$$

which gives

$$\psi(x) = \sum_{n=1}^{\infty} D_n \sin\left(\frac{n\pi x}{L}\right). \tag{10.32}$$

This is considered as a Fourier series for function $\psi(x)$. Thus, the coefficients can be determined by

$$D_n = \frac{2}{L} \int_0^L \psi(x) \sin\left(\frac{n\pi x}{L}\right) dx. \tag{10.33}$$

From (10.30), we have its derivative with respect to $t$

$$u_t = \frac{\partial u}{\partial t} = \sum_{n=1}^{\infty} \sin\left(\frac{n\pi x}{L}\right)\left[C_n \frac{n\pi v_0}{L}\cos\left(\frac{n\pi v_0 t}{L}\right) - D_n \frac{n\pi v_0}{L}\sin\left(\frac{n\pi v_0 t}{L}\right)\right]. \qquad (10.34)$$

The initial condition $u_t = 0$ at $t = 0$ means that

$$0 = \sum_{n=1}^{\infty} \sin\left(\frac{n\pi x}{L}\right)\left[C_n \frac{n\pi v_0}{L}\cos 0 - D_n \frac{n\pi v_0}{L}\sin 0\right], \qquad (10.35)$$

which leads to

$$C_n = 0, \quad (n = 1, 2, 3, ...). \qquad (10.36)$$

Therefore, the solution to the 1D wave problem becomes

$$u(x, t) = \sum_{n=1}^{\infty} D_n \sin\left(\frac{n\pi x}{L}\right)\cos\left(\frac{n\pi v_0 t}{L}\right), \quad D_n = \frac{2}{L}\int_0^L \psi(x)\sin\left(\frac{n\pi x}{L}\right)dx. \qquad (10.37)$$

## Example 10.5

As an example, let us look at the basic acoustics of classical guitar strings. For an acoustic guitar, the base length is $L = 0.6477$ metres (25.5 in). The wave speed on a string depends on the tension $\tau$ of the string

$$v_0 = \sqrt{\frac{\tau}{\mu}},$$

where $\mu$ is the mass per unit length. Thus, the total mass of a string is $M = \mu L$. Suppose a string has a tension such that the wave speed $v_0 = 142.5$ m/s, what is its basic frequency? We know from the above result that

$$\omega_n = 2\pi f = \frac{n\pi v_0}{L},$$

where $f$ is the frequency in Hz. This means the base frequency for $n = 1$ is

$$f = \frac{n v_0}{2L} = \frac{1 \times 142.5}{2 \times 0.6477} = 110.0,$$

which is the basic frequency of note A (i.e., 110 Hz). We also know that the wavelength $\Lambda$ can be calculated by

$$\Lambda = \frac{v_0}{f} = \frac{142.5}{110.0} \approx 1.2954,$$

which is exactly twice the base length. That is $\Lambda = 2L$.

Given tension $\tau$ in newtons, length $L$ and mass $M$ (thus $\mu = M/L$), we can calculate the frequency $f$ in the following manner:

$$f = \frac{v_0}{\Lambda} = \frac{\sqrt{\tau/\mu}}{2L} = \sqrt{\frac{\tau}{4\mu L^2}} = \sqrt{\frac{\tau}{4ML}}. \tag{10.38}$$

For a steel-string guitar, once we know $f$, we can calculate the tension needed for a string $\tau = 4MLf^2$.

In fact, Eq. (10.38) also applies to the vibrations of cables used for suspension bridges. The basic vibration frequency of a cable is

$$f = \frac{1}{2L} \sqrt{\frac{\tau}{\mu}}. \tag{10.39}$$

Let $w$ be the weight per unit length of the cable, then $\mu = w/g$ where $g$ is the acceleration due to gravity. Therefore, the natural frequency $\omega_n = 2\pi f$ of a vibrating cable with tension $\tau$ is given by

$$\omega_n = 2\pi f = \frac{n\pi}{L} \sqrt{\frac{\tau g}{w}}, \quad (n = 1, 2, 3, ...), \tag{10.40}$$

where $n = 1$ corresponds to the first mode (see Fig. 10.1).

As the method of separation of variables is very useful and widely used, let us try to solve a slightly more complex problem of 1D heat conduction (see Fig. 10.2).

## Example 10.6

As a classic example, we now try to solve the 1D heat conduction equation in the domain $x \in [0, L]$ and $t \geq 0$

$$u_t = ku_{xx}, \tag{10.41}$$

with the initial value and boundary conditions

$$u(0, t) = 0, \quad \frac{\partial u(x, t)}{\partial x}\bigg|_{x=L} = 0, \quad u(x, 0) = \psi(x). \tag{10.42}$$

Letting $u(x, t) = X(x)T(t)$, we have

$$\frac{X''(x)}{X} = \frac{T'(t)}{kT}. \tag{10.43}$$

As the left-hand side depends only on $x$ and the right-hand side only depends on $t$, therefore, both sides must be equal to the same constant, and the constant can be assumed to be $-\lambda^2$. The negative sign is just for convenience because we will see below that the finiteness of the solution $T(t)$ requires that eigenvalues are real and

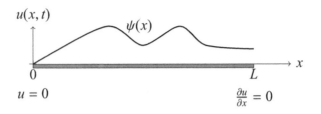

Figure 10.2: 1D heat conduction along a rod of $L$ and initial profile $\psi(x)$.

$\lambda^2 > 0$. Hence, we now get two ordinary differential equations

$$X''(x) + \lambda^2 X(x) = 0, \qquad T'(t) + k\lambda^2 T(t) = 0, \tag{10.44}$$

where $\lambda$ is the eigenvalue. The solution for $T(t)$ is

$$T = A_n e^{-\lambda^2 kt}. \tag{10.45}$$

The basic solution for $X(x)$ is simply

$$X(x) = \alpha \cos \lambda x + \beta \sin \lambda x. \tag{10.46}$$

So the fundamental solution for $u$ is

$$u(x, t) = (\alpha \cos \lambda x + \beta \sin \lambda x) e^{-\lambda^2 kt}, \tag{10.47}$$

where we have absorbed the coefficient $A_n$ into $\alpha$ and $\beta$ because they are the undetermined coefficients anyway. As the value of $\lambda$ varies with the boundary conditions,
it forms an eigenvalue problem. The general solution for $u$ should be derived by
superimposing solutions of (10.47), and we now have

$$u = \sum_{n=1}^{\infty} X_n T_n = \sum_{n=1}^{\infty} (\alpha_n \cos \lambda_n x + \beta_n \sin \lambda_n x) e^{-\lambda_n^2 kt}. \tag{10.48}$$

Obviously, we have not completed the solution yet because the unknown coefficients need to be determined. From the boundary condition $u(0, t) = 0$ at $x = 0$, we
have

$$0 = \sum_{n=1}^{\infty} \alpha_n e^{-\lambda_n^2 kt}, \tag{10.49}$$

which leads to $\alpha_n = 0$ since $\exp(-\lambda^2 kt) > 0$. From $\frac{\partial u}{\partial x}\Big|_{x=L} = 0$, we have

$$\lambda_n \cos \lambda_n L = 0, \tag{10.50}$$

which requires

$$\lambda_n L = \frac{(2n-1)\pi}{2}, \qquad (n = 1, 2, ...). \tag{10.51}$$

Therefore, $\lambda$ cannot be continuous, and it only takes an infinite number of discrete values, called eigenvalues.

Each eigenvalue $\lambda = \lambda_n = \frac{(2n-1)\pi}{2L}, (n = 1, 2, ...)$ has a corresponding eigenfunction $X_n = \sin(\lambda_n x)$. Substituting into the solution for $T(t)$, we have

$$T_n(t) = A_n e^{-\frac{[(2n-1)\pi]^2}{4L^2}kt}. \tag{10.52}$$

By expanding the initial condition into a Fourier series so as to determine the coefficients, we have

$$u(x, t) = \sum_{n=1}^{\infty} \beta_n \sin(\frac{(2n-1)\pi x}{2L}) e^{-[\frac{(2n-1)\pi}{2L}]^2 kt}, \quad \beta_n = \frac{2}{L} \int_0^L \psi(x) \sin[\frac{(2n-1)\pi x}{2L}]dx.$$

This procedure is quite lengthy, though the calculations mostly involve integrals.

## Example 10.7

In the special case when initial condition $u(x, t = 0) = \psi = u_0$ is constant, the requirement for $u = u_0$ at $t = 0$ becomes

$$u_0 = \sum_{n=1}^{\infty} \beta_n \sin \frac{(2n-1)\pi x}{2L}. \tag{10.53}$$

Using the orthogonal relationships

$$\int_0^L \sin \frac{m\pi x}{L} \sin \frac{n\pi x}{L} dx = 0, \quad (m \neq n), \qquad \int_0^L (\sin \frac{n\pi x}{L})^2 dx = \frac{L}{2}, \qquad (n = 1, 2, ...),$$

and multiplying both sides of Eq.(10.53) by $\sin[(2n-1)\pi x/2L]$, we have the integral

$$\beta_n \frac{L}{2} = \int_0^L \sin \frac{(2n-1)\pi x}{2L} u_0 dx = \frac{2u_0 L}{(2n-1)\pi}, \qquad \text{or } \beta_n = \frac{4u_0}{(2n-1)\pi}, \quad (n = 1, 2, ...),$$

and thus the solution becomes

$$u = \frac{4u_0}{\pi} \sum_{n=1}^{\infty} \frac{1}{(2n-1)} e^{-\frac{(2n-1)^2 \pi^2 kt}{4L^2}} \sin \frac{(2n-1)\pi x}{2L}. \tag{10.54}$$

This solution is essentially the same as the classic heat conduction problem discussed by Carslaw and Jaeger in 1986. This same solution can also be obtained using the Fourier series of $u_0$ in $0 < x < L$.

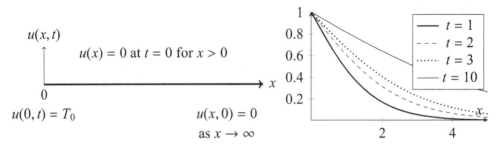

Figure 10.3: 1D heat transfer in a semi-infinite domain with a fixed temperature $u = T_0$ at $x = 0$ and initial temperature $u(x, 0) = 0$ for $x > 0$ (left) and the solution evolves with $t$ for $k = 1$ and $T_0 = 1$ (right).

## 5.2. Laplace Transform

The integral transform can reduce the number of the independent variables. For the 1D time-dependent case, it transforms a partial differential equation into an ordinary differential equation. By solving the ordinary differential equation and inverting it back, we can obtain the solution for the original partial differential equation.

When doing the transform, we have to consider the boundary conditions as well. Let us look at an example.

### Example 10.8

As an example, we now solve the heat conduction problem over a semi-infinite interval $x \in [0, \infty)$ and $t \geq 0$ (see Fig. 10.3),

$$u_t = k u_{xx}, \qquad u(x, 0) = 0, \quad u(0, t) = T_0. \tag{10.55}$$

Let $\bar{u}(x, s) = \int_0^\infty u(x, t) e^{-st} dt$ be the Laplace transform of $u(x, t)$, then Eq.(10.55) becomes

$$s\bar{u} = k \frac{d^2 \bar{u}}{dx^2}, \quad \text{and} \quad \bar{u}_{x=0} = \frac{T_0}{s},$$

which is an ordinary differential equation whose general solution can be written as

$$\bar{u} = A e^{-\sqrt{\frac{s}{k}} x} + B e^{\sqrt{\frac{s}{k}} x}.$$

The finiteness of the solution as $x \to \infty$ requires that $B = 0$, and the boundary condition at $x = 0$ leads to

$$\bar{u} = \frac{T_0}{s} e^{-\sqrt{\frac{s}{k}} x}.$$

By using the inverse Laplace transform, we have

$$u = T_0 \mathrm{erfc}\left(\frac{x}{2\sqrt{kt}}\right) = T_0\left[1 - \mathrm{erf}\left(\frac{x}{2\sqrt{kt}}\right)\right],$$

where $\mathrm{erfc}(x)=1-\mathrm{erf}(x)$ is the complementary error function. The evolution of $T$ for different times is shown in Fig. 10.3.

The Fourier transform works in a similar manner to the Laplace transform. A PDE is first transformed into an ordinary differential equation in the frequency domain, and the solution is obtained by solving its corresponding ODE. Then, the solution to the PDE in the time domain is obtained by inverse Fourier transforms.

## 5.3. Similarity Solution

Sometimes, the diffusion equation

$$u_t = \kappa u_{xx}, \tag{10.56}$$

can be solved by using the so-called similarity method by defining a similar variable

$$\eta = \frac{x}{\sqrt{\kappa t}}, \quad \text{or} \quad \zeta = \frac{x^2}{\kappa t}. \tag{10.57}$$

One can assume that the solution to the equation has the form

$$u = (\kappa t)^\alpha f\left[\frac{x^2}{(\kappa t)^\beta}\right]. \tag{10.58}$$

By substituting it into the diffusion equation, the coefficients $\alpha$ and $\beta$ can be determined. For most applications, one can assume $\alpha = 0$ so that $u = f(\zeta)$. In this case, we have

$$4\zeta u'' + 2u' + \zeta\beta(\kappa t)^{\beta-1}u' = 0, \tag{10.59}$$

where $u' = du/d\zeta$. In deriving this equation, one has to use the chain rules of differentiations

$$\frac{\partial}{\partial x} = \frac{\partial}{\partial \zeta}\frac{\partial \zeta}{\partial x}, \quad \frac{\partial}{\partial t} = \frac{\partial}{\partial \zeta}\frac{\partial \zeta}{\partial t}. \tag{10.60}$$

Since the original equation does not have time-dependent terms explicitly, this means that all the exponents for any $t$-terms must be zero. Therefore, we have $\beta = 1$. Now, the diffusion equation becomes

$$\zeta f''(\zeta) = -\left(\frac{1}{2} + \frac{\zeta}{4}\right)f'. \tag{10.61}$$

Using $(\ln f')' = f''/f'$ and integrating the above equation once, we get

$$f' = \frac{Ke^{-\zeta/4}}{\sqrt{\zeta}}. \tag{10.62}$$

Integrating it again and using the substitution $\zeta = 4\xi^2$, we obtain

$$u = A \int_0^\xi e^{-\xi^2} d\xi = C\mathrm{erf}(\frac{x}{\sqrt{4\kappa t}}) + D, \tag{10.63}$$

where $C$ and $D$ are constants that can be determined from appropriate boundary conditions.

For the same problem as (10.55), the boundary condition as $x \to \infty$ implies that $C + D = 0$, while $u(0, t) = T_0$ means that $D = -C = T_0$. Therefore, we finally have

$$u = T_0[1 - \mathrm{erf}(\frac{x}{\sqrt{4\kappa t}})] = T_0\mathrm{erfc}(\frac{x}{\sqrt{4\kappa t}}). \tag{10.64}$$

There are other important methods for solving partial differential equations. These include Green's function, series methods, asymptotic methods, approximate methods, perturbation methods and naturally the numerical methods. Interested readers can refer to more advanced literature on these topics.

## 6. Integral Equations

Differential equations are mathematical equations that relate functions with their derivatives. Ordinary differential equations concern univariate functions, while partial differential equations concern functions in terms of at least two independent variables. None of these equations explicitly include integrals in their standard forms. As many conservation laws, such as conservation of energy and mass, require integrals, integral equations can be useful and more convenient in certain applications.

An integral equation is an equation that relates a function with its integrals. For example, the following equation

$$y(x) = x + \frac{1}{2} \int_0^x y(u)du, \tag{10.65}$$

is an integral equation. In general, integral equations can be very challenging to solve, and here we only try to solve linear integral equations of two types: Fredholm and Volterra integral equations, depending on the limits of integration.

### 6.1. Fredholm and Volterra Integral Equations

If both limits of integration are fixed, the integral equation is of the Fredholm type. If one of the integration limits is a variable, the integral equation is of the Volterra type.

A linear integral equation for $y(x)$ can be written in the following generic form

$$u(x) + \lambda \int_a^b K(x, \eta)y(\eta)d\eta = v(x)y(x), \tag{10.66}$$

where $K(x, \eta)$ is referred to as the kernel of the integral equation. The parameter $\lambda$ is a known constant. If the function $u(x) = 0$, the equation is then called homogeneous. If $u(x) \neq 0$, the equation is inhomogeneous.

If the function $v(x) = 0$, then the unknown $y(x)$ appears only once in the integral equation, and it is under the integral sign only. This is called the linear Fredholm integral equation of the first kind

$$u(x) + \lambda \int_a^b K(x, \eta)y(\eta)d\eta = 0. \tag{10.67}$$

On the other hand, if $v(x) = 1$, equation (10.66) becomes the integral equation of the second kind

$$u(x) + \lambda \int_a^b K(x, \eta)y(\eta)d\eta = y(x). \tag{10.68}$$

An integral equation with fixed integration limits $a$ and $b$, is called a Fredholm equation. If the upper integration limit $b$ is not fixed, then the equation becomes a Volterra equation. The integral equation becomes singular if at least one of its integration limits approaches infinite. In addition, if derivatives also appear in the equation, they are often called integro-differential equations. For example,

$$y(x) = \exp(-x) + \lambda \int_0^1 x\eta y(\eta)d\eta, \tag{10.69}$$

is a Fredholm-type integral equation, while

$$y(x) = x^2 + \sin(x) + \lambda \int_0^x e^{x-\eta}y(\eta)d\eta, \tag{10.70}$$

is a Volterra-type integral equation.

In general, the Volterra integral equation can be written as

$$u(x) + \lambda \int_a^x K(x, \eta)y(\eta)d\eta = v(x)y(x). \tag{10.71}$$

The first kind [or $v(x) = 0$] and second kind [or $v(x) = 1$] are defined in a similar manner to the Fredholm equation.

The kernel is said to be separable or degenerate if it can be written in the finite sum form

$$K(x, \eta) = \sum_{i=1}^{N} f_i(x)g_i(\eta),$$ (10.72)

where $f_i(x)$ and $g_i(\eta)$ are functions of $x$ and $\eta$, respectively.

A kernel is called a convolution kernel (also displacement kernel) if it can be written as a function of the difference $(x - \eta)$ of its two arguments: $K(x, \eta) = K(x - \eta)$. For example,

$$y(x) = e^x + \lambda \int_0^x e^{x-\eta} y(\eta) d\eta,$$

is a Volterra-type integral equation with a convolution kernel.

## 6.2. Solutions of Integral Equations

Most integral equations do not have closed-form solutions. For linear integral equations, closed-form solutions are only possible for the special cases of separable and displacement kernels.

For a Fredholm integral equation of the second kind with a separable kernel, we can substitute the kernel (10.72) into the equation and we have

$$u(x) + \lambda \int_a^b \sum_{i=1}^{N} f_i(x)g_i(\eta)y(\eta)d\eta = y(x),$$ (10.73)

which becomes

$$u(x) + \lambda \sum_{i=1}^{N} f_i(x) \int_a^b g_i(\eta)y(\eta)d\eta = y(x).$$ (10.74)

Because the integration limits are fixed, the integrals over $\eta$ are constants that are to be determined. By defining

$$\alpha_i = \int_a^b g_i(\eta)y(\eta)d\eta,$$ (10.75)

we now have the solution in the form

$$y(x) = u(x) + \lambda \sum_{i=1}^{N} \alpha_i f_i(x),$$ (10.76)

where the $N$ coefficients $\alpha_i$ are determined by

$$\alpha_i = \int_a^b g_i(\eta)u(\eta)d\eta + \lambda \sum_{i=1}^{N} \int_a^b [\alpha_i f_i(\eta)g_i(\eta)]d\eta,$$ (10.77)

for $i = 1, 2, ..., N$. Only for a few special cases can these coefficients be written as simple explicit expressions.

## Example 10.9

Let us try to solve

$$x^3 + \int_0^1 x^2\eta^2 y(\eta)d\eta = y(x), \tag{10.78}$$

which means that $u(x) = x^3$ and $\lambda = 1$. The solution is in the form

$$y(x) = x^3 + x^2\alpha, \quad \alpha = \int_0^1 \eta^2 y(\eta)d\eta.$$

So

$$\alpha = \int_0^1 \eta^2 y(\eta)d\eta = \int_0^1 \eta^2[\eta^3 + \eta^2\alpha]d\eta = \int_0^1 \eta^5 d\eta + \alpha\int_0^1 \eta^4 d\eta = \frac{1}{6} + \alpha\frac{1}{5},$$

which gives $\alpha = 5/24$. Therefore, we finally have

$$y(x) = x^3 + \frac{5}{24}x^2.$$

A Volterra equation with separable kernels may be solved by transforming into a differential equation via direct differentiation. Let us demonstrate this by an example.

## Example 10.10

To solve an integral equation of Volterra type

$$y(x) = e^x + \int_0^x e^x \sin(\zeta)y(\zeta)d\zeta,$$

we first divide both sides by $e^x$ to get

$$\frac{y(x)}{e^x} = 1 + \int_0^x \sin(\zeta)y(\zeta)d\zeta,$$

whose differentiation with respect to $x$ leads to

$$\left[\frac{y(x)}{e^x}\right]' = y(x)\sin(x), \quad \text{or} \quad \frac{1}{e^x}y'(x) - y(x)e^{-x} = y(x)\sin(x).$$

Dividing both sides by $y(x)$ and using $[\ln y(x)]' = y'(x)/y(x)$, we have

$$[\ln y(x)]' = e^x \sin x + 1.$$

By direct integration, we have

$$\ln y(x) = x - \frac{1}{2}e^x \cos x + \frac{1}{2}e^x \sin x.$$

Thus, we finally obtain

$$y(x) = \exp\left[x - \frac{e^x}{2}(\cos x - \sin x)\right].$$

---

There are other solution techniques for integral equations such as the series expansion method, successive approximations and transform methods using Laplace transforms. Interested readers can refer to more advanced literature.

In addition, both partial differential equations and integral equations can be solved using numerical methods and there are many textbooks on numerical methods.

## Exercises

**10.1.** Discuss the type of each of the following PDEs:
- $xu_x + yu_y = xyu$
- $(u_t)^2 = u_x + xt$
- $u_t + u^2u_x = 0$
- $u_{tt} = tu_{xx}$
- $u_t = D\nabla^2 u + \gamma u(1-u)$ where $D > 0$ and $\gamma > 0$ are constants.

**10.2.** The Euler-Bernoulli beam equation can be written as

$$EI\frac{\partial^2}{\partial x^2}\left(EI\frac{\partial^2 u(x,t)}{\partial x^2}\right) = -\mu\frac{\partial^2 u(x,t)}{\partial t^2},$$

where $\mu = \rho A$ is the so-called linear mass density of the beam. $\rho$ is the density and $A$ is the area of its cross section. In addition, $E$ is the elastic modulus and $I$ is the moment of inertia of the beam. Using $u(x,t) = X(x)f(t)$, show that the above equation leads to

$$\frac{\partial^4 X}{\partial x^4} - \lambda X = 0, \quad \frac{\partial^2 f}{\partial t^2} + \omega^2 f = 0, \quad \lambda = \frac{\omega^2\mu}{EI}.$$

**10.3.** Show that $\nabla^2 u = 0$ has a solution

$$u = \frac{1}{\sqrt{x^2 + y^2 + z^2}}.$$

**10.4.** Solve $y(x) = 1 + \int_0^1 xy(t)dt$.

# CHAPTER 11

# Numerical Methods and Optimization

## Contents

---

## Key Points

- Root-finding algorithms are used to solve polynomial equations and such methods include bisection methods, Newton's method and others.
- Numerical integration and numerical solutions of ODEs are introduced with detailed discussions of Euler's method and the Runge-Kutta method.
- Basic optimization techniques are introduced briefly with discussion of Lagrange multipliers and KKT conditions.

---

There are many different numerical methods for solving a wide range of problems with different orders of accuracy and various levels of complexity. For example, for numerical solutions of ODEs, we can have a simple Euler integration scheme or higher-order

Runge-Kutta scheme. For PDEs, we can use finite difference methods, finite element methods, finite volume methods and others. As this book is an introductory textbook, we will only introduce the most basic methods that are most useful to a wide range of problems in engineering.

To demonstrate how these numerical methods work, we will use step-by-step examples to find the roots of nonlinear equations, to estimate integrals by numerical integration, and to solve ODEs by direct integration and higher-order Runge-Kutta methods.

## 1. Root-Finding Algorithms

Let us start by trying to find the square root of a number $a$. It is essentially equivalent to finding the solution of

$$x^2 - a = 0. \tag{11.1}$$

We can rearrange it as

$$x = \frac{1}{2}(x + \frac{a}{x}), \tag{11.2}$$

which makes it possible to estimate the root $x$ iteratively. If we start from a guess, say $x_0 = 1$, we can calculate the new estimate $x_{n+1}$ from any previous value $x_n$ using

$$x_{n+1} = \frac{1}{2}(x_n + \frac{a}{x_n}). \tag{11.3}$$

Let us look at an example.

### Example 11.1

In order to find $\sqrt{10}$, we have $a = 10$ with an initial guess $x_0 = 1$. The first five iterations are as follows:

$$x_1 = \frac{1}{2}(x_0 + \frac{10}{x_0}) = \frac{1}{2}(1 + \frac{10}{1}) = 5.5,$$

$$x_2 = \frac{1}{2}(x_1 + \frac{10}{x_1}) = \frac{1}{2}(5.5 + \frac{10}{5.5}) \approx 3.6590909.$$

$$x_3 = \frac{1}{2}(x_2 + \frac{10}{x_2}) \approx 3.1960051, \quad x_4 \approx 3.1624556, \quad x_5 \approx 3.16227767.$$

We know that the exact solution is $\sqrt{10} = 3.16227766017$. We can see that after only five iterations, $x_5$ is accurate to the 7th decimal place.

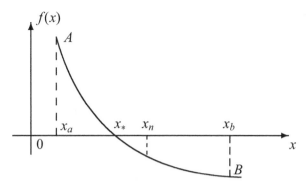

Figure 11.1: Bisection method for finding the root $x_*$ of $f(x_*) = 0$ between two bounds $x_a$ and $x_b$ in the domain $x \in [a, b]$.

Iteration is in general very efficient; however, we have to be careful about the proper design of the iteration formula and the selection of an appropriate initial guess. For example, we cannot use $x_0 = 0$ as the initial value. Similarly, if we start $x_0 = 1000$, we have to do more iterations to get the same accuracy.

There are many methods for finding the roots such as the bisection method, though the Newton-Raphson method is by far the most successful and widely used.

## Bisection Method

The above-mentioned iteration method to find $x = \sqrt{k}$ is in fact equivalent to finding the solution or the root of the function $f(x) = x^2 - k = 0$. For any function $f(x)$ in the interval $[a, b]$, the root-finding bisection method works in the following way as shown in Fig. 11.1.

The iteration procedure starts with two initial guessed bounds $x_a$ (lower bound), and $x_b$ (upper bound) so that the true root $x = x_*$ lies between these two bounds. This requires that $f(x_a)$ and $f(x_b)$ have different signs. In our case shown in Fig. 11.1, $f(x_a) > 0$ and $f(x_b) < 0$, but $f(x_a)f(x_b) < 0$. The obvious choice is $x_a = a$ and $x_b = b$. The next estimate is just the midpoint of $A$ and $B$, and we have

$$x_n = \frac{1}{2}(x_a + x_b). \tag{11.4}$$

We then have to test the sign of $f(x_n)$. If $f(x_n) < 0$ (having the same sign as $f(x_b)$), we then update the new upper bound as $x_b = x_n$. If $f(x_n) > 0$ (having the same sign as $f(x_a)$), we update the new lower bound as $x_a = x_n$. In a special case when $f(x_n) = 0$, you have found the true root. The iterations continue in the same manner until a given accuracy is achieved or the prescribed number of iterations is reached.

## Example 11.2

If we want to find $\sqrt{\pi}$, we have

$$f(x) = x^2 - \pi = 0.$$

We can use $x_a = 1$ and $x_b = 2$ since $\pi < 4$ (thus $\sqrt{\pi} < 2$). The first bisection point is

$$x_1 = \frac{1}{2}(x_a + x_b) = \frac{1}{2}(1 + 2) = 1.5.$$

Since $f(x_a) < 0$, $f(x_b) > 0$ and $f(x_1) = -0.8916 < 0$, we update the new lower bound $x_a = x_1 = 1.5$. The second bisection point is

$$x_2 = \frac{1}{2}(1.5 + 2) = 1.75,$$

and $f(x_2) = -0.0791 < 0$, so we update lower bound again $x_a = 1.75$. The third bisection point is

$$x_3 = \frac{1}{2}(1.75 + 2) = 1.875.$$

Since $f(x_3) = 0.374 > 0$, we now update the new upper bound $x_b = 1.875$. The fourth bisection point is

$$x_4 = \frac{1}{2}(1.75 + 1.875) = 1.8125.$$

It is within 2.5% of the true value of $\sqrt{\pi} \approx 1.7724538509$.

In general, the convergence of the bisection method is very slow, and Newton's method is a much better choice in most cases.

## Newton-Raphson Method

Newton's method is a widely-used classical method for finding the solution to a nonlinear univariate function of $f(x)$ on the interval $[a, b]$. It is also referred to as the Newton-Raphson method. At any given point $x_n$ shown in Fig. 11.2, we can approximate the function by a Taylor series

$$f(x_{n+1}) = f(x_n + \Delta x) \approx f(x_n) + f'(x_n)\Delta x, \tag{11.5}$$

where

$$\Delta x = x_{n+1} - x_n, \tag{11.6}$$

which leads to

$$x_{n+1} - x_n = \Delta x \approx \frac{f(x_{n+1}) - f(x_n)}{f'(x_n)}, \tag{11.7}$$

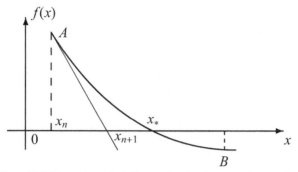

Figure 11.2: Newton's method of approximating the root $x_*$ by $x_{n+1}$ from $x_n$.

or

$$x_{n+1} \approx x_n + \frac{f(x_{n+1}) - f(x_n)}{f'(x_n)}. \tag{11.8}$$

Since we try to find an approximation to $f(x) = 0$ with $f(x_{n+1})$, we can use the approximation $f(x_{n+1}) \approx 0$ in the above expression. Thus we have the standard Newton iterative formula

$$x_{n+1} = x_n - \frac{f(x_n)}{f'(x_n)}. \tag{11.9}$$

The iteration procedure starts from an initial guess value $x_0$ and continues until a predefined criterion is met. A good initial guess will use fewer steps; however, if there is no obvious initial good starting point, you can start at any point on the interval $[a, b]$. But if the initial value is too far from the true zero, the iteration process may fail. So it is a good idea to limit the number of iterations.

## Example 11.3

To find the root of

$$f(x) = x - e^{-x} = 0,$$

we use Newton-Raphson method starting from $x_0 = 0$. We know that

$$f'(x) = \left(x - e^{-x}\right)' = 1 + e^{-x},$$

and thus the iteration formula becomes

$$x_{n+1} = x_n - \frac{x_n - e^{-x_n}}{1 + e^{-x_n}}.$$

Since $x_0 = 0$, we have

$$x_1 = 0 - \frac{0 - e^{-0}}{1 + e^{-0}} = 0.5, \quad x_2 = 0.5 - \frac{0.5 - e^{-0.5}}{1 + e^{-0.5}} \approx 0.566311003$$

$$x_3 \approx 0.56714317, \quad x_4 \approx 0.56714329.$$

We can see that $x_3$ (only three iterations) is very close (to the 6th decimal place) to the true root which is $x_* \approx 0.5671432904$, while $x_4$ is accurate to the 8th decimal place.

---

Sometimes, we can use a linear interpolation of two consecutive values to approximate $f'(x)$ and we have an alternative Newton's formula as

$$x_{n+1} = x_n - \frac{f(x_n)}{[f(x_n) - f(x_{n-1})]/(x_n - x_{n-1})} = x_n - \frac{(x_n - x_{n-1})f(x_n)}{f(x_n) - f(x_{n-1})}, \tag{11.10}$$

which can be more convenient in certain applications.

We have seen that Newton-Raphson's method is very efficient and that is why it is so widely used. Using this method, we can virtually solve almost all root-finding problems, though care should be taken when dealing with multiple roots. Obviously, this method is not applicable to carrying out integration.

## 2. Numerical Integration

For any smooth function, we can always calculate its derivatives by direct differentiation; however, integration is often difficult even for seemingly simple integrals such as the error function

$$\text{erf}(x) = \frac{2}{\sqrt{\pi}} \int_0^x e^{-u^2} du. \tag{11.11}$$

The integration of this simple integrand $\exp(-u^2)$ does not lead to any simple explicit expression, which is why it is often written as erf(), referred to as the error function. If we pick up a mathematical handbook, we know that erf(0) = 0, and erf($\infty$) = 1, while

$$\text{erf}(0.5) \approx 0.52049, \qquad \text{erf}(1) \approx 0.84270. \tag{11.12}$$

If we want to calculate such integrals, numerical integration is the best alternative.

Now if we want to numerically evaluate the following integral

$$I = \int_a^b f(x)dx, \tag{11.13}$$

where $a$ and $b$ are fixed and finite, we know that the value of the integral is exactly

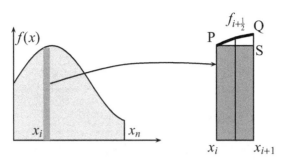

Figure 11.3: Integral as a sum of multiple thin stripes.

the total area under the curve $y = f(x)$ between $a$ and $b$. As both the integral and the area can be considered as the sum of the values over many small intervals, the simplest way of evaluating such numerical integration is to divide up the integral interval into $n$ equal small sections and split the area into $n$ thin strips of width $h$ so that $h \equiv \Delta x = (b-a)/n$, $x_0 = a$ and $x_i = ih + a(i = 1, 2, ..., n)$. The values of the functions at the dividing points $x_i$ are denoted as $y_i = f(x_i)$, and the value at the midpoint between $x_i$ and $x_{i+1}$ is labelled as $y_{i+1/2} = f_{i+1/2}$

$$y_{i+1/2} = f(x_{i+1/2}) = f_{i+1/2}, \qquad x_{i+1/2} = \frac{x_i + x_{i+1}}{2}. \qquad (11.14)$$

The accuracy of such approximations depends on the number $n$ and the way to approximate the curve in each interval.

Figure 11.3 shows such an interval $[x_i, x_{i+1}]$ which is exaggerated in the figure for clarity. The curve segment between $P$ and $Q$ is approximated by a straight line with a slope

$$\frac{\Delta y}{\Delta x} = \frac{f(x_{i+1}) - f(x_i)}{h}, \qquad (11.15)$$

which approaches $f'(x_{i+1/2})$ at the midpoint point when $h \to 0$.

The trapezium (formed by $P$, $Q$, $x_{i+1}$, and $x_i$) is a better approximation than the rectangle ($P$, $S$, $x_{i+1}$ and $x_i$) because the former has an area

$$A_i = \frac{f(x_i) + f(x_{i+1})}{2} h, \qquad (11.16)$$

which is closer to the area

$$\mathcal{I}_i = \int_{x_i}^{x_{i+1}} f(x)dx, \qquad (11.17)$$

under the curve in the small interval $x_i$ and $x_{i+1}$. If we use the area $A_i$ to approximate $\mathcal{I}_i$, we have the trapezium rule of numerical integration. Thus, the integral is simply the sum of all these small trapeziums, and we have

$$I \approx \frac{h}{2}[f_0 + 2(f_1 + f_2 + \dots + f_{n-1}) + f_n]$$

$$= h[f_1 + f_2 + \dots + f_{n-1} + \frac{(f_0 + f_n)}{2}]. \tag{11.18}$$

From the Taylor series, we know that

$$\frac{f(x_i) + f(x_{i+1})}{2} \approx \frac{1}{2}\{[f(x_{i+1/2}) - \frac{h}{2}f'(x_{i+1/2}) + \frac{1}{2!}(\frac{h}{2})^2 f''(x_{i+1/2})]$$

$$+[f(x_{i+1/2}) + \frac{h}{2}f'(x_{i+1/2}) + \frac{1}{2!}(\frac{h}{2})^2 f''(x_{i+1/2})]\}$$

$$= f(x_{i+1/2}) + \frac{h^2}{8}f''(x_{i+1/2}). \tag{11.19}$$

where $O(h^2 f'')$ means that the value is about the order of $h^2 f''$, or $O(h^2) = Kh^2 f''$ where $K$ is a constant.

The trapezium rule is just one of the simple and popular schemes for numerical integration with the error of $O(h^3 f'')$. If we want higher accuracy, we can either reduce $h$ or use a better approximation for $f(x)$. A small $h$ means a large $n$, which implies that we have to do the sum of many small sections, which may increase the computational time.

On the other hand, we can use higher-order approximations for the curve. Instead of using straight lines or linear approximations for curve segments, we can use parabolas or quadratic approximations. For any consecutive three points $x_{i-1}$, $x_i$ and $x_{i+1}$, we can construct a parabola in the form

$$f(x_i + t) = f_i + \alpha t + \beta t^2, \qquad t \in [-h, h]. \tag{11.20}$$

As this parabola must go through the three known points $(x_{i-1}, f_{i-1})$ at $t = -h$, $(x_i, f_i)$ at $t = 0$ and $x_{i+1}, f_{i+1}$ at $t = h$, we have the following equations for $\alpha$ and $\beta$

$$f_{i-1} = f_i - \alpha h + \beta h^2, \tag{11.21}$$

and

$$f_{i+1} = f_i + \alpha h + \beta h^2, \tag{11.22}$$

which lead to

$$\alpha = \frac{f_{i+1} - f_{i-1}}{2h}, \qquad \beta = \frac{f_{i-1} - 2f_i + f_{i+1}}{2h^2}. \tag{11.23}$$

In fact, $\alpha$ is the centred approximation for the first derivative $f_i'$ and $\beta$ is related to the central difference scheme for the second derivative $f_i''$. Therefore, the integral from $x_{i-1}$ to $x_{i+1}$ can be approximated by

$$I_i = \int_{x_{i-1}}^{x_{i+1}} f(x)dx \approx \int_{-h}^{h} [f_i + \alpha t + \beta t^2]dt = \frac{h}{3}[f_{i-1} + 4f_i + f_{i+1}],$$

where we have substituted the expressions for $\alpha$ and $\beta$. To ensure the whole interval $[a, b]$ can be divided up to form three-point approximations without any point left out, $n$ must be even. Therefore, the estimate of the integral becomes

$$I \approx \frac{h}{3}[f_0 + 4(f_1 + f_3 + \dots + f_{n-1}) + 2(f_2 + f_4 + \dots + f_{n-2}) + f_n], \qquad (11.24)$$

which is the standard Simpson's rule.

As the approximation for the function $f(x)$ is quadratic, an order higher than the linear form, the error estimate of Simpson's rule is thus $O(h^4)$ or $O(h^4 f''')$ to be more specific. There are many variations of Simpson's rule with higher-order accuracies such as $O(h^5 f^{(4)})$.

## Example 11.4

We know the value of the integral

$$I = \text{erf}(1) = \frac{2}{\sqrt{\pi}} \int_0^1 e^{-x^2} dx = 0.8427007929.$$

Let us now estimate

$$I = \frac{2}{\sqrt{\pi}} J, \qquad J = \int_0^1 e^{-x^2} dx,$$

using the Simpson rule with $n = 8$ and $h = (1 - 0)/8 = 0.125$. We have

$$J \approx \frac{h}{3}[f_0 + 4(f_1 + f_3 + f_5 + f_7) + 2(f_2 + f_4 + f_6) + f_8].$$

Since $f_i = e^{-x_i^2} = e^{-(i*h)^2}$, we have $f_0 = 1$, $f_1 = 0.984496$, $f_2 = 0.939413$, $f_3 = 0.868815$, $f_4 = 0.778801$, $f_5 = 0.676634$, $f_6 = 0.569783$, $f_7 = 0.465043$, and $f_8 = 0.367879$. Now the integral estimate of $J$ is

$$J \approx \frac{0.125}{3}[1 + 4 \times 2.9949885 + 2 \times 2.2879967 + 0.367879]$$

$$\approx \frac{0.125}{3} \times 17.923827 \approx 0.746826.$$

Finally, the integral estimate of $I$ is

$$I = \frac{2}{\sqrt{\pi}}J = \frac{2}{\sqrt{3.1415926}} \times 0.746826 \approx 0.842703.$$

We can see that this estimate is accurate to the 5th decimal place.

There are other even better ways for evaluating the integral more accurately using fewer points of evaluation. Such numerical integration is called the Gaussian integration or Gaussian quadrature. Interested readers can refer to more advanced books on numerical methods or numerical analysis.

## 3. Numerical Solutions of ODEs

The simplest first-order differential equation can be written as

$$\frac{dy}{dx} = f(x, y), \tag{11.25}$$

where $f(x, y)$ is a known function of $x$ and $y$. In principle, the solution can be obtained by direct integration

$$y(x) = y(x_0) + \int_{x_0}^{x} f(x, y(x))dx, \tag{11.26}$$

but in practice it is usually impossible to do the integration analytically, as it requires the solution $y(x)$ to evaluate the right-hand side. Numerical integration in this case is the most common technique for obtaining approximate solutions.

A naive approach is to use the standard techniques such as Simpson's rule for numerical integration to evaluate the integral numerically. However, since we have to use some approximation for $y(x)$, such techniques hardly work here without modification. A better approach is to start from the known initial value $y(x_0)$ at $x_0$, and try to march to the next point at $x = x_0 + h$ where $h$ is a small increment. In this manner, the solution at all other values of $x$ can be estimated numerically. This is essentially the Euler scheme.

## 3.1. Euler Scheme

The basic idea of the Euler method is to approximate the integral (11.26) using an iterative procedure

$$y_{n+1} = y_n + \int_{x_n}^{x_{n+1}} f(x, y)dx = y_n + hf(x_n, y_n), \tag{11.27}$$

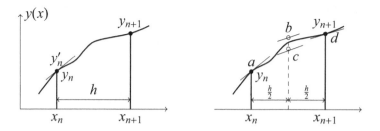

Figure 11.4: Forward Euler method and Runge-Kutta higher-order method.

where $h = \Delta x = x_{n+1} - x_n$ is a small increment. Here we use the notations $x_n = x_0 + nh$ for ($n = 0, 1, 2, ..., N$) and $y_n = y(x_n)$. This is essentially to divide the interval from $x_0$ to $x = Nh$ into $N$ small interval of width $h$, so the value of $f(x, y)$ is approximated by its value in the interval as $f(x_n, y_n)$.

The above equation can also be viewed as the approximation of the first derivative

$$\frac{dy_n}{dx} = y'_n = \frac{y_{n+1} - y_n}{h}. \tag{11.28}$$

This is a forward difference scheme for approximating the first derivative (see Fig. 11.4), and it uses only one evaluation at $x_n$.

From the differentiation and integration discussed earlier in this book, we know that such an approximation is very crude, and the accuracy is at most $O(h^2)$.

A potentially better way is to use the so-called central difference to approximate the first derivative by

$$y'_n = \frac{y_{n+1} - y_{n-1}}{2h}, \tag{11.29}$$

which uses two steps. This does not always solve the problem. In order to get a reasonable accuracy, we have to use a very small value of $h$. There is also an issue called numerical stability. If the step size $h$ is too large, there is a possibility that any error will be amplified during each iteration, and after some iterations, the value $y_n$ might be unstable and become meaningless. A far better and more stable numerical scheme is the well-known Runge-Kutta method.

## 3.2. Runge-Kutta Method

The Runge-Kutta method uses a trial step to march the solution to the midpoint of the interval by the central difference

$$y_{n+1/2} = y_n + \frac{h}{2}f(x_n, y_n), \tag{11.30}$$

then it combines with the standard forward difference as used in Euler scheme

$$y_{n+1} = y_n + hf(x_{n+1/2}, y_{n+1/2}).$$

(11.31)

This scheme can be viewed as a predictor-corrector method as it first predicts a value in the midpoint, and then corrects it to an even better estimate at the next step.

If we follow the same idea, we can devise higher-order methods with multiple steps, which will give higher accuracy. The popular classical Runge-Kutta method is a fourth-order method, which involves four steps to advance from $x_n$ to $x_{n+1}$. Unlike a single evaluation of the first derivative $a = y_n'$ at $x_n$, the Runge-Kutta method evaluates the first derivatives or gradients four times (see Fig. 11.4). First, the gradient is evaluated in the same way as in the Euler scheme at $x_n$ and we have $a = y_n' = f(x_n, y_n)$. Then, the gradient $b = f(x_n + h/2, y_n + ah/2)$ is evaluated at the middle point $x_n + h/2$ and $y_{n+1/2} = y_n + a\frac{h}{2}$ using the new information from $a$. Again, the gradient is evaluated at the middle way using the newly corrected $\tilde{y}_{n+1/2} = y_n + bh/2$ such that $c = f(x_n + h/2, \tilde{y}_{n+1/2})$. Finally, the gradient at the point $(x_{n+1}, y_{n+1})$ is evaluated, that is $d = f(x_n + h, y_n + ch)$. These four gradient values are then averaged by putting more weights on the middle point so as to approximate the overall gradient more accurately. Therefore, We have

$$a = f(x_n, y_n), \qquad b = f(x_n + h/2, y_n + ah/2),$$

$$c = f(x_n + h/2, y_n + bh/2), \qquad d = f(x_n + h, y_n + ch),$$

$$y_{n+1} = y_n + \frac{h[a + 2(b + c) + d]}{6},$$

(11.32)

which is fourth-order accurate.

## Example 11.5

Let us solve the nonlinear equation numerically

$$\frac{dy}{dx} + y^2 = -1, \qquad x \in [0, 2],$$

with the initial condition $y(0) = 1$. We know that $\frac{dy}{dx} = -(1 + y^2)$, or

$$-\int \frac{1}{1 + y^2} dy = -\tan^{-1} y = \int dx = x + K,$$

where $K$ is the constant of the integration. This gives

$$y = -\tan(x + K).$$

Using the initial condition $y = 1$ and $x = 0$, we have $1 = -\tan(K)$, or

$$K = -\tan^{-1}(1) = -\frac{\pi}{4}.$$

The analytical solution becomes

$$y(x) = -\tan(x - \frac{\pi}{4}).$$

On the interval $[0, 2]$, let us first solve the equation using the Euler scheme for $h = 0.5$. There are five points $x_i = ih$ (where $i = 0, 1, 2, 3, 4$). As $dy/dx = f(y) = -1 - y^2$, we have the Euler scheme

$$y_{n+1} = y_n + hf(y_n) = y_n - h - hy_n^2.$$

From the initial condition $y_0 = 1$, we now have

$$y_1 = y_0 - h - hy_0^2 = 1 - 0.5 - 0.5 \times 1^2 = 0,$$

$$y_2 \approx -0.5, \qquad y_3 \approx -1.125, \qquad y_4 = -2.2578.$$

These are significantly different (about 30%) from the exact solution

$$y_0* = 1, \quad y_1^* = 0.2934079, \quad y_2^* = -0.21795809,$$

$$y_3^* = -0.86756212, \quad y_4^* = -2.68770693.$$

Now let us use the Runge-Kutta method to solve the same equation to see if it is better. Since $f(x_n, y_n) = -1 - y_n^2$, we have

$$a = f(x_n, y_n) = -h(1 + y_n^2), \qquad b = -[1 + (y_n + \frac{a}{2})^2],$$

$$c = -[1 + (y_n + \frac{b}{2})^2], \qquad d = -[1 + (y_n + c)^2],$$

and

$$y_{n+1} = y_n + h\frac{[a + 2(b + c) + d]}{6}.$$

From $y_0 = 1$ and $h = 0.5$, we have

$$y_1 \approx 0.29043, \quad y_2 \approx -0.22062, \quad y_3 \approx -0.87185, \quad y_4 \approx -2.67667.$$

These values are within about 1% of the analytical solution $y_n^*$.

We can see that even with the same step size $h$, the Runge-Kutta method is more efficient and accurate than the Euler scheme. Generally speaking, higher-order schemes are better than lower-order schemes, however, higher-order methods may not always be a good choice because they may require higher computation costs. Thus, the choice of methods should balance the accuracy needed and computation ease in practice.

So far, we have only introduced the numerical method to the first-order equations. Higher-order ordinary differential equations can always be converted to a first-order system. For example, the following second-order equation

$$y''(x) + p(x)y'(x) + q(x)y(x) = r(x), \tag{11.33}$$

can be rewritten as a system if we let $u = y'(x)$

$$y' = u, \qquad u' = r(x) - q(x)y(x) - p(x)u. \tag{11.34}$$

The above system of two equations can be written as

$$\frac{d}{dx}\begin{pmatrix} y \\ u \end{pmatrix} = \begin{pmatrix} u \\ r(x) - q(x)y - p(x)u \end{pmatrix}. \tag{11.35}$$

This is essentially the same as (11.25) in the context of vectors and matrices, and it can essentially be solved using similar methods with little modifications.

## Example 11.6

Let us try to solve the following well-known Van der Pol oscillator:

$$\frac{d^2x}{dt^2} - \mu(1 - x^2)\frac{dx}{dt} + x = 0,$$

where $\mu \geq 0$ is a damping parameter and $x(t)$ is the coordinate.

By setting $y(t) = x'(t)$, we have a system of two equations:

$$\begin{cases} \frac{dx}{dt} = y, \\ \frac{dy}{dt} = \mu(1 - x^2)y - x. \end{cases}$$

This can be solved by any method discussed earlier. For example, letting $f(x, y) = \mu(1 - x^2)y - x$ and using the simple Euler method, we have

$$\begin{cases} x_{n+1} = x_n + hy_n, \\ y_{n+1} = y_n + hf(x_n, y_n) = y_n + h\left[\mu(1 - x_n^2)y_n - x_n\right], \end{cases}$$

where $h > 0$ is the step size. To actually solve this system, two initial conditions are required. For example, for $h = 0.1$, $\mu = 1$ and the initial conditions,

$$x(0) = 1, \quad y(0) = x'(0) = 0,$$

we can solve it as follows:

$$x_0 = 1, \quad y_0 = 0.$$

$$x_1 = x_0 + hy_0 = 1 + 0.1 \times 0 = 1, \quad y_1 = y_0 + hf(x_0, y_0) = 0 + 0.1\Big[1 \times (1 - 1^2) - 1\Big] = -0.1,$$

$$x_2 = x_1 + hy_1 = 1 + 0.1 \times (-0.1) = 0.99,$$

$$y_2 = y_1 + hf(x_1, y_1) = -0.1 + 0.1\Big[1 \times (1 - 0.99^2) \times (-0.1) - 0.99\Big] = 0.1992.$$

Following the same procedure, we have

$$x_5 \approx 0.9002, \quad y_5 \approx -0.4968, \quad ..., \quad x_{100} \approx -2.1199, \quad y_{100} \approx 0.48644.$$

If we plot out $x$ and $y$ or $x_n$ versus $y_n$, we will see the famous Van der Pol oscillator.


Numerical solutions of partial differential equations (PDE) are even more difficult, especially when the domain is multidimensional and irregular, and numerical methods for solving PDEs include finite different methods, finite volume methods, finite element methods, boundary element methods, spectral methods, and meshless methods. Interested readers can refer to more advanced literature.

## 4. Optimization

Optimization is everywhere, from engineering design to business planning. After all, time and resources are limited, and optimal use of such valuable resources is crucial. In addition, designs of products have to maximize the performance, sustainability, energy efficiency, and to minimize the costs. Therefore, optimization is specially important for engineering applications.

Let us start with a very simple example to design a container with a volume capacity $V_0 = 10 \text{ m}^3$. As the main cost is related to the cost of materials, thus the main aim is to minimize the total surface area $S$.

The first thing we have to decide is the shape of the container (cylinder, cubic, sphere or ellipsoid or more complex geometry). For simplicity, let us start with a cylindrical shape with a radius $r$ and height $h$. Thus, the total surface area is

$$S = 2(\pi r^2) + 2\pi rh, \tag{11.36}$$

and the volume is

$$V = \pi r^2 h. \tag{11.37}$$

There are only two design variables $r$ and $h$, and one objective function $S$ to be minimized. Obviously, if there is no capacity constraint, we can choose not to build the container, thus the cost of materials is zero for $r = 0$ and $h = 0$. However, the constraint requirement means that we have to build a container with a fixed volume

$V_0 = \pi r^2 h = 10 \text{ m}^3$. Therefore, this optimization problem can be written as

$$\text{minimize} \quad S = 2\pi r^2 + 2\pi rh, \tag{11.38}$$

subject to the equality constraint

$$\pi r^2 h = V_0 = 10. \tag{11.39}$$

To solve this problem, we can first try to use the equality constraint to reduce the number of design variables by solving $h$. So we have

$$h = \frac{V_0}{\pi r^2}. \tag{11.40}$$

Substituting it into (11.38), we get

$$S = 2\pi r^2 + 2\pi rh = 2\pi r^2 + 2\pi r \frac{V_0}{\pi r^2} = 2\pi r^2 + \frac{2V_0}{r}. \tag{11.41}$$

This is a univariate function. From basic calculus, we know that the minimum or maximum can occur at the stationary point where the first derivative is zero. That is

$$\frac{dS}{dr} = 4\pi r - \frac{2V_0}{r^2} = 0, \tag{11.42}$$

which gives

$$r^3 = \frac{V_0}{2\pi}, \quad \text{or} \quad r = \sqrt[3]{\frac{V_0}{2\pi}}. \tag{11.43}$$

Thus, the height is

$$\frac{h}{r} = \frac{V_0/(\pi r^2)}{r} = \frac{V_0}{\pi r^3} = 2. \tag{11.44}$$

This means that the height is twice the radius: $h = 2r$. Thus, the minimum surface is

$$S_* = 2\pi r^2 + 2\pi rh = 2\pi r^2 + 2\pi r(2r) = 6\pi r^2 = 6\pi \left(\frac{V_0}{2\pi}\right)^{2/3} = \frac{6\pi}{\sqrt[3]{4\pi^2}} V_0^{2/3}. \tag{11.45}$$

For $V_0 = 10$, we have

$$r = \sqrt[3]{\frac{V_0}{(2\pi)}} = \sqrt[3]{\frac{10}{2\pi}} \approx 1.1675, \quad h = 2r = 2.335,$$

and the total surface area

$$S_* = 2\pi r^2 + 2\pi rh \approx 25.69.$$

It is worth pointing out that the above optimal solution is based on the assumption or requirement to design a cylindrical container. If we decide to use a sphere with a radius $R$, we know that its volume and surface area is

$$V_0 = \frac{4\pi}{3}R^3, \quad S = 4\pi R^2. \tag{11.46}$$

We can solve $R$ directly

$$R^3 = \frac{3V_0}{4\pi}, \quad \text{or } R = \sqrt[3]{\frac{3V_0}{4\pi}}, \tag{11.47}$$

which gives the surface area

$$S = 4\pi\left(\frac{3V_0}{4\pi}\right)^{2/3} = \frac{4\pi\sqrt[3]{9}}{\sqrt[3]{16\pi^2}}V_0^{2/3}. \tag{11.48}$$

Since $6\pi/\sqrt[3]{4\pi^2} \approx 5.5358$ and $4\pi\sqrt[3]{9}/\sqrt[3]{16\pi^2} \approx 4.83598$, thus $S < S_*$. That is, the surface area of a sphere is smaller than the minimum surface area of a cylinder with the same volume. In fact, for the same $V_0 = 10$, we have

$$S(\text{sphere}) = \frac{4\pi\sqrt[3]{9}}{\sqrt[3]{16\pi^2}}V_0^{2/3} \approx 22.47, \tag{11.49}$$

which is smaller than $S_* = 25.69$ for a cylinder.

This highlights the importance of the choice of type of design (here in terms of shape) before we can do any truly useful optimization. Obviously, there are many other factors that can influence the choice of design, including the manufacturability of the design, stability of the structure, ease of installation and space availability, etc. For a container in most applications, a cylinder may be much easier to produce than a sphere, and thus the overall cost may be lower in practice. Though there are so many factors to be considered in engineering design, for the purpose of optimization, here we will only focus on the improvement and optimization of a design with well-posed mathematical formulations.

Whatever the real-world applications may be, it is usually possible to formulate an optimization problem in a generic form. All optimization problems with explicit objectives can in general be expressed as a nonlinearly constrained optimization problem

$$\text{maximize/minimize} \quad f(\mathbf{x}), \quad \mathbf{x} = (x_1, x_2, ..., x_d)^T \in \mathbb{R}^d,$$

$$\text{subject to } \phi_j(\mathbf{x}) = 0, \quad (j = 1, 2, ..., M),$$

$$\psi_k(\mathbf{x}) \leq 0, \quad (k = 1, ..., N), \tag{11.50}$$

where $f(\mathbf{x})$, $\phi_j(\mathbf{x})$ and $\psi_k(\mathbf{x})$ are scalar functions of the design vector $\mathbf{x}$. Here the components $x_i$ of $\mathbf{x} = (x_1, ..., x_d)^T$ are called design or decision variables, and they can be either continuous, discrete or a mixture of these two. The vector $\mathbf{x}$ is often called the decision vector, which varies in a $d$-dimensional space $\mathbb{R}^d$. It is worth pointing out that we use a column vector here for $\mathbf{x}$ (thus with a transpose $T$). We can also use a row vector $\mathbf{x} = (x_1, ..., x_d)$ and the results will be the same. Different textbooks may use slightly different formulations. Once we are aware of such minor variations, this causes no difficulty or confusion.

In addition, the function $f(\mathbf{x})$ is called the objective function or cost function. In addition, $\phi_j(\mathbf{x})$ are constraints in terms of $M$ equalities, and $\psi_k(\mathbf{x})$ are constraints written as $N$ inequalities. So there are $M + N$ constraints in total. The optimization problem formulated here is a nonlinear constrained problem.

The space spanned by the decision variables is called the search space $\mathbb{R}^d$, while the space formed by the values of the objective function is called the solution space. The optimization problem essentially maps the $\mathbb{R}^d$ domain or space of decision variables into a solution space $\mathbb{R}$ (or the real axis in general).

The objective function $f(\mathbf{x})$ can be either linear or nonlinear. If the constraints $\phi_j$ and $\psi_k$ are all linear, it becomes a linearly constrained problem. Furthermore, when $\phi_j$, $\psi_k$ and the objective function $f(\mathbf{x})$ are all linear, then it becomes a linear programming problem. If the objective is at most quadratic with linear constraints, then it is called quadratic programming. If all the values of the decision variables can be integers, then this type of linear programming is called integer programming or integer linear programming.

On the other hand, if no constraints are specified so that $x_i$ can take any values in the real axis (or any integers), the optimization problem is referred to as an unconstrained optimization problem.

As a very simple example of optimization problems without any constraints, we discuss the search of the maxima or minima of a univariate function.

## Example 11.7

For example, to find the maximum of a univariate function $f(x)$

$$f(x) = x^2 e^{-x^2}, \qquad -\infty < x < \infty, \tag{11.51}$$

is a simple unconstrained problem, while the following problem is a simple constrained minimization problem

$$f(x_1, x_2) = x_1^2 + x_1 x_2 + x_2^2, \qquad (x_1, x_2) \in \mathbb{R}^2, \tag{11.52}$$

subject to

$$x_1 \geq 1, \qquad x_2 - 2 = 0. \tag{11.53}$$

It is worth pointing out that the objectives are explicitly known in all the optimization problems to be discussed in this book. However, in reality, it is often difficult to quantify what we want to achieve, but we still try to optimize certain things such as the degree of enjoyment or service quality on holiday. In other cases, it might be impossible to write the objective function in any explicit form mathematically.

From basic calculus, we know that, for a given curve described by $f(x)$, its gradient $f'(x)$ describes the rate of change. When $f'(x) = 0$, the curve has a horizontal tangent at that particular point. This means that it becomes a point of special interest. In fact, the maximum or minimum of a curve can only occur at

$$f'(x_*) = 0, \tag{11.54}$$

which is a critical condition or stationary condition. The solution $x_*$ to this equation corresponds to a stationary point and there may be multiple stationary points for a given curve.

In order to see if it is a maximum or minimum at $x = x_*$, we have to use the information of its second derivative $f''(x)$. In fact, $f''(x_*) > 0$ corresponds to a minimum, while $f''(x_*) < 0$ corresponds to a maximum. Let us see a concrete example.

## Example 11.8

To find the minimum of $f(x) = x^2 e^{-x^2}$, we have the stationary condition $f'(x) = 0$ or

$$f'(x) = 2x \times e^{-x^2} + x^2 \times (-2x)e^{-x^2} = 2(x - x^3)e^{-x^2} = 0.$$

As $e^{-x^2} > 0$, we have

$$x(1 - x^2) = 0, \quad \text{or} \quad x = 0, \quad \text{and} \quad x = \pm 1.$$

The second derivative is given by

$$f''(x) = 2e^{-x^2}(1 - 5x^2 + 2x^4),$$

which is an even function with respect to $x$.

So at $x = \pm 1$, $f''(\pm 1) = 2[1 - 5(\pm 1)^2 + 2(\pm 1)^4]e^{-(\pm 1)^2} = -4e^{-1} < 0$. Thus, there are two maxima that occur at $x_* = \pm 1$ with $f_{max} = e^{-1}$. At $x = 0$, we have $f''(0) = 2 > 0$, thus the minimum of $f(x)$ occurs at $x_* = 0$ with $f_{min}(0) = 0$.

Whatever the objective is, we have to evaluate it many times. In most cases, the evaluations of the objective functions consume a substantial amount of computational power (which costs money) and design time. Any efficient algorithm that can reduce the number of objective evaluations will save both time and money.

In mathematical programming, there are many important concepts, and we will first introduce three related concepts: feasible solutions, optimality criteria, the strong local optimum and weak local optimum.

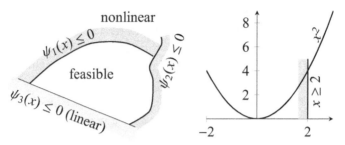

Figure 11.5: (a) Feasible domain with nonlinear inequality constraints $\psi_1(x)$ and $\psi_2(x)$ (left) as well as a linear inequality constraint $\psi_3(x)$. (b) An example with an objective of $f(x) = x^2$ subject $x \geq 2$ (right).

## 4.1. Feasible Solution

A point **x** which satisfies all the constraints is called a feasible point and thus is a feasible solution to the problem. The set of all feasible points is called the feasible region (see Fig. 11.5).

For example, we know that the domain $f(x) = x^2$ consists of all the real numbers. If we want to minimize $f(x)$ without any constraint, all solutions such as $x = -1$, $x = 1$ and $x = 0$ are feasible. In fact, the feasible region is the whole real axis. Obviously, $x = 0$ corresponds to $f(0) = 0$ as the true minimum.

However, if we want to find the minimum of $f(x) = x^2$ subject to $x \geq 2$, it becomes a constrained optimization problem. The points such as $x = 1$ and $x = 0$ are no long feasible because they do not satisfy $x \geq 2$. In this case, the feasible solutions are all the points that satisfy $x \geq 2$. So $x = 2$, $x = 100$ and $x = 10^8$ are all feasible. It is obvious that the minimum occurs at $x = 2$ with $f(2) = 2^2 = 4$. That is, the optimal solution for this problem occurs at the boundary point $x = 2$ (see Fig. 11.5).

## 4.2. Optimality Criteria

A point $\mathbf{x}_*$ is called a strong local maximum of the nonlinearly constrained optimization problem if $f(\mathbf{x})$ is defined in a $\delta$-neighbourhood $N(\mathbf{x}_*, \delta)$ and satisfies $f(\mathbf{x}_*) > f(\mathbf{u})$ for $\forall \mathbf{u} \in N(\mathbf{x}_*, \delta)$ where $\delta > 0$ and $\mathbf{u} \neq \mathbf{x}_*$. If $\mathbf{x}_*$ is not a strong local maximum, the inclusion of equality in the condition $f(\mathbf{x}_*) \geq f(\mathbf{u})$ for $\forall \mathbf{u} \in N(\mathbf{x}_*, \delta)$ defines the point $\mathbf{x}_*$ as a weak local maximum (see Fig. 11.6). The local minima can be defined in a similar manner when > and $\geq$ are replaced by < and $\leq$, respectively.

Figure 11.6 shows various local maxima and minima. Point $A$ is a strong local maximum, while point $B$ is a weak local maximum because there are many (in fact infinite) different values of **x** which will lead to the same value of $f(\mathbf{x}_*)$. Point $D$ is the global maximum, and point $E$ is the global minimum. In addition, point $F$ is a strong local minimum. However, point $C$ is a strong local minimum, but it has a discontinuity

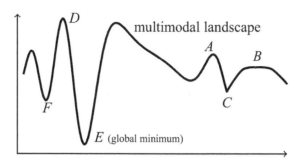

Figure 11.6: Local optima, weak optima and global optimality.

in $f'(\mathbf{x}_*)$. So the stationary condition for this point $f'(\mathbf{x}_*) = 0$ is not valid. We will not deal with these types of minima or maxima in detail.

As we briefly mentioned before, for a smooth curve $f(x)$, optimal solutions usually occur at stationary points where $f'(x) = 0$. This is not always the case because optimal solutions can also occur at the boundary, as we have seen in the previous example of minimizing $f(x) = x^2$ subject to $x \geq 2$. In our present discussion, we will assume that both $f(\mathbf{x})$ and $f'(\mathbf{x})$ are always continuous or $f(\mathbf{x})$ is everywhere twice-continuously differentiable. Obviously, the information of $f'(x)$ is not sufficient to determine whether a stationary point is a local maximum or minimum. Thus, higher-order derivatives such as $f''(x)$ are needed, but we do not make any assumption at this stage. We will discuss this further in detail in the next section.

## 5. Unconstrained Optimization

Optimization problems can be classified as either unconstrained or constrained. Unconstrained optimization problems can in turn be subdivided into univariate and multivariate problems.

### 5.1. Univariate Functions

The simplest optimization problem without any constraints is probably the search for the maxima or minima of a univariate function $f(x)$. For unconstrained optimization problems, the optimality occurs at either boundary or more often at the critical points given by the stationary condition $f'(x) = 0$.

However, this stationary condition is just a necessary condition, but it is not a sufficient condition. If $f'(x_*) = 0$ and $f''(x_*) > 0$, it is a local minimum. Conversely, if $f'(x_*) = 0$ and $f''(x_*) < 0$, then it is a local maximum. However, if $f'(x_*) = 0$ and $f''(x_*) = 0$, care should be taken because $f''(x)$ may be indefinite (both positive and negative) when $x \to x_*$, then $x_*$ corresponds to a saddle point.

For example, for $f(x) = x^3$, we have

$$f'(x) = 3x^2, \quad f''(x) = 6x. \tag{11.55}$$

The stationary condition $f'(x) = 3x^2 = 0$ gives $x_* = 0$. However, we also have

$$f''(x_*) = f''(0) = 0.$$

In fact, $f(x) = x^3$ has a saddle point $x_* = 0$ because $f'(0) = 0$ but $f''$ changes sign from $f''(0+) > 0$ to $f''(0-) < 0$.

## Example 11.9

For example, in order to find the maximum or minimum of a univariate function $f(x)$

$$f(x) = 3x^4 - 4x^3 - 12x^2 + 9, \quad -\infty < x < \infty,$$

we have to find first the stationary point $x_*$ when the first derivative $f'(x)$ is zero. That is

$$f'(x) = 12x^3 - 12x^2 - 24x = 12(x^3 - x^2 - 2x) = 0.$$

Since $f'(x) = 12(x^3 - x^2 - 2x) = 12x(x+1)(x-2) = 0$, we have

$$x_* = -1, \quad x_* = 2, \quad x_* = 0.$$

The second derivative of $f(x)$ is simply

$$f''(x) = 36x^2 - 24x - 24.$$

From the basic calculus we know that the maximum requires $f''(x_*) \le 0$ while the minimum requires $f''(x_*) \ge 0$.

At $x_* = -1$, we have

$$f''(-1) = 36(-1)^2 - 24(-1) - 24 = 36 > 0,$$

so this point corresponds to a local minimum

$$f(-1) = 3(-1)^4 - 4(-1)^3 - 12(-1)^2 + 9 = 4.$$

Similarly, at $x_* = 2$, $f''(x_*) = 72 > 0$, we have another local minimum

$$f(x_*) = -23.$$

But, at $x_* = 0$, we have $f'(0) = -24 < 0$, it has a local maximum $f(0) = 9$. However, this maximum is not a global maximum because the global maxima for $f(x)$ occur at $x = \pm\infty$.

If we plot out the graph of this function, we will see that $x_* = 2$ is the global minimum with $f(2) = -23$.

The maximization of a function $f(x)$ can be converted into the minimization of

$A - f(x)$ where $A$ is usually a large positive number (though $A = 0$ will do). For example, we know the maximum of $f(x) = e^{-x^2}$, $x \in (-\infty, \infty)$ is 1 at $x_* = 0$. This problem can be converted to a minimization problem $-f(x)$. For this reason, the optimization problems can be expressed as either minimization or maximization depending on the context and convenience of finding the solutions.

In fact, in the optimization literature, some books formulate all the optimization problems in terms of maximization, while others write these problems in terms of minimization, though they are in essence dealing with the same topics.

## 5.2. Multivariate Functions

We can extend the optimization procedure for univariate functions to multivariate functions using partial derivatives and relevant conditions. Let us start with an example

$$\text{minimize} \quad f(x, y) = x^2 + y^2, \quad x, y \in \mathcal{R}. \tag{11.56}$$

It is obvious that $x = 0$ and $y = 0$ is a minimum solution because $f(0, 0) = 0$. The question is how to solve this problem formally. We can extend the stationary condition to partial derivatives, and we have $\frac{\partial f}{\partial x} = 0$, and $\frac{\partial f}{\partial y} = 0$. In this case, we have

$$\frac{\partial f}{\partial x} = 2x + 0 = 0, \quad \frac{\partial f}{\partial y} = 0 + 2y = 0. \tag{11.57}$$

The solution is obviously $x_* = 0$ and $y_* = 0$.

Now how do we know it corresponds to a maximum or minimum? If we try to use the second derivatives, we have four different partial derivatives such as $f_{xx}$ and $f_{yy}$ and which one should we use? In fact, we need to define a Hessian matrix from these second partial derivatives and we have

$$\mathbf{H} = \begin{pmatrix} f_{xx} & f_{xy} \\ f_{yx} & f_{yy} \end{pmatrix} = \begin{pmatrix} \frac{\partial^2 f}{\partial x^2} & \frac{\partial^2 f}{\partial x \partial y} \\ \frac{\partial^2 f}{\partial y \partial x} & \frac{\partial^2 f}{\partial y^2} \end{pmatrix}. \tag{11.58}$$

Since $\partial x \partial y = \partial y \partial x$ or

$$\frac{\partial^2 f}{\partial x \partial y} = \frac{\partial^2 f}{\partial y \partial x}, \tag{11.59}$$

we can conclude that the Hessian matrix is always symmetric. In the case of $f = x^2 + y^2$, it is easy to check that the Hessian matrix is

$$\mathbf{H} = \begin{pmatrix} 2 & 0 \\ 0 & 2 \end{pmatrix}. \tag{11.60}$$

Mathematically speaking, if $\mathbf{H}$ is positive definite, then the stationary point $(x_*, y_*)$ corresponds to a local minimum. Similarly, if $\mathbf{H}$ is negative definite, the stationary point corresponds to a maximum. Since the Hessian matrix here does not involve any

$x$ or $y$, it is always positive definite in the whole search domain $(x, y) \in \mathbb{R}^2$, so we can conclude that the solution at point $(0, 0)$ is the global minimum.

Obviously, this is a special case. In general, the Hessian matrix will depend on the independent variables, but the definiteness test conditions still apply. That is, positive definiteness of a stationary point means a local minimum. Alternatively, for bivariate functions, we can define the determinant of the Hessian matrix in Eq. (11.58) as

$$\Delta = \det(\mathbf{H}) = f_{xx} f_{yy} - (f_{xy})^2. \tag{11.61}$$

At the stationary point $(x_*, y_*)$, if $\Delta > 0$ and $f_{xx} > 0$, then $(x_*, y_*)$ is a local minimum. If $\Delta > 0$ but $f_{xx} < 0$, it is a local maximum. If $\Delta = 0$, it is inconclusive and we have to use other information such as higher-order derivatives. However, if $\Delta < 0$, it is a saddle point. A saddle point is a special point where a local minimum occurs along one direction while the maximum occurs along another (orthogonal) direction.

## Example 11.10

Let us solve the minimization of $f(x, y) = (x - 1)^2 + x^2 y^2$. In this case, we have

$$\frac{\partial f}{\partial x} = 2(x - 1) + 2xy^2 = 0, \quad \frac{\partial f}{\partial y} = 0 + 2x^2 y = 0. \tag{11.62}$$

The second condition gives $y = 0$ or $x = 0$. Substituting $y = 0$ to the first condition, we have $x = 1$. However, $x = 0$ does not satisfy the first condition. Therefore, we have a solution $x_* = 1$ and $y_* = 0$.

For our example with $f = (x - 1)^2 + x^2 y^2$, we have

$$\frac{\partial^2 f}{\partial x^2} = 2y^2 + 2, \quad \frac{\partial^2 f}{\partial x \partial y} = 4xy, \quad \frac{\partial^2 f}{\partial y \partial x} = 4xy, \quad \frac{\partial^2 f}{\partial y^2} = 2x^2, \tag{11.63}$$

and thus we have

$$\mathbf{H} = \begin{pmatrix} 2y^2 + 2 & 4xy \\ 4xy & 2x^2 \end{pmatrix}. \tag{11.64}$$

At the stationary point $(x_*, y_*) = (1, 0)$, the Hessian matrix becomes

$$\mathbf{H} = \begin{pmatrix} 2 & 0 \\ 0 & 2 \end{pmatrix},$$

which is positive definite because its double eigenvalues 2 are positive. Alternatively, we have $\Delta = 4 > 0$ and $f_{xx} = 2 > 0$. Therefore, $(1, 0)$ is a local minimum.

In fact, for a multivariate function $f(x_1, x_2, ..., x_d)$ in a $d$-dimensional space, the stationary condition can be extended to

$$\mathbf{G} = \nabla f = (\frac{\partial f}{\partial x_1}, \frac{\partial f}{\partial x_2}, ..., \frac{\partial f}{\partial x_d})^T = 0, \tag{11.65}$$

where $\mathbf{G}$ is called the gradient vector. The second derivative test becomes the definiteness of the Hessian matrix

$$\mathbf{H} = \begin{pmatrix} \frac{\partial^2 f}{\partial x_1^2} & \frac{\partial^2 f}{\partial x_1 \partial x_2} & \cdots & \frac{\partial^2 f}{\partial x_1 \partial x_d} \\ \frac{\partial^2 f}{\partial x_2 \partial x_1} & \frac{\partial^2 f}{\partial x_2^2} & \cdots & \frac{\partial^2 f}{\partial x_2 \partial x_d} \\ \vdots & \vdots & \ddots & \vdots \\ \frac{\partial^2 f}{\partial x_d \partial x_1} & \frac{\partial^2 f}{\partial x_d \partial x_2} & \cdots & \frac{\partial^2 f}{\partial x_d^2} \end{pmatrix}. \tag{11.66}$$

At the stationary point defined by $\mathbf{G} = \nabla f = 0$, the positive definiteness of $\mathbf{H}$ gives a local minimum, while the negative definiteness corresponds to a local maximum. In essence, the eigenvalues of the Hessian matrix $\mathbf{H}$ determine the local behaviour of the function. As we mentioned before, if $\mathbf{H}$ is positive semi-definite, it corresponds to a local minimum.

## 6. Gradient-Based Methods

Gradient-based methods are iterative methods that extensively use the gradient information of the objective function during iterations. For the minimization of a function $f(\mathbf{x})$, the essence of this method is

$$\mathbf{x}^{(n+1)} = \mathbf{x}^{(n)} + \alpha g(\nabla f, \mathbf{x}^{(n)}), \tag{11.67}$$

where $\alpha$ is the step size, which can vary during iterations. $g(\nabla f, \mathbf{x}^{(n)})$ is a function of the gradient $\nabla f$ and the current location $\mathbf{x}^{(n)}$. Different methods use different forms of $g(\nabla f, \mathbf{x}^{(n)})$.

We know that Newton's method is a popular iterative method for finding the zeros of a nonlinear univariate function of $f(x)$ on the interval $[a, b]$. It can be modified for solving optimization problems because it is equivalent to finding the zeros of the first derivative $f'(\mathbf{x})$ once the objective function $f(\mathbf{x})$ is given.

For a given function $f(\mathbf{x})$ which is continuously differentiable, we have the Taylor expansion about a known point $\mathbf{x} = \mathbf{x}_n$ (with $\Delta \mathbf{x} = \mathbf{x} - \mathbf{x}_n$)

$$f(\mathbf{x}) = f(\mathbf{x}_n) + (\nabla f(\mathbf{x}_n))^T \Delta \mathbf{x} + \frac{1}{2} \Delta \mathbf{x}^T \nabla^2 f(\mathbf{x}_n) \Delta \mathbf{x} + ...,$$

which is minimized near a critical point when $\Delta \mathbf{x}$ is the solution of the following linear equation

$$\nabla f(\mathbf{x}_n) + \nabla^2 f(\mathbf{x}_n) \Delta \mathbf{x} = 0, \quad \text{or} \quad \mathbf{x} = \mathbf{x}_n - \mathbf{H}^{-1} \nabla f(\mathbf{x}_n), \tag{11.68}$$

where $\mathbf{H} = \nabla^2 f(\mathbf{x}_n)$ is the Hessian matrix. If the iteration procedure starts from the initial vector $\mathbf{x}^{(0)}$ (usually taken to be a guessed point in the domain), then Newton's iteration formula for the $n$th iteration is

$$\mathbf{x}^{(n+1)} = \mathbf{x}^{(n)} - \mathbf{H}^{-1}(\mathbf{x}^{(n)})\nabla f(\mathbf{x}^{(n)}). \tag{11.69}$$

It is worth pointing out that if $f(\mathbf{x})$ is quadratic, then the solution can be found exactly in a single step. However, this method may become tricky for non-quadratic functions, especially when we have to calculate the large Hessian matrix.

It can usually be time-consuming to calculate the Hessian matrix for second derivatives. A good alternative is to use an identity matrix to approximate the Hessian by using $\mathbf{H}^{-1} = \mathbf{I}$, and we have the quasi-Newton method

$$\mathbf{x}^{(n+1)} = \mathbf{x}^{(n)} - \alpha \mathbf{I} \, \nabla f(\mathbf{x}^{(n)}), \tag{11.70}$$

where $\alpha \in (0, 1)$ is a step size. In this case, the method is essentially the steepest descent method.

Though gradient-based methods can be very efficient, the final solution tends to be dependent on the starting point. If the starting point is very far away from the optimal solution, the algorithm can either reach a completely different solution for multimodal problems or simply fail in some cases. Therefore, there is no guarantee that the global optimal solution can be found.

It is worth pointing out that there are many variations of the steepest descent methods. If such optimization aim is to find the maximum, then this method becomes the *hill-climbing* method because the aim is to climb up the hill to the highest peak.

## 7. Nonlinear Optimization

As most real-world problems are nonlinear, nonlinear mathematical programming forms an important part of mathematical optimization methods. A broad class of nonlinear programming problems is about the minimization or maximization of $f(\mathbf{x})$ subject to no constraints, and another important class is the minimization of a quadratic objective function subject to nonlinear constraints. There are many other nonlinear programming problems as well.

Nonlinear programming problems are often classified according to the convexity of the defining functions. An interesting property of a convex function $f$ is that the vanishing of the gradient $\nabla f(\mathbf{x}_*) = 0$ guarantees that the point $x_*$ is a global minimum or maximum of $f$. If a function is not convex or concave, then it is much more difficult to find global minima or maxima.

## 7.1. Penalty Method

For the simple function optimization with equality and inequality constraints, a common method is the penalty method. For the optimization problem

$$\text{minimize} \quad f(\mathbf{x}), \quad \mathbf{x} = (x_1, ..., x_n)^T \in \mathbb{R}^n,$$

$$\text{subject to } \phi_i(\mathbf{x}) = 0, \ (i = 1, ..., M), \quad \psi_j(\mathbf{x}) \le 0, \ (j = 1, ..., N), \tag{11.71}$$

the idea is to define a penalty function so that the constrained problem is transformed into an unconstrained problem. Now we define $\Pi(\mathbf{x}, \mu_i, \nu_j)$

$$\Pi(\mathbf{x}, \mu_i, \nu_j) = f(\mathbf{x}) + \sum_{i=1}^{M} \mu_i \phi_i^2(\mathbf{x}) + \sum_{j=1}^{N} \nu_j \psi_j^2(\mathbf{x}), \tag{11.72}$$

where $\mu_i \gg 1$ and $\nu_j \ge 0$.

For example, let us solve the following minimization problem:

$$\text{minimize} \quad f(x) = 40(x - 1)^2, \quad x \in \mathbb{R}, \quad \text{subject to} \quad g(x) = x - a \ge 0, \tag{11.73}$$

where $a$ is a given value. Obviously, without this constraint, the minimum value occurs at $x = 1$ with $f_{\min} = 0$. If $a < 1$, the constraint will not affect the result. However, if $a > 1$, the minimum should occur at the boundary $x = a$ (which can be obtained by inspecting or visualizing the objective function and the constraint). Now we can define a penalty function $\Pi(x)$ using a penalty parameter $\mu \gg 1$. We have

$$\Pi(x, \mu) = f(x) + \mu[g(x)]^2 = 40(x - 1)^2 + \mu(x - a)^2, \tag{11.74}$$

which converts the original constrained optimization problem into an unconstrained problem. From the stationary condition $\Pi'(x) = 0$, we have

$$80(x - 1) - 2\mu(x - a) = 0, \quad \text{or} \quad x_* = \frac{40 - \mu a}{40 - \mu}. \tag{11.75}$$

For a special case $a = 1$, $x_* = 1$ and the result does not depend on $\mu$. However, in the case of $a > 1$ (say, $a = 5$), the result will depend on $\mu$. When $a = 5$ and $\mu = 100$, we have $x_* = 40 - 100 \times 5/40 - 100 = 7.6667$. If $\mu = 1000$, this gives $50 - 1000 * 5/40 - 1000$ 5.1667. Both values are far from the exact solution $x_{\text{true}} = a = 5$. If we use $\mu = 10^4$, we have $x_* \approx 5.0167$. Similarly, for $\mu = 10^5$, we have $x_* \approx 5.00167$. This clearly demonstrates that the solution will in general depend on $\mu$.

This means the solution depends on the value of $\mu$, and it is very difficult to use extremely large values without causing extra computational difficulties.

Ideally, the formulation using the penalty method should be properly designed so that the results will not depend on the penalty coefficient, or at least the dependence should be sufficiently weak.

## 7.2. Lagrange Multipliers

Another powerful method without the above limitation of using large $\mu$ is the method of Lagrange multipliers. If we want to minimize a function $f(\mathbf{x})$

$$\text{minimize}\ \ f(\mathbf{x}), \qquad \mathbf{x} = (x_1, ..., x_n)^T \in \mathbb{R}^n, \tag{11.76}$$

subject to the following nonlinear equality constraint

$$h(\mathbf{x}) = 0, \tag{11.77}$$

then we can combine the objective function $f(\mathbf{x})$ with the equality to form a new function, called the Lagrangian

$$\Pi = f(\mathbf{x}) + \lambda h(\mathbf{x}), \tag{11.78}$$

where $\lambda$ is the Lagrange multiplier, which is an unknown scalar to be determined.

This again converts the constrained optimization into an unconstrained problem for $\Pi(\mathbf{x})$, which is the beauty of this method. If we have $M$ equalities,

$$h_j(\mathbf{x}) = 0, \qquad (j = 1, ..., M), \tag{11.79}$$

then we need $M$ Lagrange multipliers $\lambda_j (j = 1, ..., M)$. We thus have

$$\Pi(x, \lambda_j) = f(\mathbf{x}) + \sum_{j=1}^{M} \lambda_j h_j(\mathbf{x}). \tag{11.80}$$

The requirement of stationary conditions leads to

$$\frac{\partial \Pi}{\partial x_i} = \frac{\partial f}{\partial x_i} + \sum_{j=1}^{M} \lambda_j \frac{\partial h_j}{\partial x_i}, \quad (i = 1, ..., n), \qquad \frac{\partial \Pi}{\partial \lambda_j} = h_j = 0, \quad (j = 1, ..., M). \tag{11.81}$$

These $M + n$ equations will determine the $n$-component $\mathbf{x}$ and $M$ Lagrange multipliers. As $\frac{\partial \Pi}{\partial g_j} = \lambda_j$, we can consider $\lambda_j$ as the rate of the change of $\Pi$ as a functional of $h_j$.

## Example 11.11

For the well-known Monkey Surface $f(x, y) = x^3 - 3xy^2$, the function does not have a unique maximum or minimum. In fact, the point $x = y = 0$ is a saddle point (see Fig. 11.7). However, if we impose an extra equality $x - y^2 = 1$, we can formulate an optimization problem as

$$\text{minimize}\ \ f(x, y) = x^3 - 3xy^2, \ \ (x, y) \in \mathbb{R}^2, \quad \text{subject to} \quad h(x, y) = x - y^2 = 1.$$

Now we can define

$$\Phi = f(x, y) + \lambda h(x, y) = x^3 - 3xy^2 + \lambda(x - y^2 - 1).$$

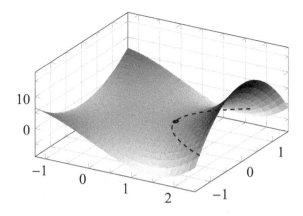

Figure 11.7: Minimization of $f(x,y) = x^3 - 3xy^2$ subject to $x = y^2 + 1$.

The stationary conditions become

$$\frac{\partial \Phi}{\partial x} = 3x^2 - 3y^2 + \lambda = 0, \quad \frac{\partial \Phi}{\partial y} = 0 - 6xy + (-2\lambda y) = 0, \quad \frac{\partial \Phi}{\partial \lambda} = x - y^2 - 1 = 0.$$

The second condition $-6xy - 2\lambda y = -2y(3x + \lambda) = 0$ implies that $y = 0$ or $\lambda = -3x$.

- If $y = 0$, the third condition $x - y^2 - 1 = 0$ gives $x = 1$. The first condition $3x^2 + 3y^2 - \lambda = 0$ leads to $\lambda = -3$. Therefore, $x = 1$ and $y = 0$ is an optimal solution with $f_{min} = 1$. As an exercise, you can verify that this solution corresponds to a minimum (not a maximum).

- If $\lambda = -3x$, the first condition becomes $3x^2 - 3y^2 - 3x = 0$. Substituting $x = y^2 + 1$ (from the third condition), we have

$$3(y^2 + 1)^2 - 3y^2 - 3(y^2 + 1) = 0, \quad \text{or} \quad 3(y^4 + 2) = 0.$$

This equation has no solution in the real domain. Therefore, the optimality occurs at $(1, 0)$ with $f_{min} = 1$. This point is marked in Fig. 11.7 where the dashed curve is the equality constraint.

## 7.3. Karush-Kuhn-Tucker Conditions

There is a counterpart of the Lagrange multipliers for nonlinear optimization with constraint inequalities. The Karush-Kuhn-Tucker (KKT) conditions concern the requirement for a solution to be optimal in nonlinear programming.

Let us know focus on the following nonlinear optimization problem

$$\text{minimize} \ \ f(\mathbf{x}), \quad \mathbf{x} \in \mathbb{R}^n,$$

subject to   $\phi_i(\mathbf{x}) = 0, \ (i = 1, ..., M), \quad \psi_j(\mathbf{x}) \le 0, \ (j = 1, ..., N).$     (11.82)

If all the functions are continuously differentiable, at a local minimum $\mathbf{x}_*$, there exist constants $\lambda_0, \lambda_1, ..., \lambda_q$ and $\mu_1, ..., \mu_p$ such that

$$\lambda_0 \nabla f(\mathbf{x}_*) + \sum_{i=1}^{M} \mu_i \nabla \phi_i(\mathbf{x}_*) + \sum_{j=1}^{N} \lambda_j \nabla \psi_j(\mathbf{x}_*) = 0,$$     (11.83)

$$\psi_j(\mathbf{x}_*) \le 0, \quad \lambda_j \psi_j(\mathbf{x}_*) = 0, \quad (j = 1, 2, ..., N),$$     (11.84)

where $\lambda_j \ge 0, (i = 0, 1, ..., N)$. The constants satisfy $\sum_{j=0}^{N} \lambda_j + \sum_{i=1}^{M} |\mu_i| \ge 0$. This is essentially a generalized method of the Lagrange multipliers. However, there is a possibility of degeneracy when $\lambda_0 = 0$ under certain conditions.

It is worth pointing out that such KKT conditions can be useful to prove theorems and sometimes useful to gain insight into certain types of problems. However, they are not really helpful in practice in the sense that they do not give any indication where the optimal solutions may lie in the search domain so as to guide the search process.

Optimization problems, especially highly nonlinear multimodal problems, are usually difficult to solve. However, if we are mainly concerned about local optimal or suboptimal solutions (not necessarily about global optimal solutions), there are relatively efficient methods such as interior-point methods, trust-region methods, the simplex method, sequential quadratic programming, and swarm intelligence-based methods. All these methods have been implemented in a diverse range of software packages. Interested readers can refer to more advanced literature.

## Exercises

**11.1.** Find the global minimum of $f(x, y) = (1 - x)^2 + 100(y - x^2)^2$.

**11.2.** Solve the constrained problem $f = x^2 + y^2 + 2xy$ subject to $y - x^2 + 2 = 0$.

**11.3.** For $f(x) = x^2 + 5y^2 + xy$, calculate first the gradients and Hessian matrix, and then find its minimum.

**11.4.** The Lorenz equations are a set of three ordinary differential equations

$$\frac{du}{dt} = \sigma(v - u), \quad \frac{dv}{dt} = u(\rho - w) - v, \quad \frac{dw}{dt} = uv - \beta w,$$

where $\sigma, \rho, \beta > 0$ are constants. Use any programming language to solve the system for $\sigma = 10, \beta = 8/3$ and $\rho = 28$ and reproduce the Lorenz attractor.[1]

**11.5.** Use any method to find the minimum of $f(x_1, ..., x_n) = \left[ \sum_{i=1}^{n} \sin^2(x_i) - e^{-\sum_{i=1}^{n} x_i^2} \right]$ $e^{\left( -\sum_{i=1}^{n} \sin^2 \sqrt{|x_i|} \right)}$ where $-10 \le x_i \le 10$ for $i = 1, 2, ..., n$ and $n \ge 2$ is an integer.

---

[1]E.N. Lorenz, Deterministic nonperiodic flow, *J. Atmospheric Sciences*, vol. 20, no. 2, 130-141 (1963).

# APPENDIX A

# Answers to Exercises

This appendix provides the brief answers to all the exercises in the book.

## Chapter 1

**1.1** The solutions are: $x = 2$, $x = 3$, $x = -\pi/7$, $x = 1$ and $y = 2$, and $x = 2$ and $y = 1$, respectively.

**1.2** The answers are: 20000, 0.0012, 160 (2 significant figures), and 6.28 (2 decimal places).

**1.3** The domains and ranges for the functions are:

- The domain of $f = (x - 1)^2$ is the whole real number axis, that is $x \in \mathbb{R}$. The range is $[0, \infty)$.
- The domain of $f = |x - 2| + (|x + 2| + 1)^2$ is $\mathbb{R}$, but its range is $[1, \infty)$.
- The domain of $f = \ln(x^2 - 1)$ is that $x^2 - 1 > 0$, which gives $x < -1$ and $x > +1$. Thus, the domain is the union of two parts: $(-\infty, -1) \cup (+1, +\infty)$. The range is $\mathbb{R}$.
- The domain of $f = 2^{-x} + 3$ is $\mathbb{R}$. For all any $x \in \mathbb{R}$, $f$ cannot be negative. In fact, the minimum value of $f$ is 3 when $x$ is very large. Thus, the range of $f$ is $(3, +\infty)$.
- Since $f = \ln 2^{-x^2} = -x^2 \ln 2$, its domain is $x \in \mathbb{R}$, but its function value cannot be positive because $-x^2$ is negative and $\ln 2 = 0.693 > 0$. So the range is $(-\infty, 0]$.

**1.4** From $\sin^2 x + \cos^2 x = 1$, we have

$$\frac{\sin^2 x + \cos^2 x}{\cos^2 x} = \frac{1}{\cos^2 x},$$

which gives

$$\tan^2 x + 1 = \frac{1}{\cos^2 x}, \quad \text{or} \quad \cos^2 x = \frac{1}{1 + \tan^2 x}.$$

Since $\sin(2x) = 2 \sin x \cos x$, we have

$$\sin(2x) = 2 \sin x \cos x = 2 \frac{\sin x}{\cos x} \cdot \cos^2 x = 2 \tan x \cdot \frac{1}{1 + \tan^2 x} = \frac{2 \tan x}{1 + \tan^2 x}.$$

## Chapter 2

**2.1** The answers are 2, 3, 7, 2 and 1, respectively.

**2.2** Since $\begin{pmatrix} 6 \\ 4 \end{pmatrix} = 15$, $\begin{pmatrix} 5 \\ 3 \end{pmatrix} = 10$, and $\begin{pmatrix} 5 \\ 4 \end{pmatrix} = 5$, the equality holds.

**2.3** You can expand the right-hand side to get $x^6 - 1$. Since

$$x^6 - 1 = (x - 1)(x + 1)(x^2 - x + 1)(x^2 + x + 1) = 0,$$

but $x^2 - x + 1 = 0$ and $x^2 + x + 1 = 0$ have no real roots, so the only two roots are $x = 1$ and $x = -1$.

**2.4** Using $a^2 - b^2 = (a + b)(a - b)$, we have

$$x^4 - 81 = (x^2)^2 - 9^2 = (x^2 + 9)(x^2 - 9) = (x^2 + 9)(x + 3)(x - 3),$$

which has two real roots ($x = -3$ and $x = 3$).

**2.5** The equation $x^x = x$ is a special case because $x^x$ is not a simple power function or exponential function. We can use the graph method plotting out $x^x$ and $x$ to find the point(s) both curves cross. By inspection, we can see that $x = 1$ is a solution. Since $(-1)^{-1} = -1$, we can see that $x = -1$ is also a solution. Can you find more solutions?

## Chapter 3

**3.1** For the three vectors

$$\mathbf{u} = \begin{pmatrix} 1 \\ 2 \\ 0 \end{pmatrix}, \quad \mathbf{v} = \begin{pmatrix} 3 \\ -4 \\ 0 \end{pmatrix}, \quad \mathbf{w} = \begin{pmatrix} 2 \\ 3 \\ 1 \end{pmatrix},$$

we have $|\mathbf{u}| = \sqrt{1^2 + 2^2 + 0^2} = \sqrt{5}$, $|\mathbf{v}| = \sqrt{3^2 + (-4)^2 + 0^2} = 5$ as well as $|\mathbf{w}| = \sqrt{2^2 + 3^2 + 1^2} = \sqrt{14}$. Since $\mathbf{u} + \mathbf{v} + \mathbf{w} = \begin{pmatrix} 1 + 3 + 2 \\ 2 - 4 + 3 \\ 0 + 0 + 1 \end{pmatrix} = \begin{pmatrix} 6 \\ 1 \\ 1 \end{pmatrix}$, we have

$$|\mathbf{u} + \mathbf{v} + \mathbf{w}| = \sqrt{6^2 + 1^2 + 1^2} = \sqrt{38} < \sqrt{5} + 5 + \sqrt{14} = |\mathbf{u}| + |\mathbf{v}| + |\mathbf{w}|.$$

**3.2** The dot product of $\mathbf{u}$ and $\mathbf{v}$ is $\mathbf{u} \cdot \mathbf{v} = 1 \times 3 + 2 \times (-4) + 0 \times 0 = -5$. The cross product of $\mathbf{u}$ and $\mathbf{v}$ is

$$\mathbf{u} \times \mathbf{v} = \begin{pmatrix} 0 \\ 0 \\ -10 \end{pmatrix},$$

and thus their triple product is

$$\mathbf{w} \cdot (\mathbf{u} \times \mathbf{v}) = \begin{pmatrix} 2 \\ 3 \\ 1 \end{pmatrix} \cdot \begin{pmatrix} 0 \\ 0 \\ -10 \end{pmatrix} = -10.$$

**3.3** For two matrices

$$\mathbf{A} = \begin{pmatrix} 2 & 3 \\ 3 & 2 \end{pmatrix}, \quad \mathbf{B} = \begin{pmatrix} 1 & -2 \\ -3 & 4 \end{pmatrix},$$

we have

$$\mathbf{AB} = \begin{pmatrix} -7 & 8 \\ -3 & 2 \end{pmatrix}, \quad \mathbf{BA} = \begin{pmatrix} -4 & -1 \\ 6 & -1 \end{pmatrix},$$

so

$$\mathbf{AB} - \mathbf{BA} = \begin{pmatrix} -3 & 9 \\ -9 & 3 \end{pmatrix}.$$

In addition, $\det(\mathbf{A}) = 2 \times 2 - 3 \times 3 = -5$ and $\det(\mathbf{B}) = -2$. Their inverses are

$$\mathbf{A}^{-1} = \begin{pmatrix} -2/5 & 3/5 \\ 3/5 & -2/5 \end{pmatrix}, \quad \mathbf{B}^{-1} = \begin{pmatrix} -2 & -1 \\ -3/2 & -1/2 \end{pmatrix}.$$

**3.4** The two eigenvalues of $\mathbf{A}$ are $\lambda_1 = -1$ and $\lambda_2 = 5$ and their corresponding eigenvectors are

$$\mathbf{v}_1 = \frac{1}{\sqrt{2}} \begin{pmatrix} -1 \\ 1 \end{pmatrix}, \quad \mathbf{v}_2 = \frac{1}{\sqrt{2}} \begin{pmatrix} 1 \\ 1 \end{pmatrix}.$$

Since $\det(\mathbf{A}) = -5$, it is clear that $\lambda_1 \cdot \lambda_2 = \det(\mathbf{A})$.

## Chapter 4

**4.1** The first and second derivatives of $f(x) = x^4 + 2x^2 + 3x$ are

$$f'(x) = 4x^3 + 4x + 3, \quad f''(x) = 12x^2 + 4.$$

Similarly, for $g(x) = xe^{-x}$, we have

$$g'(x) = e^{-x} - xe^{-x}, \quad g''(x) = xe^{-x} - 2e^{-x}.$$

For $h(x) = x^2 + \sin(x^2) + x \ln x$, we have

$$h'(x) = 2x + 2x \cos(x^2) + \ln x + 1, \quad h''(x) = 2 + 2 \cos(x^2) - 4x^2 \sin(x^2) + \frac{1}{x}.$$

**4.2** In order to calculate the first derivative of $f(x) = x^x + x$ for $x > 0$, it is better to figure out the first derivative of $g(x) = x^x$ for $x > 0$. However, from the table of differentiation of common functions, there is no entry for $x^x$. One way is to take the logarithm first (because $x > 0$), and we have

$$\ln g(x) = \ln x^x = x \ln x.$$

Differentiate both sides with respect to $x$, we have

$$[\ln g(x)]' = [x \ln x]',$$

which becomes

$$\frac{g'(x)}{g(x)} = 1 \cdot \ln x + x \cdot \frac{1}{x} = \ln x + 1, \quad \text{or} \quad g'(x) = g(x)(\ln x + 1).$$

Since $g(x) = x^x$, we have

$$g'(x) = [x^x]' = x^x(\ln x + 1).$$

Thus, the first derivative of $f(x) = x^x + x$ is

$$f'(x) = g'(x) + 1 = x^x(\ln x + 1) + 1.$$

Similarly, we have

$$f''(x) = x^x(\ln x + 1)^2 + x^{x-1},$$

and

$$f'''(x) = x^x(\ln x + 1)^3 + 3x^{x-1}(\ln x + 1) - x^{x-2}.$$

**4.3** To calculate $y'(x)$ from $y(x) - \sin(x)e^x + x^2y(x) = e^x$, we can use the implicit differentiation by differentiating both sides with respect to $x$ and we have

$$y'(x) - [\sin(x)e^x]' + [x^2y(x)]' = e^x,$$

which becomes

$$y'(x) - \cos(x)e^x - e^x \sin(x) + 2xy(x) + x^2y'(x) = e^x.$$

This means that

$$y'(x) = \frac{e^x[1 + \cos(x) + \sin(x)] - 2xy(x)}{1 + x^2}.$$

From the original equation, we can solve $y(x)$ and we have

$$y(x) = \frac{e^x + e^x \sin(x)}{1 + x^2},$$

we finally have

$$y'(x) = \frac{e^x[1 + \cos(x) + \sin(x)] - 2xe^x[1 + \sin(x)]/(1 + x^2)}{1 + x^2}$$

$$= \frac{e^x[1 + \cos(x) + \sin(x)]}{1 + x^2} - \frac{2xe^x[1 + \sin(x)]}{(1 + x^2)^2}.$$

**4.4** From $f(x) = e^{-x} + x\sin(x)$, we have

$$f'(x) = -e^{-x} + \sin(x) + x\cos(x), \quad f''(x) = e^{-x} + 2\cos(x) - x\sin(x),$$

$$f'''(x) = -e^{-x} - 3\sin(x) - x\cos(x), \quad f''''(x) = e^{-x} - 4\cos(x) + x\sin(x),$$

and

$$f'''''(x) = -e^{-x} + 5\sin(x) + x\cos(x).$$

There is some regularity in the above expressions, we can extend to get

$$f^{(99)} = -e^{-x} - 99\sin(x) - x\cos(x), \quad \text{and} \quad f^{(100)} = e^{-x} - 100\cos(x) + x\sin(x).$$

**4.5** Since the series of $\exp(x)$ is

$$\exp(x) = 1 + x + \frac{x^2}{2!} + \frac{x^3}{3!} + \dots + \frac{x^n}{n!} + \dots,$$

we can replace $x$ by $-x^2$ and we have

$$\exp(-x^2) = 1 + (-x^2) + \frac{(-x^2)^2}{2!} + \frac{(-x^2)^3}{3!} + \dots \frac{(-x^2)^n}{n!} + \dots$$

$$= 1 - x^2 + \frac{x^4}{2!} - \frac{x^6}{3!} + \dots + \frac{(-1)^n x^{2n}}{n!} + \dots,$$

which is valid for all $x \in \mathbb{R}$.

**4.6** The partial derivatives of functions can be calculated directly by using a combination of the chain and product rules. For $f(x, y) = x^2 + y^2 + xy$, we have

$$\frac{\partial f}{\partial x} = 2x + y, \quad \frac{\partial f}{\partial y} = 2y + x, \quad \frac{\partial^2 f}{\partial x \partial y} = 1, \quad \frac{\partial^2 f}{\partial y \partial x} = 1,$$

and

$$\frac{\partial^2 f}{\partial x^2} = 2, \quad \frac{\partial^2 f}{\partial y^2} = 2.$$

Similarly, for $g(x, y) = \sin(xy) + (x^2 + y^2)\exp(-x^2 - y^2)$, we have

$$\frac{\partial g}{\partial x} = y\cos(xy) + [2x - 2x(x^2 + y^2)]e^{-x^2-y^2},$$

and

$$\frac{\partial g}{\partial y} = x\cos(xy) + [2y - 2y(x^2 + y^2)]e^{-x^2-y^2}.$$

**4.7** From $\phi(r) = 1/r$ with $r = \sqrt{x^2 + y^2 + z^2}$, we have

$$\frac{\partial\psi}{\partial x} = -\frac{x}{r^3}, \quad \frac{\partial\psi}{\partial y} = -\frac{y}{r^3}, \quad \frac{\partial\psi}{\partial z} = -\frac{z}{r^3}.$$

Thus, $\nabla\phi$ is

$$\nabla\phi = \frac{\partial\psi}{\partial x}\mathbf{i} + \frac{\partial\psi}{\partial y}\mathbf{j} + \frac{\partial\psi}{\partial z}\mathbf{k} = -\frac{1}{r^3}(x\mathbf{i} + y\mathbf{j} + \mathbf{k}).$$

However, $\nabla \cdot \phi$ does not exist because $\phi$ is a scalar.
Since the second partial derivatives of $\psi$ are

$$\frac{\partial^2\psi}{\partial x^2} = \frac{3x^2}{r^5} - \frac{1}{r^3}, \quad \frac{\partial^2\psi}{\partial y^2} = \frac{3y^2}{r^5} - \frac{1}{r^3}, \quad \frac{\partial^2\psi}{\partial z^2} = \frac{3z^2}{r^5} - \frac{1}{r^3},$$

we have

$$\nabla^2 = \nabla \cdot \nabla\psi = \frac{\partial^2\psi}{\partial x^2} + \frac{\partial^2\psi}{\partial y^2} + \frac{\partial^2\psi}{\partial z^2}$$

$$= \frac{3(x^2 + y^2 + z^2)}{r^5} - \frac{3}{r^3} = \frac{3r^2}{r^5} - \frac{3}{r^3} = 3(\frac{1}{r^3} - \frac{1}{r^3}) = 0.$$

Thus $\psi(r)$ is a harmonic function.

## Chapter 5

**5.1** The indefinite integrals (omitting the integral constants) are:

$$\int(x - \frac{1}{x})dx = \frac{x^2}{2} - \ln x, \quad \int(x^2 + \cos(x) + e^{-x})dx = \frac{x^3}{3} + \sin(x) - e^{-x}.$$

$$\int \ln x\, dx = x \ln x - x, \quad \int x^2 \cos(x^3)dx = \frac{1}{3}\sin(x^3).$$

**5.2** The definite integrals are:

$$\int_0^{\pi/2} x \sin(2x)dx = \left[\frac{\sin(2x) - 2x\cos(2x)}{4}\right]\Big|_0^{\pi/2} = \frac{\pi}{4}.$$

$$\int_0^\infty xe^{-x}dx = -(1 + x)e^{-x}\Big|_0^\infty = 1, \quad \int_0^\infty xe^{-x^2}dx = -\frac{e^{-x^2}}{2}\Big|_0^\infty = \frac{1}{2}.$$

**5.3** The area under the curve $|x|\exp[-x^2]$ should be twice the area under $x\exp[-x^2]$ with $x \geq 0$ (due to symmetry). Since $\int_0^\infty xe^{-x^2}dx = 1/2$ (from the previous question), so the total area is

$$A = 2 \int_0^\infty xe^{-x^2}dx = 1.$$

**5.4** The first double integral is

$$\int_0^1 \int_0^2 xy^2 dxdy = \int_0^1 \left( x \int_0^2 y^2 dx \right)dx = \int_0^1 (x\frac{2^3}{3})dx = \frac{8}{3}\frac{x^2}{2}\Big|_0^1 = \frac{4}{3}.$$

By using the polar coordinates, $\det(J) = r$ and $dxdy = rdrd\theta$, the second double integral becomes

$$\iint_D (x^2 + y^2)^2 dxdy = \int_0^{2\pi} \left( \int_0^1 r^4\, rdr \right)d\theta = 2\pi \int_0^1 r^5 dr = 2\pi\frac{r^6}{6}\Big|_0^1 = \frac{\pi}{3}.$$

**5.5** The volume can be obtained by using the spherical coordinates by setting $x = ar \cos\theta \sin\phi$, $y = br\sin\theta \sin\phi$ and $z = cr\cos\phi$. Then, the Jacobian from $(x,y,z)$ to $(r,\theta,\phi$ is $J = abcr^2 \sin\phi$. Thus, the volume is twice the volume of the half-sphere and we have

$$V = \iiint dV = \iiint dxdydz = \int_0^1 \int_0^{2\pi} \int_0^\pi r^2(abc)\sin\phi drd\theta d\phi$$

$$= abc \int_0^1 r^2 dr \int_0^{2\pi} d\theta \int_0^\pi \sin\phi d\phi = abc\left(\frac{r^3}{3}\Big|_0^1\right) \cdot (2\pi) \cdot \left[-\cos(\phi)\Big|_0^\pi\right]$$

$$= \frac{2}{3}\pi abc[-(-1-1)] = \frac{4\pi}{3}abc.$$

## Chapter 6

**6.1** The final simplified values of the expressions are:

$$i^{1001} + i^{100} + i^{10} + i + 1 = i^{4\times250+1} + i^{4\times25} + i^{8+2} + i + 1$$

$$= i + 1 + (-1) + i + 1 = 2i + 1,$$

where we have used $i^4 = 1$. Since $\exp[\pi i] = -1$, $\exp[-\pi i] = -1$, and $\exp[-2\pi i] = 1$, so we have

$$e^{\pi i} + e^{-\pi i} + e^{-2\pi i} = -1.$$

In addition, it is easy to verify that

$$\sinh(i\pi) - \cos(i\pi) - \sinh(0) = 0 - (-1) - 0 = 1, \quad \cos(\pi) + i\sin(\pi) - i^{10} = 0.$$

Final, we have

$$(1 + 2i)(3 + 4i)/(5 + 6i) = \frac{35}{61} + \frac{80}{61}i,$$

and

$$|(3 + 4i)(1 - 2i)(2 + 4i)/(6 - 8i)| = \frac{|3 + 4i| \cdot |1 - 2i| \cdot |2 + 4i|}{|6 - 8i|} = \frac{5\sqrt{5}\sqrt{20}}{10} = 5.$$

**6.2**  To compute $\ln(1 + i)$, we now use $z = re^{i\theta}$ so that

$$\ln z = \ln(re^{i\theta}) = \ln r + \ln e^{i\theta} = \ln r + i\theta.$$

Writing $1 + i = \sqrt{2}e^{i\pi/4}$, we have

$$\ln(1 + i) = \ln \sqrt{2} + i\frac{\pi}{4} = \frac{1}{2}\ln 2 + i\frac{\pi}{4}.$$

Similarly, we have $i = e^{i\pi/2}$ and

$$\ln i = \ln e^{i\pi/2} = \frac{i\pi}{2}.$$

**6.3**  From $y = \tanh^{-1}(x)$ and the definition of tanh, we have

$$x = \tanh y = \frac{e^y - e^{-y}}{e^y + e^{-y}} = \frac{(e^y - e^{-y})e^y}{(e^y + e^{-y})e^y} = \frac{e^{2y} - 1}{e^{2y} + 1},$$

which gives

$$x(e^{2y} + 1) = e^{2y} - 1, \quad \text{or} \quad e^{2y} = \frac{1 + x}{1 - x}.$$

Thus, we have

$$\ln e^{2y} = 2y = \ln\left(\frac{1 + x}{1 - x}\right),$$

which gives

$$y = \tanh^{-1}(x) = \frac{1}{2}\ln\left(\frac{1 + x}{1 - x}\right).$$

**6.4**  For the contour integral

$$I = \oint_D \frac{e^z}{(z - \pi)^3}dz,$$

and Cauchy's integral formula

$$\oint_\Gamma \frac{f(z)}{(z - z_0)^{n+1}}dx = \frac{2\pi i}{n!}f^{(n)}(z_0),$$

we have $f(z) = e^z$, $z_0 = \pi$ and $n = 2$. Since the pole $z = z_0 = \pi$ is within the square path $D$, we can use the Cauchy integral formula and we have

$$I = \frac{2\pi i}{2!}f''(\pi) = \pi i \frac{d^2 e^z}{dz^2}\Big|_{z=\pi} = i\pi e^{\pi}.$$

## Chapter 7

**7.1**  The differential equation $y'(x) = \cos(x) - \exp(-x)$ can be solved by direct integration, so we have

$$y(x) = \int [\cos(x) - e^{-x}]dx = \sin(x) + e^{-x} + C,$$

where $C$ is the integration constant. Using $y(0) = 1$ at $x = 0$, we have $y(0) = 1 = \sin(0) + e^{-0} + C$, which gives $C = 0$. Thus, the solution is

$$y(x) = \sin(x) + e^{-x}.$$

**7.2**  The solutions to the three ODEs can be obtained by characteristics or some transformation.
- For $y''(x) + 5y'(x) + 4y(x) = 0$, we have $y(x) = e^{\lambda x}$ and $\lambda^2 + 5\lambda + 4 = 0$, so the solution is

$$y(x) = Ae^{-x} + Be^{-4x},$$

  where $A$ and $B$ are two unknown constants.
- For $y'(x) + x^2 y(x) = 0$, we have

$$\frac{y'(x)}{y(x)} = -x^2, \quad \text{or} \quad \ln y(x) = -\frac{x^3}{3} + C,$$

  where $C$ is a constant. Thus, we have

$$y(x) = e^{\ln y(x)} = Ae^{-x^3/3}, \quad A = e^C.$$

- For $y''(x) + 4y(x) = 0$, we have $y(x) = e^{\lambda x}$ and $\lambda^2 + 4 = 0$. So

$$y(x) = Ae^{+2ix} + Be^{-2x}, \quad \text{or} \quad y(x) = C\cos(2x) + D\sin(2x),$$

  where $C$ and $D$ are unknown constants.

**7.3**  From $y'''(x) + y''(x) + 4y'(x) + 4y(x) = 0$, we have $y(x) = e^{\lambda x}$ and

$$\lambda^3 + \lambda^2 + 4\lambda + 4 = (\lambda + 1)(\lambda^2 + 4) = 0,$$

which gives $\lambda = -1$, and $\lambda = \pm 2i$. Thus, the solution is

$$y(x) = Ae^{-x} + C\cos(2x) + D\sin(2x).$$

**7.4**  For $y'(x) + y(x) = 1 - e^{-2x}$ with $y(0) = 0$, as a further exercise, check that the solution is

$$y(x) = 2e^{-x}[\cosh(x) - 1] = e^{-2x}(e^x - 1)^2.$$

The solution with $y(0) = 1$ is $y(x) = 2e^{-x}\cosh(x) - e^{-x} = e^{-2x} - e^{-x} + 1$.

**7.5** For a cantilever beam of length $L$, let $x$ be the axis starting from the fixed end (A) towards the free end (B) with a load $P$. According to the Euler-Bernoulli beam theory, the deflection $u(x)$ should obey the following equation:

$$EI\frac{d^2u}{dx^2} = M,$$

where $M$ is the bending moment. Since the free end (B) does not support a moment, the moment must be maximum at $x = 0$. Thus, the moment at any $x$ is $M = (L - x)P$, so the equation for deflection becomes

$$\frac{d^2u}{dx^2} = \lambda(L - x), \quad \lambda = \frac{P}{EI}.$$

Obviously, $u = 0$ at $x = 0$ (fixed end). In addition, the fixed end also means the clamping such that $u' = du/dx = 0$ at $x = 0$.

Integrating once with respect to $x$, we have

$$\frac{du}{dx} = \lambda Lx - \frac{\lambda}{2}x^2 + K_1,$$

where $C$ is the integration constant. From $u' = 0$ at $x = 0$, we have $0 = \lambda L \times 0 - \frac{\lambda}{2}0^2 + K_1$, which gives $K_1 = 0$. Now integrating with respect to $x$ again, we have

$$u(x) = \frac{L\lambda}{2}x^2 - \frac{\lambda}{6}x^3 + K_2,$$

where $K_2$ is another integration constant. From $u = 0$ at $x = 0$, we have

$$0 = \frac{L\lambda}{2} \times 0^2 - \frac{\lambda}{6} \times 0^3 + K_2,$$

which gives $K_2 = 0$. Thus, the final solution becomes

$$u(x) = \frac{L\lambda}{2}x^2 - \frac{\lambda}{6}x^3 = \frac{\lambda x^2}{2}\left(L - \frac{x}{3}\right).$$

Since the maximum deflection occurs at $x = L$, therefore, we finally get

$$D_{max}\Big|_{x=L} = \frac{\lambda L^2}{2}\left(L - \frac{L}{3}\right) = \frac{\lambda L^3}{3} = \frac{PL^3}{3EI}.$$

## Chapter 8

**8.1** Since $f(t) = |\sin t|$ is an even function because $f(-t) = |\sin(-t)| = |-\sin t| = f(t)$ with $T = \pi$, its Fourier series only contains cosine terms with $a_n \neq 0$ (and $b_n = 0$).

First, let us calculate $a_0$ and we have

$$a_0 = \frac{1}{\pi} \int_{-\pi}^{\pi} |\sin t| dt = \frac{2}{\pi} \int_0^{\pi} \sin t dt = \frac{2}{\pi} \Big[ -\cos t \Big]\Big|_0^{\pi} = -\frac{2}{\pi}[\cos(\pi) - \cos 0)] = \frac{4}{\pi}.$$

In addition, we have

$$a_n = \frac{1}{\pi} \int_{-\pi}^{\pi} |\sin t| \cos(nt) dt.$$

Since both $|\sin(t)|$ and $\cos(nt)$ are even functions, the integrand is also even. Thus, we get

$$a_n = \frac{2}{\pi} \int_0^{\pi} \sin t \cos(nt) dt.$$

As a further exercise, you can prove that

$$\int_0^{\pi} \sin(t) \cos(nt) dt = \begin{cases} 0 & \text{for odd } n \\ -\frac{2}{n^2-1} & \text{for even } n. \end{cases}$$

As $n$ is even, we can set $n = 2k(k = 1, 2, 3, ...)$ and we have

$$a_n = \frac{2}{\pi} \cdot \frac{(-2)}{n^2 - 1} = -\frac{4}{\pi} \cdot \frac{1}{4k^2 - 1}.$$

So the Fourier series becomes

$$f(t) = |\sin t| = \frac{4}{\pi} \cdot \frac{1}{2} - \frac{4}{\pi} \sum_{k=1}^{\infty} \frac{\cos(2kt)}{4k^2 - 1} = \frac{2}{\pi} \Big[ 1 - 2 \sum_{k=1}^{\infty} \frac{\cos(2kt)}{4k^2 - 1} \Big].$$

The Fourier series for $g(t) = Af(t)$ is simply the Fourier series, multiplying each term by $A$.

**8.2** For $\sinh(at)$, its Laplace transform can be obtained by using either direct integration or the Laplace transform table. From the Laplace table, we know that $\sinh(at) = (e^{at} - e^{-at})/2$ and $\mathcal{L}[e^{at}] = \frac{1}{s+a}$, so we have

$$\mathcal{L}[\sinh(at)] = \frac{1}{2}\{\mathcal{L}[e^{at}] - \mathcal{L}[e^{-at}]\} = \frac{1}{2}[\frac{1}{s-a} - \frac{1}{s+a}]$$

$$= \frac{1}{2}[\frac{(s+a)}{(s-a)(s+a)} - \frac{(s-a)}{(s+a)(s-a)}] = \frac{1}{2} \cdot \frac{2a}{s^2 - a^2} = \frac{a}{s^2 - a^2}.$$

**8.3** For the Laplace transform of $\sin(at + b)$, we first use

$$\sin(at + b) = \sin(at) \cos b + \cos(at) \sin b.$$

Then, we have

$$\mathcal{L}[\sin(at + b)] = \mathcal{L}[\sin(at) \cos b] + \mathcal{L}[\cos(at) \sin b]$$

$$= \cos b \mathcal{L}[\sin(at)] + \sin b \mathcal{L}[\cos(at)].$$

Using

$$\mathcal{L}[\sin(at)] = \frac{a}{s^2 + a^2}, \quad \mathcal{L}[\cos(at)] = \frac{s}{s^2 + a^2},$$

we have

$$\mathcal{L}[\sin(at + b)]] = \cos b \cdot \frac{a}{s^2 + a^2} + \sin b \cdot \frac{s}{s^2 + a^2} = \frac{a \cos b + s \sin b}{s^2 + a^2}.$$

**8.4** To obtain the solution of $y'(t) - y(t) = e^{-t}$ with the initial condition $y(0) = 0$, we first apply Laplace transforms on both sides of the equation and we have

$$\mathcal{L}[y'(t)] - \mathcal{L}[y(t)] = \mathcal{L}[e^{-t}],$$

which gives

$$sY(s) - y(0) - Y(s) = \frac{1}{s+1}, \quad \text{or} \quad sY(s) - Y(s) = \frac{1}{s+1},$$

where we have used $y(0) = 0$. Solving for $Y(s)$, we get

$$Y(s) = \frac{1}{(s-1)(s+1)} = \frac{1}{2}[\frac{1}{s-1} - \frac{1}{s+1}].$$

Now we use the Laplace transform pairs to get the inverse and the above equation becomes

$$y(t) = \mathcal{L}^{-1}[Y(s)] = \frac{1}{2}\{\mathcal{L}^{-1}[\frac{1}{s-1}] - \mathcal{L}^{-1}[\frac{1}{s+1}]\} = \frac{1}{2}[e^t - e^{-t}].$$

So the final solution is $y(t) = \frac{1}{2}e^t - \frac{1}{2}e^{-t} = \sinh(t)$.

## Chapter 9

**9.1** The mean $E(p)$ of a geometrical distribution can be calculated by

$$E(p) = \sum_{k=0}^{\infty} kp(1-p)^k = p(1-p)\sum_{k=0}^{\infty}(1-p)^{k-1} \cdot k = p(1-p) \cdot \frac{1}{p^2} = \frac{1-p}{p},$$

where we have used

$$\sum_{k=0}^{\infty}(1-p)^{k-1} \cdot k = -\frac{d}{dp}[\sum_{k=1}^{\infty}(1-p)^k] = -\frac{d}{dp}(\frac{1}{p}) = \frac{1}{p^2}.$$

Obviously, if $k = 0$ is excluded or $k = 1, 2, 3, ....,$ then we have $P = p(1-p)^{k-1}$; in this case, the mean is

$$E(p) + 1 = \frac{1-p}{p} + 1 = \frac{1}{p}.$$

Further, we can show that the variance is $(1 - p)/p^2$.

**9.2** The probability distribution for $S = U + V$ is

$$P(S = n) = \sum_{k=0}^{n} P(U = k, V = n - k) = \sum_{k=0}^{n} P(U = k)P(V = n - k)$$

$$= P_k(\lambda_1)P_{n-k}(\lambda_2) = \sum_{k=0}^{n} \frac{\lambda_1^k e^{-\lambda_1}}{k!} \cdot \frac{\lambda_2^{n-k} e^{-\lambda_2}}{(n - k)!} = e^{-(\lambda_1+\lambda_2)} \frac{1}{n!} \sum_{k=0}^{n} \frac{n!}{k!(n - k)!} \lambda_1^k \lambda_2^{n-k}$$

$$= e^{-(\lambda_1+\lambda_2)} \frac{(\lambda_1 + \lambda_2)^n}{n!} = e^{-\lambda} \frac{\lambda^n}{n!} = P_n(\lambda),$$

where $\lambda = \lambda_1 + \lambda_2$. We have also used the binomial expansion

$$(\lambda_1 + \lambda_2)^n = \sum_{k=0}^{n} \frac{n!}{k!(n - k)!} \lambda_1^k \lambda_2^{n-k}.$$

**9.3** For the two students, we have $\lambda_1$ and $\lambda_2$. The joint distribution of $A + B$ obeys Poisson($\lambda_1 + \lambda_2$) with $\lambda = \lambda_1 + \lambda_2 = 3$, which gives

$$P(A + B = 4) = \frac{(\lambda_1 + \lambda_2)^4 e^{-(\lambda_1+\lambda_2)}}{4!} = \frac{3^4 e^{-3}}{4!} \approx 0.168.$$

For a period of $t$, the distribution becomes

$$P(\lambda) = \frac{(\lambda t)^n e^{-\lambda t}}{n!},$$

so the probability of receiving no emails after a 2-hour lesson is

$$P = \frac{(3 \times 2)^0 e^{-3 \times 2}}{0!} \approx 0.00248.$$

**9.4** In order to best fit a function $y = \exp[ax^2 + b\sin(x)]$, we first take the logarithm of both sides and we have

$$\ln y(x) = \ln e^{ax^2+b\sin(x)} = ax^2 + b\sin(x),$$

which can be fitted using the generalized linear regression method.

## Chapter 10

**10.1** The type of each PDE is given as follows:
- $xu_x + yu_y = xyu$ is a linear first-order PDE.
- $(u_t)^2 = u_x + xt$ is a nonlinear PDE.
- $u_t + u^2 u_x = 0$ is a nonlinear PDE.
- $u_{tt} = tu_{xx}$ is a mixed type. It is parabolic if $t > 0$ and hyperbolic for $t < 0$. This

is the well-known Euler-Tricomi equation.

- $u_t = D\nabla^2 u + \gamma u(1 - u)$ is a nonlinear parabolic equation, which is also known as a nonlinear reaction-diffusion equation.

**10.2** Since $EI$ is constant, the original beam equation becomes

$$EI\frac{\partial^4 u}{\partial x^4} = -\mu\frac{\partial^2 u}{\partial t^2}.$$

Using $u = X(x)f(t)$, we have

$$EI\frac{\partial^4 (Xf)}{\partial x^4} = -\mu\frac{\partial^2 (Xt)}{\partial t^2}, \quad \frac{EI}{\mu}\frac{1}{X}\frac{\partial^4 X}{\partial x^4} = -\frac{1}{f}\frac{\partial^2 f}{\partial t^2}.$$

As the left-hand side depends only on $x$ and the right-hand side depends only on $t$, both sides should be equal to the same constant. For convenience, we set this constant as $\omega^2$, and we have

$$-\frac{1}{f}\frac{\partial^2 f}{\partial t^2} = \omega^2 = \frac{EI}{\mu}\frac{1}{X} = \frac{\partial^4 X}{\partial x^4},$$

which gives

$$\frac{\partial^2 f}{\partial t^2} + \omega^2 f = 0, \quad \frac{\partial^4 X}{\partial x^4} - \lambda X = 0,$$

where $\lambda = \omega^2\mu/EI$. Here, $\omega$ is essentially the angular frequency of the vibrations.

**10.3** For Laplace's equation in 3D

$$\nabla^2 u = \frac{\partial^2 u}{\partial x^2} + \frac{\partial^2 u}{\partial y^2} + \frac{\partial^2 u}{\partial z^2} = 0,$$

initial conditions and boundary conditions are needed. Thus, its solution can be very lengthy and interested readers can refer to more advanced literature listed at the end of the book. However, for a given solution, it is relatively straightforward to verify the solution by direct differentiation. From $u = 1/r = 1/\sqrt{x^2 + y^2 + z^2}$, we have

$$\frac{\partial^2 u}{\partial x^2} = \frac{3x^2}{r^5} - \frac{1}{r^3}, \quad \frac{\partial^2 u}{\partial y^2} = \frac{3y^2}{r^5} - \frac{1}{r^3}, \quad \frac{\partial^2 u}{\partial z^2} = \frac{3z^2}{r^5} - \frac{1}{r^3},$$

and thus we have

$$\nabla^2 u = \frac{\partial^2 u}{\partial x^2} + \frac{\partial^2 u}{\partial y^2} + \frac{\partial^2 u}{\partial z^2} = \frac{3(x^2 + y^2 + z^2)}{(x^2 + y^2 + z^2)^{5/2}} - \frac{3}{(x^2 + y^2 + z^2)^{3/2}}$$

$$= \frac{3}{(x^2 + y^2 + z^2)^{3/2}} - \frac{3}{(x^2 + y^2 + z^2)^{3/2}} = 0.$$

**10.4** To solve $y(x) = 1 + \int_0^1 xy(t)dt$, we can use a method called successive approximation. The first approximation is $y_0(x) = 1$. Then, substituting this to the original
equation, we have

$$y_1(x) = 1 + \int_0^1 xy_0 dt = 1 + x\int_0^1 1dt = (1 + x),$$

which is again used for better approximation. Thus, we have

$$y_2(x) = 1 + \int_0^1 xy_1 dt = 1 + \int_0^1 x(1 + t)dt = 1 + x(1 + \frac{1}{2}).$$

Similarly, the next approximation is

$$y_3(x) = 1 + \int_0^1 xy_2 dt = 1 + \int_0^1 x[1 + t(1 + \frac{1}{2})]dt = 1 + x(1 + \frac{1}{2} + \frac{1}{2^2}].$$

Following the same procedure, we can get

$$y_{n+1} = 1 + x\left[1 + \frac{1}{2} + \frac{1}{2^2} + \dots + \frac{1}{2^n}\right].$$

As $n \to \infty$, the above series becomes

$$y(x) = 1 + x\sum_{i=0}^{\infty} \frac{1}{2^i} = 1 + 2x.$$

It is straightforward to check that this is indeed a solution to the integral equation.

## Chapter 11

**11.1** This function $f = (1 - x)^2 + 100(y - x^2)^2$ is called Rosenbrock's function or the
banana function because the global minimum $x = y = 1$ lies in a narrow, flat, banana-
shaped valley. The stationary conditions are

$$\frac{\partial f}{\partial x} = 2(1 - x)(-1) + 200(y - x^2)(-2x) = 0, \quad \frac{\partial f}{\partial y} = 200(y - x^2) = 0,$$

or

$$2(1 - x) + 400x(y - x^2) = 0, \quad 200(y - x^2) = 0.$$

The solution is $x = 1$ and $y = x^2 = 1$ with the minimum of $f_* = 0$.

**11.2** The constrained problem

$$f(x, y) = x^2 + y^2 + 2xy = (x + y)^2, \quad \text{subject to} \quad y - x^2 + 2 = 0,$$

has three optimal solutions $(x, y) = (1, -1)$, $(-2, 2)$ and $(-\frac{1}{2}, -\frac{7}{4})$. The solution
$(-1/2, -7/4)$ corresponds to the global maximum $f_{max} = 81/16$, while the two solutions $(1, -1)$ and $(-2, 2)$ correspond to the two minima with $f_{min} = 0$. You can

obtain the solution by the method of Lagrange multipliers and/or spotting the solution by using the fact that $x + y = 0$ for the minima.

**11.3** For $f = x^2 + 5y^2 + xy$, we have its gradient

$$\nabla f = (\frac{\partial f}{\partial x}, \frac{\partial f}{\partial y})^T = (2x + y, \ 10y + x)^T,$$

and the Hessian

$$\mathbf{H} = \begin{pmatrix} \frac{\partial^2 f}{\partial x^2} & \frac{\partial^2 f}{\partial x \partial y} \\ \frac{\partial^2 f}{\partial y \partial x} & \frac{\partial^2 f}{\partial y^2} \end{pmatrix} = \begin{pmatrix} 2 & 1 \\ 1 & 10 \end{pmatrix}.$$

Its eigenvalues are $6 + \sqrt{17}$ and $6 - \sqrt{17}$. Since both eigenvalues are positive, the Hessian is positive definite and thus the function has a minimum at $x = y = 0$.

**11.4** For Lorenz equations

$$\frac{du}{dt} = \sigma(v - u), \quad \frac{dv}{dt} = u(\rho - w) - v, \quad \frac{dw}{dt} = uv - \beta w,$$

with $\sigma = 10, \rho = 28$ and $\beta = 8/3$, we can use the simple forward Euler methods, so that we have

$$u_{n+1} = u_n + \delta t \, \sigma(v_n - u_n), \quad v_{n+1} = v_n + \delta t \, [u_n(\rho - w_n) - v_n],$$

and

$$w_{n+1} = w_n + \delta t \, [u_n v_n - \beta w_n].$$

By setting $u_0 = v_0 = w_0 = 1$ and $\delta t = 0.01$, we can program the above equations and visualize the results using any programming language such as Matlab, Python, R, C++ or Java.

**11.5** This is number 4 of Xin-She Yang's functions, and it takes the following form

$$f(x_1, ..., x_n) = \Big[ \sum_{i=1}^{n} \sin^2(x_i) - e^{-\sum_{i=1}^{n} x_i^2} \Big] e^{-\sum_{i=1}^{n} \sin^2 \sqrt{|x_i|}},$$

where $-10 \le x_i \le 10$ for $i = 1, 2, ..., n$, and $n \ge 2$ is an integer. It has a global minimum at $(0, 0, ..., 0)$ with $f_{min} = -1$. However, it is not straightforward to find this optimal solution. The gradient is not easy to calculate due to $\sin^2(x_i)$ terms and $\sqrt{|x_i|}$ terms. You can try to solve it using Newton's method for the case of $n = 2$ to see if it works.

# Bibliography

1. Abramowitz M and Stegun IA, (1972). *Handbook of Mathematical Functions*. National Bureau of Standards, Applied Mathematics Series 55, 10th ed., Washington DC: US Government Printing Office.
2. Arfken GB, Weber HJ, Harris FE, (2012). *Mathematical Methods for Physicists: A Comprehensive Guide*. 7th ed., Waltham: Academic Press.
3. Armstrong M., (1998). *Basic Linear Geostatistics*. Berlin: Springer.
4. Atluri SN, (2005). *Methods of Computer Modeling in Engineering and the Sciences*. Vol. I, Forsyth: Tech Science Press.
5. Bird J, (2014). *Engineering Mathematics*. 7th ed., New York: Routledge.
6. Bird J, (2014). *Higher Engineering Mathematics*. 7th ed. New York: Routledge.
7. Carrrier GF and Pearson CE, (1988). *Partial Differential Equations: Theory and Technique*. 2nd ed. Waltham: Academic Press.
8. Carslaw HS and Jaeger JC, (1986). *Conduction of Heat in Solids*. 2nd ed. Oxford: Oxford University Press.
9. Courant R and Hilbert D, (1962). *Methods of Mathematical Physics*. 2 volumes, New York: Wiley-Interscience.
10. Crank J, (1970). *Mathematics of Diffusion*. Oxford: Clarendon Press.
11. Croft A, Davison R, Hargreaves M and Flint J, (2012). *Engineering Mathematics: A Foundation for Electronic, Electrical, Communications and Systems Engineers*. 4th ed. Essex: Pearson Education.
12. Deb K, (1995). *Optimisation for Engineering Design: Algorithms and Examples*. New Delhi: Prentice-Hall.
13. Farlow SJ, (1982). *Partial Differential Equations for Scientists and Engineers*. New York: John Wiley and Sons.
14. Fowler AC, (1997). *Mathematical Models in the Applied Sciences*. Cambridge: Cambridge University Press.
15. Gershenfeld N, (1998). *The Nature of Mathematical Modeling*. Cambridge: Cambridge University Press.
16. Gill PE, Murray W and Wright MH, (1982). *Practical optimization*. Bingley: Emerald Group Publishing.
17. Goodman R, (1957). *Teach Yourself Statistics*. London: Teach Yourself Books.
18. Hicks TG, (2010). *Civil Engineering Formulas*, 2nd Edition, New York: McGraw-Hill.
19. Jeffrey A, (2002). *Advanced Engineering Mathematics*. San Diego: Academic Press.
20. James G, (2015). *Modern Engineering Mathematics*. 5th ed. Essex: Pearson Education.
21. Korn GA and Korn TM, (1968). *Mathematical Handbook for Scientists and Engineers*. New York: McGraw-Hill.
22. Koziel S and Yang XS, (2011). *Computational Optimization, Methods and Algorithms*. Studies in Computational Intelligence, vol. 356, Heidelberg: Springer.
23. Kreyszig E, (1988). *Advanced Engineering Mathematics*. 6th ed. New York: John Wiley & Sons.
24. Kreyszig E, (2011). *Advanced Engineering Mathematics: International Student Version*. 10th ed. New York: John Wiley and Sons.
25. Miersemann E, (2014). *Partial Differential Equations*, Amazon CreateSpace Independent Publishing Platform.
26. Moler CB, (2004). *Numerical Computing with MATLAB*. Philadelphia: SIAM.
27. Ockendon J, Howison S, Lacey A and Movchan A, (2003). *Applied Partial Differential Equations*. Oxford: Oxford University Press.
28. O'Neil PV, (2012). *Advanced Engineering Mathematics*. 7th ed. Stamford: Cengage Learning.
29. Pallour JD and Meadows DS, (1990). *Complex Variables for Scientists and Engineers*. London:

Macmillan Publishing Co.

30. Papoulis A, (1990). *Probability and statistics*. Cliffs: Englewood.
31. Pearson CE, (1983). *Handbook of Applied Mathematics*, 2nd ed. New York: Van Nostrand Reinhold.
32. Peterson JC, (2004). *Technical Mathematics with Calculus*. 3rd ed. New York: Thomson Learning.
33. Press WH, Teukolsky SA, Vetterling WT and Flannery BP, (2002). *Numerical Recipes in C++: The Art of Scientific Computing*. 2nd ed. Cambridge: Cambridge University Press.
34. Puckett EG and Colella P, (2005). *Finite Difference Methods for Computational Fluid Dynamics*. Cambridge: Cambridge University Press.
35. Rahman M, (2007). *Integral Equations and Their Applications*. Southampton: WIT Press.
36. Riley KF, Hobson MP and Bence SJ, (2006). *Mathematical Methods for Physics and Engineering*. 3rd ed. Cambridge: Cambridge University Press.
37. Singh K, (2011). *Engineering Mathematics Through Applications*. Basingstoke: Palgrave Macmillan.
38. Stroud KA and Booth DJ, (2007). *Engineering Mathematics*, 6th ed. New York: Palgrave Macmillan.
39. Topliss S, Hurst M and Skarratt G, (2007). *BTEC National Construction, Building Services Engineering and Civil Engineering*, Harlow: Heinemann.
40. Weisstein EW, http://mathworld.wolfram.com. [Accessed 15 Sept 2017]
41. Wikipedia, http://en.wikipedia.com. [Accessed 15 Sept 2017]
42. Yang XS, (2008). *Nature-Inspired Metaheuristic Algorithms*. Frome: Luniver Press.
43. Yang XS, (2010). *Engineering Optimization: An Introduction with Metaheuristic Applications*. New Jersey: John Wiley and Sons.
44. Yang XS, (2013). *Mathematical Modelling with Multidisciplinary Applications*, Hoboken, NJ: John Wiley and Sons.
45. Yang XS, (2014). *Nature-Inspired Optimization Algorithms*. Waltham: Elsevier.
46. Yang XS, (2015). *Introduction to Computational Mathematics*, 2nd ed. Singapore: World Scientific Publishing.
47. Yang XS, (2017)/ *Engineering Mathematics with Examples and Applications*, London: Academic Press/Elsevier.
48. Zill DG and Wright WS, (2009). *Advanced Engineering Mathematics*, 4th ed. Sudbury, MA: Jones & Bartlett Learning.
49. Zwillinger D, (2012). *CRC Standard Mathematical Tables and Formulae*. 32nd Ed. Boca Raton, FL: CRC Press.

# Index

*e*, 25
*i*, 140

area, 118
Argand diagram, 141
argument, 141
array, 56
auxiliary equation, 168

base, 23
binomial coefficient, 39
binomial distribution, 218
bisection method, 267

Cartesian, 48
Cartesian coordinate, 141
central difference, 275
central limit theorem, 227
chain rule, 93
characteristic equation, 168
cofactor, 66
complementary function, 168, 173
complete square, 43
complex conjugate, 142
complex number, 139
    division, 142
    function, 148
    imaginary part, 141
    real part, 141
complex root, 168, 170
computational complexity, 204
constant, 44
constrained optimization, 282
contour integral, 158
correlation coefficient, 236
cross product, 53
cumulative probability function, 222
curl, 110
curvature, 101

curve, 87
    family, 91

damped pendulum, 173
degree, 44
dependent variable, 14
derivative, 89, 92, 209, 275
    first, 97
determinant, 66
DFT, 203
differential equation, 161
    homogenous, 167
differentiation, 124
    implicit function, 96
    rule, 93
    table, 97
    vector, 107
Dirac $\delta$-function, 205
Dirichlet condition, 194
discontinuity, 190
displacement, 48
distributive law, 53
divergence, 110
dot product, 52

eigenvalue, 70
eigenvector, 70
elasticity
    Hooke's law, 82
    strain tensor, 82
    stress tensor, 82
error function, 137, 270
Euler scheme, 274
Euler's formula, 144
even function, 29–31, 120
explicit formula, 45
exponent, 35
exponential distribution, 224